ALGORITHMS
for the COMPUTATION
of MATHEMATICAL
FUNCTIONS

Academic Press Rapid Manuscript Reproduction

ALGORITHMS FOR THE COMPUTATION OF MATHEMATICAL FUNCTIONS

YUDELL L. LUKE

Department of Mathematics
University of Missouri
Kansas City, Missouri

Academic Press

NEW YORK   SAN FRANCISCO   LONDON   1977

A Subsidiary of Harcourt Brace Jovanovich, Publishers

ACADEMIC PRESS, INC.
111 Fifth Avenue, New York, New York 10003

United Kingdom Edition published by
ACADEMIC PRESS, INC. (LONDON) LTD.
24/28 Oval Road, London NW1

Luke, Yudell L.
      Algorithms for the computation of mathematical
functions.

      (Computer science and applied mathematics)
      Bibliography: p.
      Includes indexes.
      1.   Functions, Special—Data processing.
2.   Approximaton theory—Data processing.   I.   Title.
QA351.L796   515'.5'0285   77-8502
ISBN 0-12-459940-6

PRINTED IN THE UNITED STATES OF AMERICA

To
My Grandchildren

Michelle Ilene
Matthew Douglas
Melissa Jill
Staci Jo
Sara Kumiko
and
. . .

CONTENTS

PREFACE

In my previous volumes, "The Special Functions and Their Approximations," Volumes 1 and 2 (Academic Press, New York, 1969), and "Mathematical Functions and Their Approximations" (Academic Press, New York, 1975), the following topics especially pertinent to the nature of the present volume were examined in considerable detail:

1. expansion of functions in series of the Chebyshev polynomials of the first kind;
2. rational approximations of the Padé class;
3. rational approximations of a certain type not of the Padé class;
4. coefficients for approximation of many functions by forms of the above three types.

For the most part, the functions treated were of the hypergeometric type or were closely related to this type. These include, for example, the elementary functions, Debye functions, incomplete gamma functions, elliptic integrals, Bessel functions, Struve functions and their integrals, and the $_pF_q$ and G functions.

Classically, the ultimate mathematical approximations for functions were in the form of numerical values of the functions such as tables of square roots, cosines, or sines. With the advent of high-speed computers, numerical tables, though very important for teaching and research, are primarily for the occasional computer. To impress these tables into the memory of a computer and then program for table lookup and interpolation is not economical. A computer requires efficient algorithms and schemes for the evaluation of functions on demand. My philosophy of approximation is global in character. Numerical values of functions are but a facet of the overall problem. We desire approximations to compute functions and their zeros, to simplify mathematical expressions like integrals and transforms, and to facilitate directly the mathematical solution of a wide variety of functional equations such as differential equations and integral equations. So the main thrusts of my volumes were on the development of analytical expansions and approximations for universal use.

In the volumes noted, coefficients for approximating numerous functions were provided. Nonetheless, there is a large void in the literature because machine programs to generate Chebyshev coefficients, to compute rational approximations, and to evaluate the coefficients in the numerator and denominator polynomials that define these rational approximations are generally not available. Such data would serve to considerably extend and enhance the previous volumes. The production of these programs is certainly in the spirit of my philosophy of approximation. So the principal thrust of the present volume is the presentation of programs for machine use.

For approximations of the first two categories noted in the opening paragraph, complete a priori asymptotic error analyses were provided. For approximations of the third category, convergence of the approximations was assured, but complete a priori asymptotic error analyses were lacking. A further mission of the present volume is to correct this deficiency.

The machine programs are written for rather general types of functions, while the coefficients noted in my previous volumes were for special cases of the general types. In illustration, we now give general programs to compute the coefficients in the expansion of the confluent functions $_1F_1(a;c;z)$ and $z^a U(a;c;z)$ in series of Chebyshev polynomials of the first kind. For example, we write

$$_1F_1(a;c;z) = \sum_{n=0}^{\infty} C_n(w) T_n^*(z/w), \qquad 0 \le z/w \le 1$$

and call the $C_n(w)$ Chebyshev coefficients. For special cases of the parameters, we then have expansions for Bessel functions, parabolic cylinder functions, Coulomb wave functions, incomplete gamma functions, and much more. We also give programs for a certain kind of rational approximation for the confluent functions that are not of the Padé class. When $a = 1$, the confluent functions are incomplete gamma functions; in this case since rational approximations of the Padé class are known in closed form, programs for the Padé approximations are developed. There are two different programs for the rational approximations. One program generates approximations as a number by evaluating the numerator and denominator polynomials with the aid of the recursion formula satisfied by these polynomials. The second program generates the coefficients that define the numerator and denominator polynomials.

It takes several chapters to cover the confluent functions. For instance, there is a chapter dealing with the Chebyshev coefficients $C_n(w)$ noted in the preceding paragraph. Here the analytical form for $C_n(w)$ and its asymptotic representation for large n are stated. Also presented is the recursion formula satisfied by $C_n(w)$ to enable the evaluation of the Chebyshev coefficients by use of this formula in the backward direction. Machine programs are provided to generate the $C_n(w)$ once w, a, and c are specified. The ideas are illustrated with numerical examples. There are no proofs. We do, however, provide the basic machinery to understand the functions treated, to identify their relation with the named special functions, and to evaluate the coefficients. Though the above discussion bears on the Chebyshev expansions, the style of the chapters for rational approximations follows a similar pattern.

For the confluent functions noted, we give programs for Chebyshev expansions and rational approximations valid in neighborhoods both of the origin and of infinity. Besides these functions, we also treat in a similar fashion $_2F_1(a,b;c;z)$, Bessel functions, $_1F_2(a;b,c;z)$, the G functions $G_{1,3}^{m,1}(z^2/4|_{a,b,c}^1)$, $m = 3$ or $m = 2$, and much more. In effect, we cover virtually all the functions for which explicit coefficients were provided in my previous volumes.

The programming of the routines in this collection was done for use on an IBM 370/168 operating under OS/VS Release 1.7 with the FORTRAN IV H-extended compiler, Release 2.1. Thus to use the programs on other types of equipment, certain adjustments must be made as required by that equipment. In this connection, it should be realized that it is impractical to prepare programs to cover every type of machine and compiler on the market. Most certainly we cannot anticipate the characteristics of future machines and compilers. Each program is written for quadruple precision and for the variable and all parameters real. The small changes necessary to get double or single precision or to get any of the three precision types with complex arithmetic are carefully detailed. Note that quadruple precision is available only on the FORTRAN H-extended compiler.

To assist the reader in the location of material, notation and subject indexes are provided. There is also a bibliography. It is quite short, since my previous volumes provide most of the required background.

I should like to thank John Lott, John Lowry III, and Rick Whitaker for their valuable suggestions and expert assistance in the preparation of the machine programs. Finally, it is a pleasure to acknowledge the aid of Teresa Anderson in the preparation of the manuscript for camera copy production.

YUDELL L. LUKE
Kansas City, Missouri
March, 1977

I BASIC FORMULAS

1.1. Introduction

In this chapter, we present some results to facilitate use of this volume. No proofs are furnished. For further details, see Luke (1969, 1975a).

1.2. The Generalized Hypergeometric Function and the G-Function

The generalized hypergeometric function can be defined by the series

$$
{}_pF_q(\alpha_1,\alpha_2,\ldots,\alpha_p;\rho_1,\rho_2,\ldots,\rho_q;z) = {}_pF_q\left(\begin{array}{c}\alpha_1,\alpha_2,\ldots,\alpha_p \\ \rho_1,\rho_2,\ldots,\rho_q\end{array}\bigg| z\right)
$$

$$
= \sum_{k=0}^{\infty}\left[\prod_{k=1}^{p}(\alpha_h)_k z^k \bigg/ \prod_{h=1}^{q}(\rho_h)_k k!\right] ,
$$

$$
(\alpha)_k = \Gamma(\alpha+k)/\Gamma(\alpha) ,
$$

$$
p \leq q \quad \text{or} \quad p = q+1 \quad \text{and} \quad |z| < 1 , \tag{1}
$$

where $\Gamma(z)$ is the gamma function. When $p = q+1$ with certain restrictions on the parameters, we also have convergence for $|z| = 1$. Where no confusion can arise, we simply refer to (1) as ${}_pF_q(z)$ or as ${}_pF_q$. The ${}_pF_q$ can also be defined by integrals, so that when $p = q+1$, the domain of validity is $|\arg(1-z)| < \pi$. This enables the analytic continuation of (1), and we use the same notation for both (1) and its analytic continuation.

We suppose that none of the ρ_h's is a negative integer or zero. It is often convenient to employ a contracted notation and write (1) in the abbreviated form

1

$$_pF_q(\alpha_p;\rho_q;z) = {}_pF_q\left(\begin{array}{c}\alpha_p\\[4pt]\rho_q\end{array}\Bigg|z\right) = \sum_{k=0}^{\infty}\left[(\alpha_p)_k z^k/(\rho_q)_k k!\right] . \qquad (2)$$

Thus, $\Gamma(\alpha_p)$ stands for $\prod\limits_{h=1}^{p}\Gamma(\alpha_h)$ and $(\alpha_p+m)_k$ stands for $\prod\limits_{h=1}^{p}(\alpha_h+m)_k$, m independent of p . Also, empty terms which appear in products should be replaced by unity, and empty terms which appear in sums should be put to zero. All of this should be clear from the context.

The α_h's and ρ_h's are called numerator and denominator parameters, respectively. If, in a ${}_pF_q$, a numerator and denominator parameter coalesce, these parameters can be omitted, whence the ${}_pF_q$ becomes a ${}_{p-1}F_{q-1}$.

We often have need for a truncated hypergeometric series. Thus,

$$_pF_q^m\left(\begin{array}{c}\alpha_p\\[4pt]\rho_q\end{array}\Bigg|z\right) = \sum_{k=0}^{m}\frac{(\alpha_p)_k z^k}{(\rho_q)_k k!}$$

$$= \frac{(\alpha_p)_m z^m}{(\rho_q)_m m!}\ {}_{q+2}F_p^m\left(\begin{array}{c}-m,1-m-\rho_q,1\\[4pt]1-m-\alpha_p\end{array}\Bigg|\frac{(-)^{p+q+1}}{z}\right) , \qquad (3)$$

where we suppose that if any α_h is a negative integer, say $\alpha_h = -r$, then $r > m$; and if any ρ_h is a negative integer, say $\rho_h = -s$, then $m < s$.

If m is a positive integer or zero, then

$$_{p+1}F_{q+1}\left(\begin{array}{c}-m,\alpha_p\\[4pt]c,\rho_q\end{array}\Bigg|z\right) = \frac{(\alpha_p)_m(-)^m z^m}{(c)_m(\rho_q)_m}$$

$$\times\ {}_{q+2}F_p^m\left(\begin{array}{c}-m,1-m-c,1-m-\rho_q\\[4pt]1-m-\alpha_p\end{array}\Bigg|\frac{(-)^{p+q+1}}{z}\right) . \qquad (4)$$

In (4), we usually suppose that none of the α_h's, ρ_h's or c is a negative integer or zero. However, this can be relaxed in view of the comments following (3).

A more general function is the so-called G-function. It is defined by a Mellin-Barnes integral from which many useful properties can be deduced. Only a few of these are recorded below. We have

$$
G_{p,q}^{m,n}\left(z\left|\begin{array}{c}a_p\\b_q\end{array}\right.\right) = \sum_{h=1}^{m} \frac{\prod\limits_{j=1}^{m}\Gamma(b_j-b_h)^* \prod\limits_{j=1}^{n}\Gamma(1+b_h-a_j)z^{b_h}}{\prod\limits_{j=m+1}^{q}\Gamma(1+b_h-b_j)\prod\limits_{j=n+1}^{p}\Gamma(a_j-b_h)}
$$

$$
\times \ _pF_{q-1}\left(\begin{array}{c}1+b_h-a_p\\1+b_h-b_q^*\end{array}\left|(-)^{p-m-n}z\right.\right) ,
$$

$$
p < q \quad \text{or} \quad p = q \quad \text{and} \quad |\arg(1-(-)^{p-m-n}z)| < \pi . \qquad (5)
$$

Here the asterisk (*) sign means that the terms involving b_j-b_h are omitted when $j = h$. Some important properties of the G-function are

$$
G_{p,q}^{m,n}\left(z\left|\begin{array}{c}a_p\\b_q\end{array}\right.\right) = G_{q,p}^{n,m}\left(z^{-1}\left|\begin{array}{c}1-b_q\\1-a_p\end{array}\right.\right) , \qquad (6)
$$

$$
z^\sigma G_{p,q}^{m,n}\left(z\left|\begin{array}{c}a_p\\b_q\end{array}\right.\right) = G_{p,q}^{m,n}\left(z\left|\begin{array}{c}a_p+\sigma\\b_q+\sigma\end{array}\right.\right) , \qquad (7)
$$

$$
G_{p,q}^{m,n}\left(z\left|\begin{array}{c}a_p\\b_q\end{array}\right.\right) = (2\pi i)^{-1}\left\{\exp(i\pi b_{m+1})G_{p,q}^{m+1,n}\left(ze^{-i\pi}\left|\begin{array}{c}a_p\\b_q\end{array}\right.\right)\right.
$$

$$
\left.-\exp(-i\pi b_{m+1})G_{p,q}^{m+1,n}\left(ze^{i\pi}\left|\begin{array}{c}a_p\\b_q\end{array}\right.\right)\right\} ,
$$

$$
m \leq q-1 . \qquad (8)
$$

The G-function has meaning in terms of its Mellin-Barnes integral representation even when $p > q+1$. For complete details on this and the asymptotic expansion for $G_{p,q}^{m,n}(z)$ for large z, see the sources cited. For numerous applications, it is convenient to define

$$H(z) = \{\Gamma(\rho_q)/\Gamma(\alpha_p)\}\ G_{p,q+1}^{1,p}\left(-z\ \Big|\ \begin{matrix} 1-\alpha_p \\ 0, 1-\rho_q \end{matrix}\right).$$ (9)

Then

$$H(z) = {}_pF_q(\alpha_p;\rho_q;z) ,$$

$$p \le q \quad \text{or} \quad p = q+1 \quad \text{and} \quad |\arg(1-z)| < \pi ,$$ (10)

$$H(z) \sim {}_pF_q(\alpha_p;\rho_q;z) , \quad p \ge q+2 ,$$

$$z \to 0 , \quad |\arg(-z)| \le \tfrac{1}{2}(p+1-q)\pi-\varepsilon, \quad \varepsilon > 0 .$$ (11)

In the applications, when $p \ge q+2$, we normally deal with $H(-1/z)$.

1.3. Expansion of ${}_pF_q(z)$ and $G_{p+1,q}^{q-r,1}(z)$, $r = 0$ or $r = 1$, in Series of Chebyshev Polynomials of the First Kind

We have

$$_pF_q\left(\begin{matrix} a_p \\ b_q \end{matrix}\ \Big|\ z\right) = \sum_{n=0}^{\infty} C_n(w)\,T_n^*(z/w), \quad 0 \le z/w \le 1 ,$$ (1)

$$C_n(w) = E_n(w)F_{n,p,q}(w), \quad E_n(w) = \frac{\varepsilon_n\,(a_p)_n\,(w/4)^n}{(b_q)_n n!} ,$$

$$F_{n,p,q}(w) = {}_{p+1}F_{q+1}\left(\begin{matrix} \tfrac{1}{2}+n, a_p+n \\ 1+2n, b_q+n \end{matrix}\ \Big|\ w\right) ,$$

$$p \le q, \ |w| < \infty; \ p = q+1, \ w \ne 1, \ |\arg(1-w)| < \pi ,$$ (2)

where $T_n^*(x)$ is the shifted Chebyshev polynomial of the first kind, see (5) below. Also,

$$\varepsilon_0 = 1 \quad \text{and} \quad \varepsilon_n = 2 \quad \text{for} \quad n > 0 \ . \tag{3}$$

If z is a complex number, say $z = xe^{i\theta}$, then we must take
$w = \rho e^{i\theta}$, and $0 \le x/\rho \le 1$. If z is replaced by $g^2 z^2$
and w is replaced by $g^2 w^2$, then

$$_pF_q \left(\begin{matrix} a_p \\ b_q \end{matrix} \middle| g^2 z^2 \right) = \sum_{n=0}^{\infty} C_n (g^2 w^2) T_{2n}(z/w), \ 0 \le z/w \le 1 \tag{4}$$

where $T_n(x)$ is the Chebyshev polynomial of the first kind.
We have

$$T_n(x) = \cos n\theta \ , \ x = \cos \theta \ ,$$

$$T_n(x) = {}_2F_1 \left(\begin{matrix} -n,n \\ \tfrac{1}{2} \end{matrix} \middle| \frac{1-x}{2} \right) = (-)^n \ {}_2F_1 \left(\begin{matrix} -n,n \\ \tfrac{1}{2} \end{matrix} \middle| \frac{1+x}{2} \right) ,$$

$$T_n^*(x) = T_n(2x-1) \ , \ T_n^*(x^2) = T_{2n}(x) \ . \tag{5}$$

With appropriate restrictions on the parameters a_h and
b_h , the coefficients $C_n(w)$ satisfy a recurrence formula in
n of order $t = \max(q+2, p+1)$. The $C_n(w)$'s can be found
by use of the recursion formula in the backward direction,
together with the normalization relation

$$\sum_{n=0}^{\infty} (-)^n C_n(w) = 1 \ . \tag{6}$$

The general recursion formula is not presented here, but recurrence formulas for the evaluation of $C_n(w)$ are given in
later chapters for the special cases treated there. The general scheme for evaluation of $C_n(w)$ by use of the recursion
formula in the backward direction is taken up later in this
section.

The relation (6) follows from (1) by setting $z = 0$. From another point of view, (6) emerges from (1) upon expanding $T_n^*(z/w)$ in powers of z/w and equating the constant terms. By equating the coefficients of z and z^2, we get the respective relations

$$\sum_{n=1}^{\infty} (-)^n n^2 C_n(w) = - a_p w/2b_q , \tag{7}$$

$$\sum_{n=2}^{\infty} (-)^n n^2 (n^2-1) C_n(w) = 3a_p(a_p+1)w^2/4b_q(b_q+1) \tag{8}$$

which are useful for check purposes. Let $C_{n,N}(w)$ be the approximate values of $C_n(w)$ determined by the backward recursion process. Then, in the computer programs,

$$\text{TEST 1} = \sum_{n=1}^{N+1} (-)^n n^2 C_{n,N}(w) + a_p w/2b_q , \tag{9}$$

$$\text{TEST 2} = \sum_{n=2}^{N+1} (-)^n n^2 (n^2-1) C_{n,N}(w) - 3a_p(a_p+1)w^2/4b_q(b_q+1) . \tag{10}$$

Truncation and round-off errors aside, the values of the test functions should be zero.

If $p < q$, the form for $C_n(w)$ is an asymptotic representation for n large and w fixed. If $p = q$, we have

$$F_{n,p,p}(w) = \left\{ \exp \frac{(a_p+n)w}{2(b_p+n)} \right\} \left[1 + \frac{(a_p+n)w^2}{16n(b_p+n)} + 0(n^{-2}) \right] \tag{11}$$

or

$$F_{n,p,p}(w) = e^{w/2} \left[1 + \frac{w^2+8w \sum_{h=1}^{p} (a_h-b_h)}{16n} + 0(n^{-2}) \right] . \tag{12}$$

If $p = q+1$, then

$$F_{n,q+1,q}(w) = \frac{(2n)!(n\pi)^{\frac{1}{2}} e^{-n\delta} A(w)}{(n!)^2 w^n} [1 + 0(n^{-1})] ,$$

$$A(w) = \{(1+e^{-\delta})/(1-e^{-\delta})\}^u, \quad u = \sum_{h=1}^{q+1} a_h - \sum_{j=1}^{q} b_h ,$$

$$e^{-\delta} = [2-w\mp 2(1-w)^{\frac{1}{2}}]/w , \tag{13}$$

where the sign in the latter is chosen so that $|e^{-\delta}| < 1$.
This is possible for all w except $w \geq 1$. The function
$e^{-\delta}$ is the same as $-e^{-\zeta}$ of 1.8(2) if there we put $z = we^{i\pi}$.

The asymptotic forms can often be simplified by using
the formula

$$\frac{\Gamma(\alpha+n)}{\Gamma(\beta+n)} = n^{\alpha-\beta}\left[1 + \frac{\rho(\eta-1)}{2n} + \frac{\rho(\rho-1)\{3(\eta-1)^2-\rho-1\}}{24n^2}\right.$$

$$\left. + O(n^{-3})\right] , \quad \rho = \alpha-\beta , \quad \eta = \alpha+\beta . \tag{14}$$

For the expansion of a special class of G-functions in
series of Chebyshev polynomials, we have

$$BG_{p+1,q}^{q,1}\left(z\left|\begin{array}{c}1,a_p\\b_q\end{array}\right.\right) = \sum_{n=0}^{\infty} G_n(w)T_n^*(w/z) , \quad z/w \geq 1 ,$$

$$G_n(w) = \varepsilon_n(-)^n\pi^{-\frac{1}{2}} BG_{p+2,q+1}^{q+1,1}\left(w\left|\begin{array}{c}1-n,a_p,n+1\\\frac{1}{2}, b_q\end{array}\right.\right) ,$$

$$B = \frac{\Gamma(a_p)}{\Gamma(b_q)} , \tag{15}$$

under the conditions

    (1) $p \geq 0, q > p+1$,

    (2) b_j is not a negative integer or zero,
             $j = 1,2,\ldots,q$,

    (3) $|\arg w| < (q+1-p)\pi/2, w \neq 0$. $\tag{16}$

It is convenient to record the asymptotic representation

$$BG_{p+1,q}^{q,1}\left(z\left|\begin{array}{c}1,a_p\\b_q\end{array}\right.\right) \sim {}_qF_p(b_q;a_p;-1/z) ,$$

$$|z| \to \infty, \quad |arg\ z| \leq (q+1-p)\pi/2-\varepsilon, \quad \varepsilon > 0 \ . \tag{17}$$

With appropriate restrictions on the parameters a_h and b_h , $G_n(w)$ satisfies a recurrence formula in n of order $t = max(q+1,p+2)$. The $G_n(w)$'s can be computed by use of this recurrence formula in the backward direction, together with the normalization relation

$$\sum_{n=0}^{\infty} (-)^n G_n(w) = 1 \ . \tag{18}$$

The general recursion formula is not given here, but recurrence formulas are given for two special cases of (15) in Chapters 8 and 9. The backward recurrence scheme for the evaluation of $G_n(w)$ is treated later in this section.

Equation (18) follows from (15) and (17) by letting $z \to \infty$. As in the derivation of (7) and (8), we find

$$\sum_{n=1}^{\infty} (-)^n n^2 G_n(w) = \frac{b_q}{2a_p w} \ , \tag{19}$$

$$\sum_{n=2}^{\infty} (-)^n n^2 (n^2-1) G_n(w) = \frac{3b_q(b_q+1)}{4a_p(a_p+1)w^2} \ , \tag{20}$$

which are useful for check purposes. If $G_{n,N}(w)$ are the approximate values of $G_n(w)$ found by use of the backward recursion process, then in the computer programs,

$$\text{TEST 1} = \sum_{n=1}^{N+1} (-)^n n^2 G_{n,N}(w) - \frac{b_q}{2a_p w} \ , \tag{21}$$

$$\text{TEST 2} = \sum_{n=2}^{N+1} (-)^n n^2 (n^2-1) G_{n,N}(w) - \frac{3b_q(b_q+1)}{4a_p(a_p+1)w^2} \ , \tag{22}$$

and aside from truncation and round-off errors, these values should be zero.

From the work of Fields (1972,1973), we deduce the following asymptotic formula for $G_n(w)$.

$$G_n(w) = \frac{2(-)^n B}{\pi^{\frac{1}{2}} n} \left\{ \frac{(2\pi)^{\beta-1}}{\beta} \right\}^{\frac{1}{2}} \Omega^{\beta\gamma} \exp\{-\beta\Omega + 2\Omega P(y) + S/\Omega\}$$

$$\times \ [1 + 0(\Omega^{-2}(1+|w|^{4\Delta}))] \ , \quad \beta = q+1-p \ , \quad \beta \geq 3 \ ,$$

$$\Omega^\beta = n^2 w \ , \quad n^2 w \to \infty \ , \quad |\arg \Omega^\beta| \leq \pi(\beta+1)-\varepsilon, \quad \varepsilon > 0,$$

$$w = o(n^{2\mu}) \ , \quad (n^2 w)^{-1} = o(1) \ , \quad \mu = \max(\tfrac{1}{3}, \tfrac{2}{5}\beta-2) \ ,$$

$$\Delta = \min(\tfrac{2}{3}, \tfrac{1}{2\beta-5}) \ , \quad 2\beta\gamma = 1-\beta+2B_1-2A_1 \ , \tag{23}$$

$$B \quad \text{as in (15),}$$

$$\prod_{j=1}^{p}(x+a_j) = \sum_{j=0}^{p} A_j x^{p-j}, \quad (x+\tfrac{1}{2})\prod_{j=1}^{q}(x+b_j) = \sum_{j=0}^{q+1} B_j x^{q-j} \ ,$$

$$P(y) = \frac{\beta-2}{2}\left[1 - (\frac{\sinh y}{y})^{-\frac{2}{\beta-2}} \right], \quad y = (w\Omega^{2-\beta})^{\frac{1}{2}} \ ,$$

$$P(y) = \frac{y^2}{6}\left[1 - \frac{(\beta+3)y^2}{30(\beta-2)} + \frac{(4\beta^2+5\beta+9)y^4}{1890(\beta-2)^2} \right] + 0(y^8) \ ,$$

$$y \to 0 \ ,$$

$$S = \frac{[3+2(A_1-B_1)]w}{6\Gamma(4-\beta)} + A_2 - B_2 + \frac{(B_1-A_1)}{2\beta}$$

$$\times \ \{(\beta-1)B_1 + (\beta+1)A_1\} - \frac{\beta^2-1}{24 \ \beta} \ .$$

Clearly, (15) is convergent, which is quite remarkable since the right hand side of (15) can be viewed as a rearrangement of the right hand side of (17), which is divergent.

With the aid of 1.2(8), we can deduce the expansion

$$\frac{\pi B}{\sin \pi b_h} \ G_{p+1,q}^{q-1,1}\left(z \left|\begin{array}{l}1, a_p \\ b_q \end{array}\right.\right) = \sum_{n=0}^{\infty} D_n(w) T_n^*(w/z) \ , \tag{24}$$

$$D_n(w) = \frac{\pi B (2\pi i)^{-1}}{\sin \pi b_h} \{e^{i\pi b_h} G_n(we^{-i\pi}) - e^{-i\pi b_h} G_n(we^{i\pi})\} ,$$

$$h = q , \qquad\qquad (25)$$

valid under the conditions (1) and (2) of (16) and the further
condition that

$$|arg \ w| < (q-1-p)\pi/2 . \qquad\qquad (26)$$

We also have

$$\frac{\pi B}{\sin \pi b_h} \ G_{p+1,q}^{q-1,1} \left(z \left| \begin{matrix} 1, a_p \\ b_q \end{matrix} \right.\right) \sim \ _q F_p(b_q; a_p; 1/z) ,$$

$$|z| \to \infty , \quad |arg \ z| \le (q-1-p)\pi/2-\varepsilon , \quad \varepsilon > 0 . \qquad (27)$$

An asymptotic formula for $D_n(w)$ follows from (23). We omit
this result. The special cases $q = 2, \ p = 0$ and $q = 3,$
$p = 0$ are thoroughly detailed in Chapters 8 and 9, respectively.

It can be shown that $D_n(w)$ satisfies the same recurrence formula as does $G_n(w)$ with w replaced by $-w$. Further, (18), (20) and (22) are satisfied by $D_n(w)$; and so
also are (19) and (21), provided in the right hand sides of
these equations w is replaced by $-w$.

We now present the general scheme for evaluation of the
Chebyshev coefficients $C_n(w)$, $G_n(w)$ and $D_n(w)$. It is convenient to treat $C_n(w)$ first. Let N be a large positive
integer. Let Q be an arbitrary number and put

$$E_{N+1,N}(w) = Q , \ E_{N+r,N}(w) = 0 , \ r = 2,3,\ldots,t ,$$

$$t = max(q+2,p+1) . \qquad\qquad (28)$$

Though Q is arbitrary, it is often set equal to unity.

However, in practice, to minimize the risk of overflow, take
Q equal to, or near, the asymptotic estimate of $C_{N+1}(w)$ as
deduced from the discussion surrounding (11)-(13). In the
examples treated in the later chapters, we always put
$Q = 10^{-20}$ and no difficulties were encountered. Next, evaluate $E_{n,N}(w)$ from the recurrence formula for $C_n(w)$ with
$C_n(w)$ replaced by $E_{n,N}(w)$ for $n = N,N-1,\ldots,0$. Let

$$U_N(w) = \sum_{n=0}^{N+1} (-)^n E_{n,N}(w) ,$$
(29)

$$C_{n,N}(w) = E_{n,N}(w)/U_n(w) .$$
(30)

Then for w fixed,

$$\lim_{N\to\infty} C_{n,N}(w) = C_n(w) , \quad n = 0,1,\ldots .$$
(31)

The treatment for $G_n(w)$ and $D_n(w)$ is analogous to
that for $C_n(w)$, but there are important differences. The
general discussion around (28), (29) holds for $G_n(w)$ and
$D_n(w)$ with the understanding that the recurrence formulas
for $G_n(w)$ or $D_n(w)$ are used as appropriate, that
$t = \max(q+1,p+2)$, and that, though Q is arbitrary, it
ideally should be selected equal to, or near, the asymptotic
estimate of $G_{N+1}(w)$ or $D_{N+1}(w)$. Instead of (30), we use
the notation

$$G_{n,N}(w) = E_{n,N}(w)/U_n(w) ,$$
(32)

or

$$D_{n,N}(w) = E_{n,N}(w)/U_n(w) ,$$
(33)

as appropriate.

Then for w fixed,

$$\lim_{N\to\infty} G_{n,N}(w) = G_n(w) , \quad n = 0,1,\dots ,$$

$$|\arg w| < \pi , \tag{34}$$

$$\lim_{N\to\infty} D_{n,N}(w) = D_n(w) , \quad n = 0,1,\dots ,$$

$$0 < |\arg w| < \pi . \tag{35}$$

As a function of arg w , convergence of the backward
recursion schemes for evaluation of $G_n(w)$ and $D_n(w)$
weakens as $|\arg w| \to \pi$ and arg w → 0 , respectively, and
fails when $|\arg w| = \pi$ and arg w = 0 , respectively. In
these situations, the coefficients can still be efficiently
found provided the backward recurrence schemes are modified
as outlined by Wimp [1969,1970] and Luke [1969,1975a]. It is
sufficient to treat the case for $G_n(w)$ as that for $D_n(w)$
is analogous. The technique is as follows. Let N,M be two
large but distinct positive integers. Evaluate $G_{n,N}(w)$ and
$G_{n,M}(w)$ as outlined above. Let $W = W(G_n)$ be an operator
acting on the $G_n(w)$'s which produces a known numerical
value. For example, if we know the value of $BG_{p+1,q}^{q,1}(z)$ in
(15) when z = w , then

$$W = BG_{p+1,q}^{q,1}\left(w \left|\begin{matrix}1,a_p\\b_q\end{matrix}\right.\right) = \sum_{n=0}^{\infty} G_n(w) . \tag{36}$$

We determine μ from

$$\mu W_1 + (1-\mu)W_2 = W ,$$

$$W_1 = W(G_{n,N}) , \quad W_2 = W(G_{n,M}) \tag{37}$$

and evaluate

$$G_{n,N,M}(w) = \mu G_{n,N}(w) + (1-\mu)G_{n,M}(w) \ . \qquad (38)$$

Then, if the hypotheses of some results given by Wimp (1969)
apply, we have

$$\lim_{\substack{N,M\to\infty \\ N\neq M}} G_{n,N,M}(w) = G_n(w) \ , \ n = 0,1,\dots \ . \qquad (39)$$

This was the procedure used to get numerous coefficients in
Luke (1969,1975a). Other possible choices of W could be
based on the relations (19) or (20). Another possibility is
to evaluate $BG_{p+1,q}^{q,1}(z)$ for z sufficiently large with the
aid of (17). In practice, the hypotheses noted in the statement (39) might be difficult to confirm and some experiments
might be necessary. We have always succeeded when W was
evaluated by use of (36), but failed when W was evaluated
by use of (19) or (20) or the other possibilities as noted in
the discussion after 8(8). In the unsuccessful situations,
we found that W_1 and W_2 were much nearer to each other
than $C_{n,N}(w)$ and $C_{n,M}(w)$. Notice that if $W_1 = W_2$, any
value of μ satisfies (37) and this appears to be part of the
difficulty. In practice, it is usually best to circumvent the
evaluation of μ per se, since considerable loss of figures
occurs if W_1 is near W_2 . Alternative and preferred forms
are

$$C_{n,N,M}(w) = C_{n,N}(w) + (W-W_1)\{C_{n,N}(w) - C_{n,M}(w)\}/(W_1-W_2)$$

$$(40)$$

or the right hand side of this equation with the subscripts

N and M interchanged.

In the ensuing discussion, unless noted to the contrary, $C_n(w)$ is a generic term for any of the Chebyshev coefficients treated in this section. Since $|T_n^*(x)| \le 1$ when $0 \le x \le 1$, if $R_m(z)$ is the error when a Chebyshev expansion like (1) is truncated after m terms, then

$$|R_m(z)| \le \sum_{n=m}^{\infty} |C_n(w)| = R_m^*(z) , \qquad (41)$$

and if the $C_n(w)$'s decay with sufficient rapidity, which is usually the case, then for m sufficiently large

$$R_m^*(z) \sim |C_m(w)| [1 + \delta_m] ,$$

$$\delta_m = \left| \frac{C_{m+1}(w)}{C_m(w)} \right| . \qquad (42)$$

If indeed $C_n(w)$ is given by (2), then

$$\frac{C_{n+1}(w)}{C_n(w)} = \frac{(a_{p+n})(w/4)F}{(b_{q+n})(n+1)} [1 + 0(n^{-j})]$$

$$p < q, \; j = q+1-p, \; F = 1 \; ; \; p = q, \; j = 2, \; F = 1 \; ;$$

$$p = q+1, \; j = 1, \; F = e^{-\delta} . \qquad (43)$$

We also have

$$\frac{G_{n+1}(w)}{G_n(w)} = \left(\frac{n+1}{n}\right)^{2\gamma-1} \left\{ \exp\left(\frac{-2\Omega}{n}\right) \right\} [1 + 0(\Omega^{-2}\{1+|w|^{4\Delta}\})] . \qquad (44)$$

Clearly, the generic symbol $C_{n+1}(w)/C_n(w)$ can be conceived as a measure of the rapidity that a Chebyshev series like (1) converges. Let w be fixed. Then from (43),

convergence improves as   (q-p)   increases.   If   p = q , convergence is good.   The same is true when   p = q+1   over a
large sector of the complex   w-plane.   In this connection,
see the table giving numerical values of $|e^{-\delta}|$ noted in the
discussion following (13).   A better appreciation of these
remarks can be gleaned from the numerical examples given in
the later chapters.

It is readily seen from (44) and (23) that if $\beta = 3$
and   w   is fixed, positive and sufficiently large, say
$w \geq 5$, convergence of (15) is quite rapid.   The situation
deteriorates as β increases and as $|\arg w| \to \pi$.   We do
not give $D_{n+1}(w)/D_n(w)$, but for $\beta = 3$, convergence of (24)
is acceptable for most practical purposes when $w \geq 8$.   The
situation rapidly deteriorates as β increases, unless at the
same time we can afford to increase   w .   This discussion is
rather qualitative, but an illuminating picture of the state
of affairs is afforded by the numerical examples treated in
Chapters 8 and 9.

The size of   N   needed to achieve a given level of accuracy in the Chebyshev coefficients is rather difficult to
determine a priori.   In the absence of more exact criteria,
we suggest the following approach.   Again let $C_n(w)$ be a
generic symbol for a Chebyshev coefficient.   Choose $N \geq N^*$
such that the asymptotic estimate for $C_{N^*+2}(w)$ as found
from (2), (11)-(13), (23) or the combination (23) and (25) as
appropriate, is less than the accuracy desired by truncation
of a Chebyshev series like (1).   Thus, if we desire the magnitude of the truncation error to be less than $0.5 \cdot 10^{-r}$,
then choose N^* such that the asymptotic estimate of

$$|C_{N*+2}(w)| < 0.5 \cdot 10^{-r} \, . \tag{45}$$

Then N* is a first estimate for N . Actually, one should
choose N considerably larger than N* since the extra cost
in computation is trivial. Indeed, for the real $_2F_1$ example
treated in Chapter 4, computations were made for N = 30(5)50.
The CPU time varied from 0.60 to 0.65 seconds, and the cost
varied from $0.70 to $0.73. We recommend that coefficients be
evaluated for at least two different values of N .[†] It is
our heuristic experience that the number of significant figures the coefficients have in common for the two, or possibly
more, N values used is correct. This is especially true
for all the coefficients treated in this volume.

We recognize that acceptance of the number of common decimals in $C_n(w)$ for different N values does not guarantee
the accuracy of the coefficient. Such a procedure can be conceived as a measure of the accuracy. The procedure is valuable, and when supplemented by the additional checks noted
above, more credence is given to the validity of the coefficients. Further measures of accuracy for the errors in the
coefficients are afforded by TESTS 1 and 2. For suppose it is
believed that the error in the magnitude of each coefficient
does not exceed $\sigma \cdot 10^{-r}$ where $0 < \sigma < 1$. Then, from (9)
and (10), if ET m is the magnitude of the error in TEST m ,
m = 1,2, then

$$|ET\ 1| \le \sigma \cdot 10^{-r} \sum_{n=1}^{N+1} n^2 = (\sigma \cdot 10^{-r})(N+1)(N+2)(2N+3)/6 \, ,$$

[†] Here we assume that the backward recursion process as described by (28)-(35) need not be modified as in the discussion
surrounding (36)-(40). In the modified situation, we suggest
use of at least three different N values which provides at
least three sets of coefficients like $G_{n,N,M}(w)$.

$$|ET\ 2| \leq \sigma \cdot 10^{-r} \sum_{n=2}^{N+1} n^2 (n^2-1) = (\sigma \cdot 10^{-r}) N(N+1)(N+2)(N+3)(2N+3)/10.$$

$$(46)$$

For further comments on checks of the Chebyshev coefficients and the accuracy of the Chebyshev expansions, see 1.7.

1.4.   Efficient Evaluation of Series of Chebyshev Polynomials

We consider a more general situation. Let

$$f_n(x) = \sum_{k=0}^{n} a_k p_k(x) \ , \tag{1}$$

where $p_k(x)$ is an arbitrary function which satisfies

$$p_{n+1}(x) + \alpha_n p_n(x) + \beta_n p_{n-1}(x) = 0 \ , \ n \geq 1 \ . \tag{2}$$

Here α_n and β_n may depend on both n and x . Consider the backward recursion system

$$B_k = -\alpha_k B_{k+1} - \beta_{k+1} B_{k+2} + a_k, \ k = n, n-1, \ldots, 1, 0,$$

$$B_{n+1} = B_{n+2} = 0 \ . \tag{3}$$

Then

$$f_n(x) = B_0 p_0(x) + B_1 \{p_1(x) + \alpha_0 p_0(x)\} \ , \tag{4}$$

$$f'_n(x) = (B_0 + B_1 \alpha_0) p'_0(x) + B_1 p'_1(x)$$

$$+ B_1 \alpha'_0 p_0(x) - g_n(x) \ ,$$

$$g_n(x) = \sum_{k=0}^{n-1} b_k p_k(x) \ , \ b_k = B_{k+1} \alpha'_k + B_{k+2} \beta'_{k+1} \tag{5}$$

and $g_n(x)$ can be evaluated after the manner of evaluating $f_n(x)$. Thus, let

$$C_k = -\alpha_k C_{k+1} - \beta_{k+1} C_{k+2} + b_k \ , \quad k = n-1, n-2, \ldots, 1, 0 \ ,$$

$$C_n = C_{n+1} = 0 \ . \tag{6}$$

Then

$$f_n'(x) = (B_0 + B_1 \alpha_0) p_0'(x) + B_1 p_1'(x) + B_1 \alpha_0' p_0(x)$$

$$- C_0 p_0(x) - C_1 \{p_1(x) + \alpha_0 p_0(x)\} \ . \tag{7}$$

Computations for the case of Chebyshev polynomials of the first kind can be summarized as follows. Let $f_n(x)$ be given by (1) and the recursion formula for $p_n(x)$ be given by (2) with $\beta_k = 1$ for all k . Evaluate B_k and C_k as indicated in (3) and (6), respectively, with $\beta_k = 1$ for all k . Here $b_k = B_{k+1} \alpha_k'$. Other pertinent data are described in the following table.

$p_n(x)$	α_n	b_n	$f_n(x)$	$f_n'(x)$
$T_n(x)$	$-2x$	$-2B_{n+1}$	$B_0 - xB_1$	$-B_1 - C_0 + xC_1$
$T_{2n}(x)$	$-2(2x^2-1)$	$-8xB_{n+1}$	$B_0 - (2x^2-1)B_1$	$-4xB_1 - C_0 + (2x^2-1)C_1$
$T_{2n+1}(x)$	$-2(2x^2-1)$	$-8xB_{n+1}$	$x(B_0 - B_1)$	$B_0 - B_1 - x(C_0 - C_1)$
$T_n^*(x)$	$-2(2x-1)$	$-4B_{n+1}$	$B_0 - (2x-1)B_1$	$-2B_1 - C_0 + (2x-1)C_1$

$$\tag{8}$$

As remarked, the general formulation only requires that $p_k(x)$ satisfy (2). However, in practice, the scheme cannot always be used indiscriminately because of the possible growth of round-off errors.

Let $p_k(x)$ be a Chebyshev polynomial of the first kind, and let $f_n^*(x)$ be the computed value of f_n . Suppose the numerics are done in fixed point arithmetic to t decimal places. Suppose that all round-off errors which occur at any stage of the computation do not exceed $\frac{1}{2} \cdot 10^{-t}$ in magnitude. Then it can be shown that

$$|f_n - f_n^*| \leq E_n = \frac{1}{2} \cdot 10^{-t} \left((2n+3) + \sum_{k=1}^{n} k^2 |a_k| \right) \quad . \tag{9}$$

If the a_k's decrease with sufficient rapidity, which is certainly the case for all the Chebyshev series considered in this tome, and if n is sufficiently large, then

$$E_n \sim \frac{1}{2} \cdot 10^{-t}(2n+3) \tag{10}$$

and round-off is not a serious problem. If (1) converges slowly, for example, if $a_k = O(k^{-2})$, and if there is a constant A such that $|a_k| \leq A/k^2$ for all $k \geq 1$, then

$$E_n \leq \frac{1}{2} \cdot 10^{-t}\{(A+2)n+3\} \tag{11}$$

and again round-off effects are easily controlled. For the functions treated in this volume, the a_k's decay exponentially and (10) applies. Suppose the a_k's are accurate to 20 decimals and all round-off errors in magnitude do not exceed $\frac{1}{2} \cdot 10^{-20}$. Let $n = 30$. Then the final result f_n should be accurate to better than 18 decimals.

```
C    ----------- IBM S/370 ----------- MULTIPLE PRECISION -------------
C
C    ***************************************************************
C    *THIS FUNCTION SUBPROGRAM EVALUATES SERIES IN VARIOUS CHEBYSHEV  *
C    *POLYNOMIALS.                                                    *
C    *                                                               *
C    *DESCRIPTION OF VARIABLES.                                       *
C    *                                                               *
C    *N      -INPUT  - THE NUMBER OF COEFFICIENTS IN THE SERIES.      *
C    *                                                               *
C    *A      -INPUT  - A VECTOR CONTAINING THE COEFFICIENTS IN THE    *
C    *                 SERIES.  DIMENSION OF THE VECTOR IS SET AT 100. *
C    *                                                               *
C    *X      -INPUT  - THE ARGUMENT OF THE SERIES.                    *
C    *                                                               *
C    *LOG1,  -INPUT  - LOGICAL VARIABLES USED TO SPECIFY THE TYPE OF   *
C    *LOG2               CHEBYSHEV POLYNOMIAL.  LOG1 AND LOG2 ARE TO BE *
C    *                   ENTERED ACCORDING TO THE FOLLOWING TABLE.     *
C    *                                                               *
C    *                   TYPE       LOG1     LOG2                     *
C    *                   --------------------------                   *
C    *                   REGULAR    .TRUE.   .TRUE.                    *
C    *                   SHIFTED    .TRUE.   .FALSE.                   *
C    *                   EVEN       .FALSE.  .TRUE.                    *
C    *                   ODD        .FALSE.  .FALSE.                   *
C    *EVAL   -OUTPUT - THE VALUE OF THE SERIES.                        *
C    *                                                               *
C    *ALL OTHER VARIABLES ARE FOR INTERNAL USE.                       *
C    ***************************************************************
C          FUNCTION EVAL(N,A,X,LOG1,LOG2)
           IMPLICIT REAL*16 (A-H,O-Z)
           LOGICAL*4 LOG1,LOG2
           DIMENSION A(100),B(100)
           DATA ZERO,ONE,TWO/0.Q0,1.Q0,2.Q0/
           XFAC=TWO*(TWO*X-ONE)
           IF(LOG1.AND.LOG2) XFAC=TWO*X
           IF(.NOT.LOG1) XFAC=TWO*(TWO*X*X-ONE)
           N1=N+1
           B(N1)=ZERO
           B(N1+1)=ZERO
           DO 3 I=1,N
    3      B(N1-I)=XFAC*B(N1+1-I)-B(N1+2-I)+A(N1-I)
           EVAL=B(1)-XFAC*B(2)/TWO
           IF(.NOT.LOG1.AND..NOT.LOG2) EVAL=X*(B(1)-B(2))
           RETURN
           END
```

1.5. Rational Approximations for Generalized Hypergeometric Functions

For convenience, we prefer to deal with $H(-z)$ as defined in 1.2(9), which we now notate as $F(z)$. We write

$$F(z) = {}_pF_q(\alpha_p; \rho_q; -z) = F_n(z) + Q_n(z) , \tag{1}$$

$$F_n(z) = A_n(z)/B_n(z) , \tag{2}$$

$$B_n(z) = L_n z^n \; {}_{q+f+3}F_{p+g+1}\left(\begin{array}{c}-n,n+\lambda,\rho_q-a,c_f,1\\ \beta+1,\alpha_p+1-a,d_g\end{array}\middle| -1/z\right) \;, \qquad (3)$$

$$A_n(z) = \left[\frac{n(n+\lambda)(\rho_q-1)c_f}{(\beta+1)\alpha_p d_g}\right]^a L_n z^{n-a}$$

$$\times \sum_{k=0}^{n-a} \frac{(a-n)_k(n+\lambda+a)_k(\alpha_p)_k(c_f+a)_k}{(\beta+1+a)_k(\alpha_p+1)_k(d_g+a)_k k!}$$

$$\times \; {}_{q+f+3}F_{p+g+1}\left(\begin{array}{c}-n+a+k,n+\lambda+a+k,\rho_q+k,c_f+a+k,1\\ \beta+1+a+k,\alpha_p+1+k,d_g+a+k\end{array}\middle| -1/z\right) \;, \quad (4)$$

or

$$A_n(z) = \left[\frac{n(n+\lambda)(\rho_q-1)c_f}{(\beta+1)\alpha_p d_g}\right]^a L_n z^{n-a}$$

$$\times \sum_{k=0}^{n-a} \frac{(a-n)_k(n+\lambda+a)_k(\rho_q)_k(c_f+a)_k(-)^k z^{-k}}{(\beta+1+a)_k(\alpha_p+1)_k(d_g+a)_k}$$

$$\times \; {}_{p+q+f+2}F_{p+q+g+1}\left(\begin{array}{c}-n+a+k,n+\lambda+a+k,\rho_q+k,c_f+a+k,\alpha_p\\ \beta+1+a+k,\alpha_p+1+k,d_g+a+k,\rho_q\end{array}\middle| 1\right) \;,$$

$$(5)$$

where

$$L_n = \frac{(\beta+1)_n(\alpha_p+1-a)_n(d_g)_n}{(n+\lambda)_n n!(\rho_q-a)_n(c_f)_n} \;, \quad a = 0 \quad \text{or} \quad a = 1, \; \lambda = \alpha+\beta+1 \;.$$

$$(6)$$

Here $Q_n(z)$ is the remainder or the error in the rational approximation, and is usually expressed in the form

$$Q_n(z) = S_n(z)/B_n(z) \;. \qquad (7)$$

The parameters α and β are arbitrary so long as each is > -1 . We usually take $\alpha = \beta = 0$, for in this instance both

$A_n(z)$ and $B_n(z)$ satisfy the same finite homogeneous recursion formula. Notice that if $\beta = 0$, the $_{q+f+3}F_{p+g+1}$ in (3) becomes a $_{q+f+2}F_{p+g}$. Also, we usually put $f = g = 0$.

We suppose that if a numerator parameter α_h and a denominator parameter ρ_h are equal, then these parameters are cancelled before equations (1)-(7) are written. This will insure that all hypergeometric forms are the lowest order possible. If $\alpha_h = \rho_h$ and we let $\alpha_h \to \infty$, then our forms will be of the lowest order possible. In effect, we can assume that no numerator parameter coaslesces with a denominator parameter.

For proofs of convergence of the rational approximations, error estimates, recursion formulas for $A_n(z)$ and $B_n(z)$, and other remarks, see Luke (1969,1975a) and Fields (1972). See also the remarks following equation (21) of this section.

The notation $A_n(z)$ and $B_n(z)$ is not the same as that employed by Luke (1969,1975a). In Luke (1975a) replace z by $-z$ in $A_n(z)$ and multiply the result by z^n to get the $A_n(z)$ defined here. The same applies with $A_n(z)$ replaced by $B_n(z)$.

Recursion formulas for the numerator and denominator polynomials can be deduced from Luke (1969,1975a). We omit the general formulas in this section, but we do cite the recursion formulas for the explicit functions treated in later chapters. The recursion formula can be used to generate the coefficients which define the polynomials, though we have found this less efficient than a scheme to be shortly considered. Given an explicit value of z , the recursion formula is valuable to generate higher order approximants. In a number of cases, we can prove that such a computational scheme is

stable.   In those instances where proofs are not available,
extensive heuristic tests serve to confirm stability.

In the absence of or in lieu of asymptotic error analyses, generation of approximants by use of the recursion formulas for the polynomials can be used to determine heuristi-
cally the error as a function of the order of the approximation.   The ideas are as follows. Suppose that $A_n(z)$ and
$B_n(z)$ satisfy the same homogeneous recursion formula of order
t , that is one composed of (t+1) terms. We evaluate $A_n(z)$
and $B_n(z)$ from (3) or (4) (or by equivalent representations)
for n = 0,1,...,t-1 , and then compute $A_n(z)$ and $B_n(z)$
for n = t,t+1,...,N by use of the recursion formula. For
each n = 0,1,...,N , we compute $F_n(z) = A_n(z)/B_n(z)$. In
all of our examples, for z fixed and suitably restricted,
we know that

$$\lim_{n \to \infty} F_n(z) = F(z) . \qquad (8)$$

Thus for N sufficiently large, we accept the number of common decimals in $F_N(z)$ and $F_{N-1}(z)$ as correct, assume
$F_N(z)$ is correct to the number of common places, and approximate the error in $F_n(z)$ by $F_N(z) - F_n(z)$. Thus, in the
programs given in the later chapters, the machine prints
$A_n(z)$, $B_n(z)$ and $F_n(z)$ for n = 0,1,...,N, $F_n(z) - F_{n-1}(z)$,
that is first differences, for n = 1,2,...,N and $F_N(z)$
- $F_n(z)$, n = 0,1,...,N-1 . We recognize that acceptance of
common decimals does not guarantee such accuracy, but it may
be viewed as a measure of the accuracy. Nonetheless, the procedure is valuable, and when supplemented by other checks as
noted in 1.7, more credence is given to the error analysis.
Fortunately, in all the cases treated in the later chapters,

asymptotic error analyses are always available.

Suppose that an approximant of order n will give the desired accuracy for a needed range of z values, and that functional values are needed for many z values. Then it may be more economical to get the functional values by direct evaluation of the numerator and denominator polynomials. Thus in the later chapters, machine programs are developed to give the coefficients which define these polynomials. In some applications, as in the approximation of integral transforms, the polynomials themselves are needed rather than their numerical values for given values of z . We now turn to evaluation of the coefficients of the powers of z in (3).

It is readily verified by use of 1.2(4) that

$$B_n(z) = {}_{p+g+1}F^n_{q+f+1}\left(\begin{matrix} -n-\beta,-n-\alpha_p+a,1-n-d_g \\ 1-\lambda-2n,1-n-\rho_q+a,1-n-c_f \end{matrix}\middle| (-)^r z \right)$$

$$= \sum_{k=0}^{n} u_k z^k, \quad u_0 = 1, \quad r = p+q+f+g . \tag{9}$$

Then the u_k's are readily evaluated by use of the recurrence formula

$$u_{k+1} = \frac{(-)^r(-n-\beta+k)(-n-\alpha_p+a+k)(1-n-d_g+k)u_k}{(1-\lambda-2n+k)(1-n-\rho_q+a+k)(1-n-c_f+k)(k+1)} . \tag{10}$$

For machine computation, neither of the forms (4) or (5) is satisfactory. From (5), we can show that

$$A_n(z) = \sum_{k=0}^{n-a} v_k z^k, \quad v_k = \sum_{m=0}^{k} t_m u_{k-m}, \quad t_m = \frac{(-)^m(\alpha_p)_m}{(\rho_q)_m m!} , \tag{11}$$

where the u_k's are defined by (9) and are easily evaluated by (10). Since

$$t_{m+1} = \frac{-(\alpha_p+m)}{(\rho_q+m)(m+1)}\, t_m \ , \quad t_0 = 1 \ , \tag{12}$$

we see that computation of v_k is direct.

Next we turn to rational approximations for $F(1/z)$, or in the notation of 1.2(9) for $H(-1/z)$. We write

$$H(-1/z) = \{\Gamma(\rho_q)/\Gamma(\alpha_p)\}\ G_{p,q+1}^{1,p}\left(\frac{1}{z}\ \middle|\ \begin{matrix}1-\alpha_p\\0,1-\rho_q\end{matrix}\right) \tag{13}$$

$$= \{\Gamma(\rho_q)/\Gamma(\alpha_p)\}\ G_{q+1,p}^{p,1}\left(z\ \middle|\ \begin{matrix}1,\rho_q\\\alpha_p\end{matrix}\right)$$

$$= H_n(-1/z) + Q_n^*(z), \quad H_n(-1/z) = A_n^*(z)/B_n^*(z), \tag{14}$$

$$B_n^*(z) = {}_{q+f+3}F_{p+g+1}\left(\begin{matrix}-n,n+\lambda,\rho_q-a,c_f,1\\\beta+1,\alpha_p+1-a,d_g\end{matrix}\ \middle|\ -z\right) \ , \tag{15}$$

$$A_n^*(z) = \left[\frac{n(n+\lambda)(\rho_q-1)c_f z}{(\beta+1)\alpha_p d_g}\right]^a$$

$$\times \sum_{k=0}^{n-a}\frac{(a-n)_k(n+\lambda+a)_k(\alpha_p)_k(c_f+a)_k}{(\beta+1+a)_k(\alpha_p+1)_k(d_g+a)_k k!}$$

$$\times {}_{q+f+3}F_{p+g+1}\left(\begin{matrix}-n+a+k,n+\lambda+a+k,\rho_q+k,c_f+a+k,1\\\beta+1+a+k,\alpha_p+1+k,d_g+a+k\end{matrix}\ \middle|\ -z\right) . \tag{16}$$

Notice that

$$B_n^*(z) = (z^n/L_n)B_n(1/z), \quad A_n^*(z) = (z^n/L_n)A_n(1/z) \tag{17}$$

where $B_n(z)$ and $A_n(z)$ are given by (3) and (4), respectively. Another form for $A_n^*(z)$ follows from (5) and (17).

This we omit.   From (7), we have

$$Q_n^*(z) = Q_n(1/z) .\tag{18}$$

If we write

$$B_n^*(z) = \sum_{k=0}^{n} u_k^* z^k, \; u_0^* = 1 ,\tag{19}$$

then

$$u_{k+1}^* = - \frac{(k-n)(n+\lambda+k)(\rho_q-a+k)(c_f+k)u_k^*}{(\beta+1+k)(\alpha_p+1-a+k)(d_g+k)} ,\tag{20}$$

whence the u_k^*'s are readily found.   For the evaluation of
the coefficients in $A_n^*(z)$, we have

$$A_n^*(z) = \sum_{k=0}^{n-a} v_k^* z^{k+a}, \; v_k^* = \sum_{m=0}^{n-a-k} t_m u_{m+k+a}^* ,\tag{21}$$

where t_m is given by (11).   See also (12).

We now give some further remarks on the nature of the
term $S_n(z)$ which enters in the remainder, see (7).   If in the
formulation (13-16), $\beta = 0$, $f = g = 0$, $\lambda + a$ is any one of the
numbers $0,1,...,p$ and $p \geq q + 2$, a closed form expression for
$S_n(1/z)$ has been given by Fields (1972).   We do not develop
this here, though such is given for the special case treated
in Chapter 20.   It would appear that a similar situation
should prevail for the formulation (1-5) which allows for $p <$
$q + 2$.   Indeed, this so, and such has been given by Fields
(1976) in a communique to the author.   This we present under
the same assumptions noted above for (13-16) except that $p \leq q$
or $p = q + 1$ and $|\arg(1+z)| < \pi$.   Note that the assumptions on
a and λ are more liberal than the usual cases of interest,
namely $a = 0$ or $a = 1$ and $\lambda = 1$.   Under the general conditions
noted, and with $\nu = q + 2 - p \geq 1$, we have

$$S_n(z) = \frac{(-)^{a\nu}\Gamma(\rho_q)\Gamma(1-\rho_q)\Gamma(\alpha_p+1-a)}{\Gamma(\rho_q-a)\Gamma(-\alpha_p)\Gamma(\alpha_p+1)} \quad {}_pF_q\left(\begin{matrix}\alpha_p\\ \rho_q\end{matrix}\middle| -z\right)$$

$$\times\frac{n!\,\Gamma(n+\lambda+a-\alpha_p)\{(-)^{\nu+1}z\}^{n+\lambda}}{\Gamma(n+\lambda+a+1-\rho_q)(2n+\lambda)!} \quad {}_{p+2}F_{q+1}\left(\begin{matrix}n+\lambda,n+\lambda+a-\alpha_p\\ 2n+\lambda+1,n+\lambda+a+1-\rho_q\end{matrix}\middle| -z\right)$$

$$\qquad\qquad\qquad\qquad\qquad\qquad\qquad\qquad\qquad\qquad\qquad\qquad (22)$$

$$+\sum_{k=1}^{q}\frac{\Gamma(\alpha_p+1-a)\Gamma(\rho_q)\Gamma(\rho_k-a)(-z)^{1-a}}{\Gamma(\alpha_p)(\rho_q-\rho_k)^*\Gamma(2-\rho_k)\Gamma(\rho_q-a)} \quad {}_pF_q\left(\begin{matrix}\alpha_p+1-\rho_k\\ \rho_q+1-\rho_k^*,2-\rho_k\end{matrix}\middle| -z\right)$$

$$\times\frac{n!\,\Gamma(n+\lambda+a-\rho_k)}{(n+\lambda-1)!\,\Gamma(n+1+\rho_k-a)} \quad {}_{p+1}F_{q+1}\left(\begin{matrix}\rho_k-a,\rho_k-\alpha_p\\ 1+\rho_k-\rho_q^*,1-n-\lambda-a+\rho_k,n+1-a+\rho_k\end{matrix}\middle| -z\right).$$

Here the asterisk (*) sign has the same meaning as in 1.2(5). Asymptotic representations for $S_n(z)$, $B_n(z)$ and so for the error $Q_n(z)$, see (7), can be deduced from data in Luke (1969).

1.6.   The Padé Table

Consider the at least formal power series

$$F(z) = \sum_{r=0}^{\infty} c_r z^r, \quad c_0 \neq 0 , \qquad\qquad (1)$$

and the doubly infinite array of rational functions

$$F_{mn}(z) = A_{mn}(z)/B_{mn}(z) , \qquad\qquad (2)$$

$$A_{mn}(z) = \sum_{r=0}^{m} a_r z^r, \quad B_{mn}(z) = \sum_{r=0}^{n} b_r z^r , \qquad\qquad (3)$$

determined in such a way that the power series representation of $F_{mn}(z)$ agrees with that of $F(z)$ for as high a power of z as possible. This uniquely determined rational function is called a Padé approximant to $F(z)$ and is said to occupy the (n,m) position in the Padé table. The Padé table can be

conceived as a doubly infinite matrix (F_{mn}), $m,n = 0,1,\ldots$.
Here m and n designate columns and rows, respectively.
Evidently the elements in the first row $(n = 0)$ are the partial sums of the series for $F(z)$, while the elements in the
first column $(m = 0)$ are the partial sums of the series for
$1/F(z)$. The positions (n,n) fill the main diagonal, while
the positions $(n,n-r)$ and $(n,n+r)$ fill the r-th subdiagonal and r-th superdiagonal, respectively.

Let us write

$$B_{mn}(z)F(z) - A_{mn}(z) = z^{m+n+1} G(z), \; G(0) \neq 0 . \qquad (4)$$

We have the system of linear equations

$$\sum_{r=0}^{k} b_r c_{k-r} = a_k, \; k = 0(1)m , \qquad (5)$$

$$\sum_{r=0}^{n} b_r c_{k-r} = 0, \; k = m+1(1)m+n \qquad (6)$$

where it is understood that $c_r = 0$ for $r < 0$. The polynomials $A_{mn}(z)$ and $B_{mn}(z)$ can be taken as relatively
prime. Also, without loss of generality, we can take $b_0 = 1$
whence $a_0 = c_0$. In this event, the approximation F_{mn} is
said to be reduced. Clearly, a_r and b_r depend on m and
n, but we suppress this notation-wise in the interests of
simplicity. Under appropriate conditions on the c_r's, the
b_r's can be found by solution of the linear equation system
(6) and the a_k's follow from (5). In only a few cases are
the coefficients known in closed form. For further details
on Padé approximants, see Luke (1969,1975a) and the references
given there. In these references, the roles of n and m as
in (n,m) are interchanged. Also in the latter reference

there are a few (obvious) typographical errors.

1.7. Computations of and Checks on Coefficients and Tables

Some discussion on checks for Chebyshev coefficients and
the accuracy of Chebyshev expansions have already been given
in 1.3. The matter is further explored here. Though many of
our comments are stated in terms of Chebyshev expansions, it
must be borne in mind that the ideas can be frequently applied
to check rational approximations as well.

For all the functions treated in later chapters, asymptotic estimates are provided for the Chebyshev coefficients
and the errors in the rational approximations. These are
quite valuable for purposes of planning and checking. The
asymptotic estimates are always given with order terms, but
pragmatic bounds for these terms are not given. Of course, in
the numerical examples, the order terms are neglected. These
examples show that the asymptotic forms are remarkably accurate even for small values of n . Even so, the values are
accurate to at best several, say three or four, significant
figures. In the material which follows, we concentrate on
ideas to check coefficients and errors to the full number of
significant figures provided.

If a function is of hypergeometric type, the coefficients
are also of hypergeometric type. In this event, we usually
recommend that the coefficients be produced by use of the
known closed form recurrence relations as previously described. Where possible, particular coefficients should be
checked by evaluation of their hypergeometric series representations. For many functions, coefficients can also be developed by using the results in Chapter 10. In illustration,

from the coefficients for $e^{\pm z}$ valid for $0 \le z/w \le 1$, coefficients for the functions

$$\int_0^z t^{-1}(1-e^{\pm t})dt = \mp z \; {}_2F_2\left(\begin{array}{c}1,1\\2,2\end{array}\middle|\pm z\right) \tag{1}$$

follow from 10.1(8). More generally, coefficients for

$$\int_0^z t^{-\nu}(1-e^{\pm t})dt = \mp z^{2-\nu}(1-\nu) \; {}_2F_2\left(\begin{array}{c}1,1-\nu\\2,2-\nu\end{array}\middle|\pm z\right) \; ,$$

$$R(\nu) < 2 \; , \tag{2}$$

$$\int_0^z t^{-\nu}e^{\pm t}dt = (1-\nu)^{-1}z^{1-\nu} \; {}_1F_1(a;a+1;\pm z)$$

$$= (1-\nu)^{-1}z^{1-\nu}e^{\pm z} \; {}_1F_1(1;a+1;\mp z) \; ,$$

$$R(\nu) < 1 \; , \tag{3}$$

follow from results in 11.2. The scheme outlined in 11.3 is
very valuable for applications. For example, from 2.2(50)
and Chapter 8, we can determine the coefficients in

$$K_\nu(z) = (\pi/2z)^{\frac{1}{2}}e^{-z} \sum_{n=0}^{\infty} G_n(w)T_n^*(w/z), \; w/z \ge 1 \; . \tag{4}$$

Then, by application of 11.3, we can readily deduce the coefficients in

$$(2/\pi)^{\frac{1}{2}}e^{bz}z^{-u} \int_z^{\infty} e^{-(b-1)t}t^{u+\frac{1}{2}}K_\nu(t)dt$$

$$= \sum_{n=0}^{\infty} F_n(w)T_n^*(w/z), \; w \ne 0, \; w/z \ge 1 \; ,$$

$$R(b) > 0 \quad \text{or} \quad R(b) = 0 \quad \text{and} \quad R(u) < -1 \; . \tag{5}$$

For nonhypergeometric functions, evaluation of the Chebyshev coefficients can often be accomplished from a knowledge
of the Taylor series expansions for these functions with the

aid of the basic series or rearrangement analysis developed
in Chapter 12.   For example, Chebyshev coefficients for
$\{\Gamma(z+1)\}^{-1}$ in Luke (1969,1975a) were found in this manner.
This technique was also applied in the cited references to
get coefficients for those portions of the ascending series
for $Y_n(z)$ and $K_n(z)$ not attached to $J_n(z)$ and $I_n(z)$,
respectively, n = 0,1 .   For   n = 0 , these coefficients can
also be found with the help of a recursion formula in the
backward direction.   The idea is much akin to that used in
11.4.   From 2.2(30) and Chapter 6, we can write

$$I_\nu(z) = \frac{(z/2)^\nu}{\Gamma(\nu+1)} \sum_{n=0}^{\infty} C_n(w^2/4)T_{2n}(z/w), \ 0 < z/w \le 1 , \quad (6)$$

where $C_n(w^2/4)$, of course, depends on ν .   Now

$$\partial I_\nu(z)/\partial\nu = [\ln(z/2) - \psi(\nu+1)]I_\nu(z)$$

$$+ \frac{(z/2)^\nu}{\Gamma(\nu+1)} \sum_{n=0}^{\infty} A_n(w^2/4)T_{2n}(z/w) ,$$

$$A_n(w^2/4) = \partial C_n(w^2/4)/\partial\nu . \quad (7)$$

Differentiate the known recursion formula for $C_n(w^2/4)$ as
given by 6(4) with respect to ν .   This yields an inhomo-
geneous recursion formula for the coefficients $A_n(w^2/4)$
which depends on the coefficients $C_n(w^2/4)$.   The backward
recurrence scheme for the determination of $A_n(w^2/4)$ can be
developed after the manner of the discussion in 11.4.   For
complete details, see Luke (1969,1975a).   To complete our
remarks concerning the case of Bessel functions for   n = 0 ,
note that

$$K_0(z) = -\{\partial I_\nu(z)/\partial \nu\}_{\nu=0} \ , \ Y_0(z) = (2/\pi)\{\partial J_\nu(z)/\partial \nu\}_{\nu=0} \ .$$

$$(8)$$

For certain functions, special techniques are called for. This is difficult to discuss in general. For some examples, see the developments in Luke (1969) concerning $\ln \Gamma(x+3)$ and its derivatives, $\tan \pi x/4$ and $\cot \pi x/2$.

For each set of coefficients, checks should be made by evaluating the Chebyshev series for specific arguments and comparing with known tabular values or by computing tabular values anew from power series and asymptotic series. If a function has both a Chebyshev series expansion and a rational function representation, then tabular data deduced from each representation afford an excellent appraisal of the accuracy.

Frequently, coefficients, sums of coefficients, and rational approximations can be related to transcendents known to many decimal places. In illustration, if

$$\cos x = \sum_{n=0}^{\infty} c_n T_{2n}(4x/\pi), \ 0 \le x \le \pi/4 \ , \tag{9}$$

then

$$\sum_{n=0}^{\infty} c_n = 2^{-\frac{1}{2}} \ . \tag{10}$$

Now integrate (9) from 0 to $\pi/4$. Then, with the aid of 10.2(6), we find

$$- \sum_{n=0}^{\infty} \frac{c_n}{4n^2-1} = 2^{3/2}/\pi \ . \tag{11}$$

For another example, let

$$x^{-1} \ln(1+x)dx = \sum_{n=0}^{\infty} b_n T_n^*(x), \ 0 \le x \le 1 \ . \tag{12}$$

Then

$$\sum_{n=0}^{\infty} b_n = \ln 2, \quad b_0 = 2(2^{\frac{1}{2}}-1) \ . \tag{13}$$

We also have the dilogarithm function

$$L(x) = \int_0^x t^{-1} \ln(1+t)dt = \sum_{n=0}^{\infty} a_n T_n^*(x), \quad 0 \le x \le 1 \tag{14}$$

and the a_n's are readily expressed in terms of the b_n's in view of 10.1(7). It is known that $L(1) = \pi^2/12$. So we must have

$$-\sum_{n=0}^{\infty} \frac{b_{2n}}{4n^2-1} = \pi^2/12 \ . \tag{15}$$

Another type of check can be described as follows. Suppose we develop coefficients for

$$I_\nu(z) = \frac{(z/2)^\nu}{\Gamma(\nu+1)} \sum_{n=0}^{\infty} C_n(w_1,\nu)T_{2n}(z/w_1), \quad 0 < z/w_1 \le 1, \tag{16}$$

and

$$K_\nu(z) = (\pi z/2)^{\frac{1}{2}}e^{-z} \sum_{n=0}^{\infty} G_n(w_2,\nu)T_n^*(w_2/z), \quad z/w_2 \ge 1 \ . \tag{17}$$

Now

$$K_\nu(z) = (\pi/2)(\csc \nu\pi) [I_{-\nu}(z)-I_\nu(z)] \ . \tag{18}$$

Take, for example, $w_1 = 8$, $w_2 = 5$. Then calculations in the region $5 \le z \le 8$ provide checks on the coefficients, $C_n(w_1,\pm\nu)$, $G_n(w_2,\nu)$.

We now consider yet another technique for checking coefficients. From 1.3(1,2), we have

$$_pF_q\left(\begin{matrix} a_p \\ b_q \end{matrix} \middle| z\right) = \sum_{n=0}^{\infty} C_n(w)T_n^*(z/w), \quad z/w \le 1 \ , \tag{19}$$

and in particular

$$C_0(w) = {}_{p+1}F_{q+1} \left(\begin{array}{c} \frac{1}{2}, a_p \\ 1, b_q \end{array} \middle| w \right) \quad , \tag{20}$$

$$C_0(w) = {}_pF_q \left(\begin{array}{c} \frac{1}{2}, a_{p-1} \\ b_q \end{array} \middle| w \right) \quad \text{if} \quad a_h = 1 \quad \text{for} \quad h = p \ ,$$

$$C_0(w) = {}_pF_q \left(\begin{array}{c} a_p \\ 1, b_{q-1} \end{array} \middle| w \right) \quad \text{if} \quad b_h = \frac{1}{2} \quad \text{for} \quad h = q \ .$$

$$\tag{21}$$

For either of these last two ${}_pF_q$'s , by use of 1.3(1,2), we can generate coefficients which are analogous to $C_n(w)$, call them $C_n^*(w)$. Then

$$C_0(w) = \sum_{n=0}^{\infty} C_n^*(w) \ . \tag{22}$$

Notice that the recursion formulas for $C_n(w)$ and $C_n^*(w)$ are of the same length and have the same parent.

The same idea also applies to the expansion of a G-function in series of Chebyshev polynomials as in 1.3(15). Thus

$$BG_{p+1,q}^{q,1} \left(z \middle| \begin{array}{c} 1, a_p \\ b_q \end{array} \right) = \sum_{n=0}^{\infty} G_n(w) T_n^*(w/z), \quad z/w \geq 1 \ , \tag{23}$$

and in particular

$$G_0(w) = \pi^{-\frac{1}{2}} B \ G_{p+2,q+1}^{q+1,1} \left(w \middle| \begin{array}{c} 1, a_p, 1 \\ \frac{1}{2}, b_q \end{array} \right) \quad , \tag{24}$$

$$G_0(w) = \frac{\Gamma(a_{p-1})}{\Gamma(b_q)} \ G_{p+1,q}^{q,1} \left(w \middle| \begin{array}{c} 1, a_{p-1}, 1 \\ b_q \end{array} \right) \quad ,$$

$$\text{if} \quad a_h = \frac{1}{2} \quad \text{for} \quad h = p \ ,$$

$$G_0(w) = \pi^{-\frac{1}{2}} B \; G_{p+1,q}^{q,1} \left(w \left| \begin{matrix} 1, a_p \\ \frac{1}{2}, b_{q-1} \end{matrix} \right. \right) \; ,$$

$$\text{if} \quad b_h = 1 \quad \text{for} \quad h = q \; . \tag{25}$$

For either of these last two G-functions, we can generate co-efficients which correspond to $G_n(w)$, call them $G_n^*(w)$. Then

$$G_0(w) = \sum_{n=0}^{\infty} G_n^*(w) \; . \tag{26}$$

Again the recursion formulas for $G_n(w)$ and $G_n^*(w)$ are of the same length and have the same parent.

Further checks can often be developed by use of special properties of the functions treated. In illustration, for Bessel functions, one can employ recurrence relations and Wronskians.

1.8.   Tables of the Functions $e^{-\xi}$, $e^{-\zeta}$

In a number of approximations developed in this volume, estimation of the error is facilitated by having a numerical table of absolute values for two exponential functions defined as follows.

$$e^{-\xi} = 2z + 1 \mp 2(z^2 + z)^{\frac{1}{2}} \tag{1}$$

where the sign is chosen so that $|e^{-\xi}| < 1$, which is possible for all $z, z \neq -1$, $|\arg(1 + 1/z)| < \pi$. If $-1 \leq z \leq 0$, $|e^{-\xi}| = 1$. We also have need for the form (1) with z replaced by $1/z$. We define

$$e^{-\zeta} = [z + 2 \mp 2(z+1)^{\frac{1}{2}}]/z \; , \tag{2}$$

where the sign is chosen so that $|e^{-\zeta}| < 1$. This is possible for all $z, z \neq -1$, $|\arg(1+z)| < \pi$. If $z \leq -1$, $|e^{-\zeta}| = 1$.

VALUES OF EXP(-XI) FOR Z EQUAL R*EXP(I*THETA)

THETA(DEGREES)

R	0.0	10.0	20.0	30.0	40.0	45.0	50.0	60.0	70.0	80.0	90.0	100.0	110.0	120.0
0.10	0.5367	0.5376	0.5412	0.5469	0.5549	0.5598	0.5654	0.5785	0.5943	0.6130	0.6349	0.6600	0.6887	0.7211
0.20	0.4202	0.4213	0.4246	0.4302	0.4381	0.4430	0.4486	0.4618	0.4781	0.4978	0.5213	0.5491	0.5817	0.6199
0.30	0.3510	0.3520	0.3550	0.3601	0.3673	0.3716	0.3770	0.3893	0.4046	0.4233	0.4460	0.4734	0.5064	0.5459
0.40	0.3033	0.3042	0.3069	0.3114	0.3180	0.3220	0.3267	0.3378	0.3518	0.3689	0.3900	0.4158	0.4473	0.4860
0.50	0.2679	0.2687	0.2711	0.2752	0.2810	0.2846	0.2888	0.2988	0.3113	0.3269	0.3460	0.3696	0.3989	0.4354
0.60	0.2404	0.2411	0.2432	0.2469	0.2521	0.2553	0.2590	0.2679	0.2792	0.2931	0.3104	0.3317	0.3583	0.3919
0.70	0.2183	0.2189	0.2208	0.2240	0.2267	0.2316	0.2349	0.2429	0.2530	0.2654	0.2809	0.3000	0.3239	0.3542
0.80	0.2000	0.2006	0.2023	0.2052	0.2094	0.2120	0.2150	0.2221	0.2311	0.2423	0.2561	0.2731	0.2944	0.3214
0.90	0.1847	0.1852	0.1867	0.1893	0.1931	0.1955	0.1981	0.2046	0.2127	0.2227	0.2350	0.2502	0.2691	0.2928
1.00	0.1716	0.1720	0.1734	0.1758	0.1792	0.1814	0.1838	0.1896	0.1969	0.2058	0.2168	0.2304	0.2471	0.2679
1.10	0.1603	0.1607	0.1620	0.1641	0.1672	0.1692	0.1713	0.1766	0.1832	0.1913	0.2011	0.2132	0.2280	0.2463
1.20	0.1504	0.1508	0.1519	0.1539	0.1567	0.1585	0.1605	0.1653	0.1712	0.1785	0.1874	0.1982	0.2113	0.2273
1.30	0.1417	0.1420	0.1431	0.1449	0.1475	0.1491	0.1509	0.1553	0.1607	0.1673	0.1753	0.1849	0.1966	0.2106
1.40	0.1339	0.1343	0.1352	0.1369	0.1393	0.1408	0.1424	0.1464	0.1513	0.1573	0.1646	0.1732	0.1836	0.1960
1.50	0.1270	0.1273	0.1282	0.1298	0.1319	0.1333	0.1348	0.1385	0.1430	0.1485	0.1550	0.1628	0.1721	0.1831
1.60	0.1208	0.1211	0.1219	0.1233	0.1253	0.1266	0.1280	0.1314	0.1355	0.1405	0.1465	0.1535	0.1619	0.1716
1.70	0.1151	0.1154	0.1162	0.1175	0.1194	0.1205	0.1218	0.1249	0.1288	0.1333	0.1388	0.1452	0.1527	0.1613
1.80	0.1100	0.1103	0.1110	0.1122	0.1139	0.1150	0.1162	0.1191	0.1226	0.1268	0.1318	0.1376	0.1444	0.1522
1.90	0.1053	0.1055	0.1062	0.1074	0.1090	0.1100	0.1111	0.1138	0.1170	0.1209	0.1255	0.1308	0.1370	0.1439
2.00	0.1010	0.1012	0.1019	0.1029	0.1044	0.1054	0.1064	0.1089	0.1119	0.1155	0.1197	0.1246	0.1302	0.1365
2.20	0.0934	0.0936	0.0941	0.0951	0.0964	0.0972	0.0981	0.1003	0.1029	0.1060	0.1096	0.1137	0.1184	0.1236
2.40	0.0869	0.0870	0.0875	0.0883	0.0895	0.0902	0.0910	0.0929	0.0952	0.0979	0.1010	0.1046	0.1085	0.1128
2.60	0.0812	0.0813	0.0818	0.0825	0.0835	0.0842	0.0849	0.0866	0.0886	0.0909	0.0937	0.0967	0.1001	0.1037
2.80	0.0762	0.0763	0.0767	0.0774	0.0783	0.0789	0.0795	0.0810	0.0828	0.0849	0.0873	0.0899	0.0928	0.0960
3.00	0.0718	0.0719	0.0723	0.0729	0.0737	0.0742	0.0748	0.0761	0.0777	0.0796	0.0817	0.0840	0.0866	0.0893
3.20	0.0679	0.0680	0.0683	0.0688	0.0696	0.0701	0.0706	0.0718	0.0732	0.0749	0.0768	0.0788	0.0811	0.0834
3.40	0.0644	0.0645	0.0648	0.0653	0.0659	0.0664	0.0668	0.0679	0.0692	0.0707	0.0724	0.0742	0.0762	0.0783
3.60	0.0612	0.0613	0.0616	0.0620	0.0626	0.0630	0.0635	0.0644	0.0656	0.0670	0.0685	0.0701	0.0719	0.0737
3.80	0.0583	0.0584	0.0587	0.0591	0.0597	0.0600	0.0604	0.0613	0.0624	0.0636	0.0650	0.0664	0.0680	0.0697
4.00	0.0557	0.0558	0.0560	0.0564	0.0569	0.0573	0.0576	0.0585	0.0594	0.0605	0.0618	0.0631	0.0646	0.0660
4.50	0.0501	0.0502	0.0504	0.0507	0.0511	0.0514	0.0517	0.0524	0.0532	0.0541	0.0551	0.0561	0.0572	0.0584
5.00	0.0455	0.0456	0.0458	0.0460	0.0464	0.0466	0.0469	0.0474	0.0481	0.0488	0.0496	0.0505	0.0514	0.0523
5.50	0.0417	0.0418	0.0419	0.0421	0.0425	0.0426	0.0428	0.0433	0.0439	0.0445	0.0452	0.0459	0.0466	0.0474
6.00	0.0385	0.0386	0.0387	0.0389	0.0391	0.0393	0.0395	0.0399	0.0404	0.0409	0.0415	0.0421	0.0427	0.0433
7.00	0.0334	0.0334	0.0335	0.0336	0.0338	0.0340	0.0341	0.0344	0.0348	0.0352	0.0356	0.0360	0.0365	0.0369
8.00	0.0294	0.0295	0.0295	0.0296	0.0298	0.0299	0.0300	0.0303	0.0305	0.0308	0.0312	0.0315	0.0318	0.0322
9.00	0.0263	0.0264	0.0264	0.0265	0.0266	0.0267	0.0263	0.0270	0.0272	0.0275	0.0277	0.0280	0.0283	0.0285
10.00	0.0238	0.0238	0.0239	0.0240	0.0241	0.0241	0.0242	0.0244	0.0245	0.0247	0.0250	0.0252	0.0254	0.0256

VALUES OF EXP(-XI) FOR Z EQUAL R*EXP(I*THETA)

R				THETA(DEGREES)			
	130.0	135.0	140.0	150.0	160.0	170.0	180.0
0.10	0.7574	0.7771	0.7978	0.8423	0.8909	0.9436	1.0000
0.20	0.6643	0.6891	0.7156	0.7745	0.8414	0.9167	1.0000
0.30	0.5935	0.6207	0.6504	0.7183	0.7986	0.8924	1.0000
0.40	0.5338	0.5618	0.5930	0.6664	0.7572	0.8680	1.0000
0.50	0.4815	0.5090	0.5402	0.6160	0.7145	0.8413	1.0000
0.60	0.4349	0.4610	0.4910	0.5660	0.6687	0.8104	1.0000
0.70	0.3932	0.4172	0.4450	0.5161	0.6185	0.7723	1.0000
0.80	0.3561	0.3775	0.4024	0.4669	0.5631	0.7219	1.0000
0.90	0.3233	0.3420	0.3637	0.4197	0.5038	0.6514	1.0000
1.00	0.2945	0.3105	0.3291	0.3760	0.4444	0.5587	1.0000
1.10	0.2692	0.2829	0.2985	0.3370	0.3898	0.4647	0.5367
1.20	0.2471	0.2587	0.2718	0.3031	0.3428	0.3902	0.4202
1.30	0.2277	0.2376	0.2486	0.2740	0.3040	0.3350	0.3510
1.40	0.2108	0.2193	0.2285	0.2491	0.2721	0.2936	0.3033
1.50	0.1960	0.2032	0.2110	0.2280	0.2459	0.2614	0.2679
1.60	0.1828	0.1891	0.1957	0.2098	0.2241	0.2357	0.2404
1.70	0.1712	0.1766	0.1823	0.1942	0.2057	0.2147	0.2183
1.80	0.1609	0.1656	0.1705	0.1806	0.1901	0.1972	0.2000
1.90	0.1517	0.1558	0.1601	0.1687	0.1766	0.1825	0.1847
2.00	0.1434	0.1471	0.1508	0.1582	0.1649	0.1698	0.1716
2.20	0.1292	0.1321	0.1350	0.1407	0.1456	0.1491	0.1504
2.40	0.1174	0.1198	0.1221	0.1266	0.1304	0.1330	0.1339
2.60	0.1076	0.1095	0.1114	0.1150	0.1180	0.1201	0.1208
2.80	0.0992	0.1008	0.1024	0.1054	0.1078	0.1094	0.1100
3.00	0.0920	0.0934	0.0947	0.0972	0.0992	0.1006	0.1010
3.20	0.0858	0.0870	0.0881	0.0902	0.0919	0.0930	0.0934
3.40	0.0804	0.0814	0.0824	0.0842	0.0856	0.0865	0.0869
3.60	0.0756	0.0765	0.0773	0.0789	0.0801	0.0809	0.0812
3.80	0.0713	0.0721	0.0728	0.0742	0.0753	0.0760	0.0762
4.00	0.0675	0.0682	0.0688	0.0700	0.0710	0.0716	0.0718
4.50	0.0595	0.0600	0.0605	0.0614	0.0621	0.0626	0.0627
5.00	0.0532	0.0536	0.0540	0.0547	0.0553	0.0556	0.0557
5.50	0.0481	0.0484	0.0488	0.0493	0.0498	0.0500	0.0501
6.00	0.0439	0.0442	0.0444	0.0449	0.0453	0.0455	0.0455
7.00	0.0374	0.0376	0.0377	0.0381	0.0383	0.0385	0.0385
8.00	0.0325	0.0327	0.0328	0.0330	0.0332	0.0333	0.0334
9.00	0.0288	0.0289	0.0290	0.0292	0.0293	0.0294	0.0294
10.00	0.0258	0.0259	0.0260	0.0261	0.0262	0.0263	0.0263

VALUES OF EXP(-ZETA) FOR Z EQUAL R*EXP(I*THETA)

THETA (DEGREES)

R	0.0	10.0	20.0	30.0	40.0	45.0	50.0	60.0	70.0	80.0	90.0	100.0	110.0	120.0
0.10	0.0236	0.0236	0.0239	0.0240	0.0241	0.0241	0.0242	0.0244	0.0245	0.0247	0.0250	0.0252	0.0254	0.0256
0.20	0.0455	0.0456	0.0458	0.0460	0.0464	0.0466	0.0469	0.0474	0.0481	0.0488	0.0496	0.0505	0.0514	0.0523
0.30	0.0655	0.0656	0.0659	0.0664	0.0671	0.0676	0.0680	0.0692	0.0705	0.0720	0.0738	0.0757	0.0778	0.0799
0.40	0.0838	0.0841	0.0845	0.0853	0.0864	0.0871	0.0879	0.0896	0.0918	0.0943	0.0972	0.1005	0.1041	0.1081
0.50	0.1010	0.1012	0.1019	0.1029	0.1044	0.1054	0.1064	0.1089	0.1119	0.1155	0.1197	0.1246	0.1302	0.1365
0.60	0.1170	0.1172	0.1180	0.1194	0.1213	0.1225	0.1238	0.1270	0.1309	0.1356	0.1413	0.1479	0.1556	0.1646
0.70	0.1319	0.1322	0.1332	0.1348	0.1371	0.1385	0.1402	0.1441	0.1489	0.1547	0.1617	0.1701	0.1802	0.1921
0.80	0.1459	0.1463	0.1474	0.1493	0.1520	0.1537	0.1556	0.1601	0.1658	0.1727	0.1811	0.1913	0.2037	0.2187
0.90	0.1591	0.1595	0.1606	0.1629	0.1660	0.1679	0.1701	0.1753	0.1818	0.1898	0.1995	0.2114	0.2260	0.2440
1.00	0.1716	0.1720	0.1734	0.1758	0.1792	0.1814	0.1838	0.1896	0.1969	0.2058	0.2168	0.2304	0.2471	0.2679
1.10	0.1834	0.1839	0.1854	0.1880	0.1918	0.1941	0.1967	0.2031	0.2111	0.2210	0.2332	0.2483	0.2669	0.2904
1.20	0.1946	0.1951	0.1968	0.1996	0.2036	0.2062	0.2090	0.2160	0.2246	0.2354	0.2487	0.2651	0.2856	0.3114
1.30	0.2053	0.2059	0.2076	0.2106	0.2149	0.2176	0.2207	0.2281	0.2374	0.2490	0.2633	0.2810	0.3030	0.3310
1.40	0.2154	0.2161	0.2179	0.2211	0.2257	0.2286	0.2318	0.2397	0.2496	0.2619	0.2771	0.2959	0.3194	0.3492
1.50	0.2251	0.2258	0.2278	0.2311	0.2360	0.2390	0.2424	0.2507	0.2611	0.2741	0.2901	0.3100	0.3348	0.3662
1.60	0.2344	0.2351	0.2372	0.2407	0.2458	0.2489	0.2525	0.2612	0.2721	0.2857	0.3025	0.3233	0.3492	0.3820
1.70	0.2433	0.2440	0.2462	0.2499	0.2551	0.2584	0.2622	0.2712	0.2826	0.2967	0.3142	0.3358	0.3628	0.3967
1.80	0.2519	0.2526	0.2548	0.2586	0.2641	0.2675	0.2714	0.2808	0.2926	0.3073	0.3253	0.3477	0.3755	0.4105
1.90	0.2601	0.2608	0.2631	0.2671	0.2727	0.2763	0.2803	0.2900	0.3022	0.3173	0.3359	0.3590	0.3876	0.4233
2.00	0.2679	0.2687	0.2711	0.2752	0.2810	0.2846	0.2888	0.2988	0.3113	0.3269	0.3460	0.3696	0.3989	0.4354
2.20	0.2829	0.2837	0.2862	0.2905	0.2966	0.3004	0.3048	0.3153	0.3285	0.3448	0.3648	0.3894	0.4198	0.4574
2.40	0.2967	0.2976	0.3002	0.3047	0.3111	0.3151	0.3196	0.3306	0.3443	0.3612	0.3820	0.4074	0.4386	0.4770
2.60	0.3097	0.3106	0.3133	0.3180	0.3246	0.3287	0.3335	0.3448	0.3589	0.3764	0.3977	0.4238	0.4556	0.4946
2.80	0.3219	0.3228	0.3256	0.3304	0.3372	0.3415	0.3464	0.3580	0.3725	0.3904	0.4122	0.4388	0.4711	0.5104
3.00	0.3333	0.3343	0.3372	0.3421	0.3491	0.3535	0.3584	0.3704	0.3852	0.4035	0.4257	0.4526	0.4852	0.5247
3.20	0.3441	0.3451	0.3481	0.3531	0.3602	0.3647	0.3698	0.3820	0.3971	0.4156	0.4382	0.4654	0.4982	0.5378
3.40	0.3543	0.3553	0.3584	0.3635	0.3708	0.3753	0.3805	0.3929	0.4082	0.4270	0.4498	0.4773	0.5103	0.5498
3.60	0.3640	0.3650	0.3681	0.3733	0.3807	0.3853	0.3906	0.4032	0.4187	0.4377	0.4607	0.4883	0.5214	0.5609
3.80	0.3732	0.3742	0.3774	0.3826	0.3902	0.3948	0.4002	0.4129	0.4286	0.4478	0.4709	0.4987	0.5318	0.5712
4.00	0.3820	0.3830	0.3862	0.3915	0.3991	0.4039	0.4092	0.4221	0.4380	0.4573	0.4805	0.5083	0.5414	0.5807
4.50	0.4021	0.4032	0.4064	0.4119	0.4197	0.4246	0.4301	0.4431	0.4593	0.4788	0.5023	0.5301	0.5631	0.6018
5.00	0.4202	0.4213	0.4246	0.4302	0.4381	0.4430	0.4486	0.4618	0.4781	0.4978	0.5213	0.5491	0.5817	0.6199
5.50	0.4365	0.4376	0.4410	0.4466	0.4546	0.4596	0.4652	0.4786	0.4950	0.5147	0.5381	0.5657	0.5981	0.6356
6.00	0.4514	0.4525	0.4559	0.4616	0.4697	0.4747	0.4803	0.4937	0.5101	0.5298	0.5532	0.5806	0.6125	0.6494
7.00	0.4776	0.4787	0.4821	0.4878	0.4960	0.5010	0.5067	0.5201	0.5365	0.5560	0.5790	0.6059	0.6370	0.6727
8.00	0.5000	0.5011	0.5045	0.5103	0.5184	0.5234	0.5291	0.5424	0.5586	0.5780	0.6006	0.6269	0.6572	0.6917
9.00	0.5195	0.5206	0.5240	0.5297	0.5378	0.5428	0.5484	0.5617	0.5777	0.5967	0.6190	0.6447	0.6741	0.7075
10.00	0.5367	0.5378	0.5412	0.5469	0.5549	0.5598	0.5654	0.5785	0.5943	0.6130	0.6349	0.6600	0.6887	0.7211

VALUES OF EXP(-ZETA) FOR Z EQUAL R*EXP(I*THETA)

R	THETA(DEGREES)						
	130.0	135.0	140.0	150.0	160.0	170.0	180.0
0.10	0.0258	0.0259	0.0260	0.0261	0.0262	0.0263	0.0263
0.20	0.0532	0.0536	0.0540	0.0547	0.0553	0.0556	0.0557
0.30	0.0821	0.0832	0.0842	0.0861	0.0876	0.0886	0.0889
0.40	0.1123	0.1144	0.1165	0.1205	0.1239	0.1262	0.1270
0.50	0.1434	0.1471	0.1508	0.1582	0.1649	0.1698	0.1716
0.60	0.1750	0.1806	0.1866	0.1991	0.2115	0.2213	0.2251
0.70	0.2064	0.2144	0.2232	0.2427	0.2641	0.2836	0.2922
0.80	0.2371	0.2478	0.2598	0.2879	0.3225	0.3606	0.3820
0.90	0.2666	0.2800	0.2953	0.3330	0.3841	0.4553	0.5195
1.00	0.2945	0.3105	0.3291	0.3760	0.4444	0.5587	1.0000
1.10	0.3205	0.3390	0.3604	0.4156	0.4984	0.6438	1.0000
1.20	0.3447	0.3652	0.3891	0.4509	0.5437	0.7010	1.0000
1.30	0.3671	0.3893	0.4151	0.4819	0.5807	0.7391	1.0000
1.40	0.3876	0.4113	0.4387	0.5090	0.6109	0.7660	1.0000
1.50	0.4066	0.4313	0.4599	0.5327	0.6358	0.7861	1.0000
1.60	0.4240	0.4496	0.4792	0.5535	0.6566	0.8017	1.0000
1.70	0.4401	0.4664	0.4966	0.5719	0.6743	0.8144	1.0000
1.80	0.4549	0.4818	0.5125	0.5882	0.6895	0.8248	1.0000
1.90	0.4687	0.4959	0.5269	0.6029	0.7028	0.8337	1.0000
2.00	0.4815	0.5090	0.5402	0.6160	0.7145	0.8413	1.0000
2.20	0.5045	0.5324	0.5637	0.6388	0.7342	0.8538	1.0000
2.40	0.5246	0.5526	0.5839	0.6579	0.7502	0.8637	1.0000
2.60	0.5424	0.5704	0.6014	0.6742	0.7636	0.8718	1.0000
2.80	0.5583	0.5861	0.6169	0.6883	0.7750	0.8786	1.0000
3.00	0.5726	0.6002	0.6306	0.7007	0.7848	0.8844	1.0000
3.20	0.5855	0.6129	0.6429	0.7117	0.7934	0.8894	1.0000
3.40	0.5973	0.6244	0.6540	0.7214	0.8011	0.8939	1.0000
3.60	0.6080	0.6348	0.6640	0.7303	0.8079	0.8978	1.0000
3.80	0.6179	0.6445	0.6733	0.7383	0.8141	0.9013	1.0000
4.00	0.6271	0.6533	0.6817	0.7456	0.8197	0.9045	1.0000
4.50	0.6473	0.6727	0.7002	0.7614	0.8316	0.9112	1.0000
5.00	0.6643	0.6891	0.7156	0.7745	0.8414	0.9167	1.0000
5.50	0.6790	0.7031	0.7289	0.7856	0.8497	0.9212	1.0000
6.00	0.6919	0.7153	0.7403	0.7952	0.8568	0.9251	1.0000
7.00	0.7133	0.7356	0.7593	0.8109	0.8683	0.9314	1.0000
8.00	0.7307	0.7520	0.7746	0.8234	0.8773	0.9363	1.0000
9.00	0.7451	0.7656	0.7871	0.8337	0.8847	0.9403	1.0000
10.00	0.7574	0.7771	0.7978	0.8423	0.8909	0.9436	1.0000

We have the expansions

$$|e^{-\xi}| = 1 + 2r \cos^2\theta/2 - r^{\frac{1}{2}}(r+2)\cos\theta/2 + O(r^{5/2}) ,$$

$$z = re^{i\theta} , \tag{3}$$

$$e^{-\xi} = 1/4z - 1/8z^2 + 5/64z^3 + O(z^{-4}) , \quad |z| > 1 . \tag{4}$$

The following approximations for $e^{-\xi}$, designated as $f(z)$, are based on second order Padé approximations for the square root. The domain of applicability for each formula is given. As a practical guide, formulas (5), (6), (7) and (8) work best when $|z| \geq 1$, $|z| \leq 1$, $|v| \leq 1$, and $|u| \leq 1$, respectively. This should be sufficient for practical purposes since further tests of reliability are easily determined by comparison with the absolute numerical values given in the tables.

$$f(z) = \frac{4z+1}{16z^2+12z+1} , \quad |\arg(1+1/z)| < \pi . \tag{5}$$

$$f(z) = 1 + 2z - 2z^{\frac{1}{2}} \frac{(16+20z+5z^2)}{16+12z+z^2} ,$$

$$|\arg(1+z)| < \pi , \quad -\pi < \arg z \leq \pi . \tag{6}$$

$$f(z) = -1 - 2v + 2v^{\frac{1}{2}} \frac{(16+20v+5v^2)}{16+12v+v^2} ,$$

$$z = -1 - v , \quad |\arg v| \leq \pi/2 . \tag{7}$$

$$f(z) = -1 + 2u - 2iu^{\frac{1}{2}} \frac{(16-20u+5u^2)}{16-12u+u^2} ,$$

$$z = -1 + u , \quad |\arg u| \leq \pi/2 . \tag{8}$$

II IDENTIFICATION OF FUNCTIONS

2.1. Introduction

The present chapter gives a complete list of functions for which algorithms and programs are presented in this volume along with a reference to where the material is located. Functions usually defined by the $_pF_q(z)$ and G-function symbols are noted in 2.2 and 2.3, respectively. Here we also note the names of the functions. In some cases, the names used are those normally employed except for some multiplication factor. This is done for simplicity and should cause no difficulty. Some miscellaneous functions not necessarily of the above type are noted in 2.4. For complete descriptions of the functions presented, see Erdélyi, et al (1953,1954), Abramowitz and Stegun (1964) and Luke (1962,1969,1975a).

2.2. The Generalized Hypergeometric Function $_pF_q(z)$

For some general remarks on this function, see 1.2(1-4). Some special cases follow.

The Gaussian Hypergeometric Function $_2F_1(a,b;c;z)$

See Chapters 4, 13 and 14.

Only a skeleton list of formulas are provided below. Further representations follow by use of Kummer's formulas, quadratic transformation formulas and other formulas for analytic continuation. For details, see Erdélyi, et al (1953) and Luke (1969,1975a).

a	b	c	z	$_2F_1(a,b;c;z)$	
1	1	2	$-z$	$z^{-1}\ln(1+z)$	(1)
½	1	3/2	z^2	$\ln[(1+z)/(1-z)]$	(2)
½	½	3/2	z^2	$z^{-1}\arcsin z$	(3)
½	1	3/2	$-z^2$	$z^{-1}\arctan z$	(4)
$-\nu$	$\nu+1$	$1-\mu$	$(1-z)/2$	$\Gamma(1-\mu)[(z-1)/(z+1)]^{\mu/2}P_\nu^\mu(z)$	(5)

$$\tfrac{1}{2}(\mu+\nu+1)\quad \tfrac{1}{2}(\mu+\nu+2)\quad \nu+3/2 \quad z^{-2} \qquad \frac{e^{-i\pi\mu}2^{\nu+1}\Gamma(\nu+3/2)z^{\mu+\nu+1}Q_\nu^\mu(z)}{\pi^{\frac{1}{2}}\Gamma(\mu+\nu+1)(z^2-1)^{\mu/2}}$$

$$(6)$$

$P_\nu^\mu(z)$ and $Q_\nu^\mu(z)$ are the Legendre functions of the first and second kind, respectively. Here the complex z-plane is cut along the z-axis from -1 to 1 . The representations (7,8) which follow hold for $-1 < x < 1$.

a	b	c	z	$_2F_1(a,b;c;z)$	
$-\nu$	$\nu+1$	$1-\mu$	$(1-x)/2$	$\Gamma(1-\mu)[(1-x)/(1+x)]^{\mu/2}P_\nu^\mu(x)$	(7)
				$Q_\nu^\mu(x)=\tfrac{1}{2}e^{-i\pi\mu}[e^{-i\pi\mu/2}Q_\nu^\mu(x+i0)+e^{i\pi\mu/2}Q_\nu^\mu(x-i0)]$	(8)
½	½	1	k^2	$(2/\pi)K(k)$	(9)
$-\frac{1}{2}$	½	1	k^2	$(2/\pi)E(k)$	(10)

$K(k)$ and $E(k)$ are the complete elliptic integrals of the first and second kind, respectively.

1	$\alpha+\beta$	$\alpha+1$	x	$\alpha x^{-\alpha}(1-x)^{\beta}\int_0^x t^{\alpha-1}(1-t)^{\beta-1}dt,$	(11)

the incomplete Beta function.

1	v	1	$-x$	$_1F_0(v;-x)=(1+x)^{-v},$	(12)

the binomial function.

The Kummer transformation formulae are

$$_2F_1(a,b;c;z)$$
$$= (1-z)^{c-a-b}{}_2F_1(c-a,c-b;c;z) \qquad (13)$$

(continued on next page)

$$= (1-z)^{-a} \,_2F_1(a,c-b;c;z/(z-1)) \tag{14}$$

$$= (1-z)^{-b} \,_2F_1(c-a,b;c;z/(z-1)) \ . \tag{15}$$

The following equations are useful for analytic continuation

of the $_2F_1$.

$$_2F_1\left(\begin{matrix}a,b\\c\end{matrix}\middle|z\right) = \frac{\Gamma(b-a)\Gamma(c)}{\Gamma(b)\Gamma(c-a)}\,(z^{-1}e^{i\pi})^a\ _2F_1\left(\begin{matrix}a,a+1-c\\a+1-b\end{matrix}\middle|1/z\right)$$

+ the same expression with   a   and   b   interchanged. (16)

$$_2F_1\left(\begin{matrix}a,b\\c\end{matrix}\middle|z\right) = \frac{\Gamma(c-a-b)\Gamma(c)}{\Gamma(c-a)\Gamma(c-b)}\ _2F_1\left(\begin{matrix}a,b\\a+b+1-c\end{matrix}\middle|1-z\right)$$

$$+ \frac{\Gamma(a+b-c)\Gamma(c)(1-z)^{c-a-b}}{\Gamma(a)\Gamma(b)}\ _2F_1\left(\begin{matrix}c-a,c-b\\c+1-a-b\end{matrix}\middle|1-z\right)\ . \tag{17}$$

Further expressions can be deduced by applying the Kummer

formulas.

The Confluent Hypergeometric Function

$$_1F_1(a;c;z) = e^z\,_1F_1(c-a;c;z) \tag{18}$$

See Chapters 5, 8, 15 and 16.

Equation (18) is the Kummer transformation formula for the

$_1F_1$.

a	c	z	$_1F_1(a;c;z)$	
a	a	z	e^z	(19)
a	a+1	-z	$az^{-a}\gamma(a,z)$, the	(20)
			incomplete gamma function.	
1	a+1	z	$az^{-a}e^z\gamma(a,z)$	(21)
1	3/2	z^2	$z^{-1}e^{z^2}\mathrm{Erf}(z)$,	(22)
			error function.	
½	3/2	$-z^2$	$z^{-1}\mathrm{Erf}(z)$	(23)
½+ν	1+2ν	-2z	$e^{-z}(z/2)^{-\nu}\Gamma(\nu+1)I_\nu(z)$,	(24)

(continued on next page)

the modified Bessel function
of the first kind.

See also $_0F_1(c;z)$.

Bessel Functions $_0F_1(c;z)$

See Chapters 6, 17, 18 and 19.

c	z	$_0F_1(c;z)$	
½	$-z^2/4$	$\cos z$	(25)
3/2	$-z^2/4$	$z^{-1}\sin z$	(26)
½	$z^2/4$	$\cosh z$	(27)
3/2	$z^2/4$	$z^{-1}\sinh z$	(28)
$\nu+1$	$-z^2/4$	$\Gamma(\nu+1)(z/2)^{-\nu}J_\nu(z)$	(29)
$\nu+1$	$z^2/4$	$\Gamma(\nu+1)(z/2)^{-\nu}I_\nu(z)$	(30)

$J_\nu(z)\{I_\nu(z)\}$ is the

{modified} Bessel function of the first kind.

The Hypergeometric Function $_1F_2(a;b,c;z)$

See Chapters 7 and 9.

a	b	c	z	$_1F_2(^a_{b,c}|z)$	
1	½$(\mu-\nu+3)$	½$(\mu+\nu+3)$	$-z^2/4$	$(\mu-\nu+1)(\mu+\nu+1)z^{-\mu-1}s_{\mu,\nu}(z)$	

$s_{\mu,\nu}(z)$ is Lommel's function. (31)

1	3/2	$\nu+3/2$	$-z^2/4$	$\Gamma(3/2)\Gamma(\nu+3/2)(z/2)^{-\nu-1}$	
				$\times\ H_\nu(z)$	(32)
1	3/2	$\nu+3/2$	$z^2/4$	$\Gamma(3/2)\Gamma(\nu+3/2)(z/2)^{-\nu-1}$	
				$\times\ L_\nu(z)$	(33)

$H_\nu(z)\{L_\nu(z)\}$ is the {modified} Struve function.

| $\nu+3/2$ | $\nu+2$ | $2\nu+2$ | $-z^2$ | $\Gamma(\nu+1)\Gamma(\nu+2)(z/2)^{-2\nu-1}$ | |
| | | | | $\times\ J_\nu(z)J_{\nu+1}(z)$ | (34) |

$J_\nu(z)$ is the Bessel function of the first kind.

(continued on next page)

a	b	c	z	$_1F_2\left(\genfrac{}{}{0pt}{}{a}{b,c}\middle\vert z\right)$
$\nu+\tfrac{1}{2}$	$\nu+1$	$2\nu+1$	$-z^2$	$\{\Gamma(\nu+1)\}^2 (z/2)^{-2\nu} J_\nu^2(z)$ (35)
$\tfrac{1}{2}$	$\nu+1$	$1-\nu$	$-z^2$	$\dfrac{\nu\pi}{\sin\,\nu\pi}\,J_{-\nu}(z)J_\nu(z)$ (36)
$\tfrac{1}{2}(\mu+\nu+1)$	$\tfrac{1}{2}(\mu+\nu+3)$	$\nu+1$	$-z^2/4$	$2^\nu(\mu+\nu+1)\Gamma(\nu+1)z^{-\mu-\nu-1}$

$$\times \int_0^z t^\mu J_\nu(t)\,dt \qquad (37)$$

The Confluent Hypergeometric Function

$$U(a;c;z) = \frac{\Gamma(1-c)}{\Gamma(b)}\,_1F_1(a;c;z) + \frac{\Gamma(c-1)}{\Gamma(a)}\,_1F_1(b;2-c;z)\ ,$$

$$b = 1+a-c\ . \qquad (38)$$

The Kummer transformation formula for the U-function is

$$U(a;c;z) = z^{1-c}U(1+a-c;2-c;z)\ . \qquad (39)$$

See Chapters 8, 15, 16, 20 and 21.

a	c	z	$U(a;c;z)$
$1-a$	$1-a$	z	$e^z\Gamma(a,z) = e^z\displaystyle\int_z^\infty t^{a-1}e^{-t}dt$

$\Gamma(a,z)$ is the complementary incomplete gamma function. (40)

1	$1+a$	z	$e^z z^{-a}\Gamma(a,z)$ (41)
1	1	z	$e^z E_1(z) = e^z\displaystyle\int_z^\infty t^{-1}e^{-t}dt$

$E_1(z) = -Ei(-z)$ is the exponential integral. (42)

1	1	$-x$	$e^{-x}Ei(x) = (e^{-x})P.V.\displaystyle\int_{-\infty}^x t^{-1}e^t dt,$

$$x > 0$$

$Ei(x)$ is Cauchy principal value of the integral. (43)

$$E_1(z) + (\gamma + \ln z) = \int_0^z t^{-1}(1-e^{-t})dt\ , \qquad (44)$$

(continued on next page)

a	c	z	U(a;c;z)

$$Ei(z) - (\gamma + \ln z) = \int_0^z t^{-1}(e^t-1)dt \; ,$$

$$Ei(z) = Ei(-ze^{\pm i\pi}) \mp i\pi \; . \tag{45}$$

| $\tfrac{1}{2}$ | $\tfrac{1}{2}$ | z^2 | $2 \exp(z^2)Erfc(z)$ |

$$= 2 \exp(z^2) \int_z^\infty \exp(-t^2)dt$$

Erfc(z) is the complementary error function.   (46)

| $\tfrac{1}{2}(n+1)$ | $\tfrac{1}{2}$ | z^2 | $2^n \pi^{\tfrac{1}{2}} \exp(z^2) i^n erfc(z)$ |

$i^n erfc(z)$ is the repeated integral

of the complementary error function.

$$i^0 erfc(z) = erfc(z) = 2\pi^{-\tfrac{1}{2}} Erfc(z) . \tag{47}$$

| $L+1-i\eta$ | $2L+2$ | $2i\rho$ | Coulomb Wave Functions, |

see Luke (1969,1975a)   (48)

| $(1-\nu)/2$ | $3/2$ | $z^2/2$ | Parabolic Cylinder Functions, |

see Luke (1969,1975a)   (49)

| $\tfrac{1}{2}+\nu$ | $1+2\nu$ | $2z$ | $\pi^{-\tfrac{1}{2}} e^z (2z)^{-\nu} K_\nu(z)$ |

$K_\nu(z)$ is the modified Bessel function of the second kind.

$$K_\nu(z) = \tfrac{1}{2}\pi (\csc \nu\pi)[I_{-\nu}(z) - I_\nu(z)] . \tag{50}$$

| $\tfrac{1}{2}+\nu$ | $1+2\nu$ | $2ze^{-i\varepsilon\pi/2}$ | $\tfrac{1}{2}\pi^{\tfrac{1}{2}} i\varepsilon e^{i\varepsilon\nu\pi}(2z)^{-\nu} e^{-iz} H_\nu^{(3-\varepsilon)/2}(z)$, |

$$\varepsilon = \pm 1, \; H_\nu^{(3-\varepsilon)/2}(z) = J_\nu(z) + i\varepsilon Y_\nu(z) \; ,$$

$$Y_\nu(z) = (\csc \nu\pi)[\cos \nu\pi J_\nu(z) - J_{-\nu}(z)] \; ,$$

$$K_\nu(z) = \tfrac{1}{2}\pi i\varepsilon e^{i\varepsilon\nu\pi/2} H_\nu^{(3-\varepsilon)/2}(ze^{i\varepsilon\pi/2}) \; . \tag{51}$$

If $\varepsilon = 1\{\varepsilon=-1\}$, $H_\nu^{(3-\varepsilon)/2}(z)$ is the

Hankel function of the first {second} kind.

$Y_\nu(z)$ is the Bessel function of the second kind.

2.3. The G-Function

For some general remarks on this function, see the comments surrounding 1.2(5-11). Some special cases of the type $G_{1,q}^{q-r,1}(z|_{b_q}^1)$, r = 0 or r = 1 are detailed below.

$G_{1,2}^{2,1}(z|_{a,b}^1)$, see 8(1) and Chapters 20 and 21.

$G_{1,2}^{1,1}(z|_{a,b}^1)$, see 8(7).

$G_{1,3}^{3,1}(z^2/4|_{a,b,c}^1)$, see 9(1).

a	b	c	z^2	$\dfrac{G_{1,3}^{3,1}(\frac{z^2}{4}|_{a,b,c}^1)}{}$
1	$\frac{1}{2}(1-\mu+\nu)$	$\frac{1}{2}(1-\mu-\nu)$	z^2	$\Gamma(\frac{1-\mu+\nu}{2})\Gamma(\frac{1-\mu-\nu}{2})z^{1-\mu}S_{\mu,\nu}(z)$ (1)

$S_{\mu,\nu}(z)$ is Lommel's function.

1	$\frac{1}{2}$	$\frac{1}{2}-\nu$	z^2	$\dfrac{\pi^2(z/2)^{1-\nu}}{\cos\,\nu\pi}[H_\nu(z) - Y_\nu(z)]$ (2)

$H_\nu(z)$ is the Struve function.

$Y_\nu(z)$ is the Bessel function of the second kind.

$\frac{1}{2}+\nu$	$\frac{1}{2}-\nu$	$\frac{1}{2}$	z^2	$\dfrac{\pi^{5/2}z}{4\cos\,\nu\pi}[J_\nu^2(z/2)+Y_\nu^2(z/2)]$ (3)

$J_\nu(z)$ is the Bessel function of the first kind.

$G_{1,3}^{2,1}(z^2/4|_{a,b,c}^1)$, see 9(8).

a	b	c	z^2	$\dfrac{G_{1,3}^{2,1}(\frac{z^2}{4}|_{a,b,c}^1)}{}$
1	$\frac{1}{2}$	$\frac{1}{2}-\nu$	z^2	$\pi(z/2)^{1-\nu}[I_\nu(z)-L_\nu(z)]$ (4)
1	$\frac{1}{2}-\nu$	$\frac{1}{2}$	z^2	$\dfrac{\pi(z/2)^{1-\nu}}{\cos\,\nu\pi}[I_{-\nu}(z)-L_\nu(z)]$ (5)

$I_\nu(z)$ is the modified Bessel function of the first kind.

(continued on next page)

a	b	c	z^2	$G^{2,1}_{1,3}(\frac{z^2}{4} \mid \genfrac{}{}{0pt}{}{1}{a,b,c})$

$L_\nu(z)$ is the modified Struve function.

| $\tfrac{1}{2}$ | $\tfrac{1}{2}+\nu$ | $\tfrac{1}{2}-\nu$ | z^2 | $\pi^{\tfrac{1}{2}} z I_\nu(z/2) K_\nu(z/2)$ | (6) |

$K_\nu(z)$ is the modified Bessel function of the second kind.

| $\tfrac{1}{2}+\nu$ | $\tfrac{1}{2}-\nu$ | $\tfrac{1}{2}$ | z^2 | $\dfrac{\pi^{3/2} z}{2 \sin 2\pi\nu} [I^2_{-\nu}(z/2) - I^2_\nu(z/2)].$ | (7) |

2.4. Miscellaneous Functions

If the Chebyshev coefficients for $f(x)$ are known, schemata for development of Chebyshev coefficients for functions related to

$$\int_0^X e^{at} t^u f(t)\,dt, \quad \int_X^\infty e^{-bt} t^u f(t)\,dt , \tag{1}$$

$$\int_0^X e^{at} t^u (\ln t) f(t)\,dt, \quad \int_X^\infty e^{-bt} t^u (\ln t) f(t)\,dt \tag{2}$$

are detailed in Chapter 11.

3 GENERAL REMARKS ON THE ALGORITHMS AND PROGRAMS

3.1. Introduction

In this chapter we discuss in a very general way the
nature of the programs presented in this volume. It should
be realized that it is impractical to prepare programs to
cover every type of machine and compiler now on the market.
Most certainly, we cannot anticipate the characteristics of
future machines and compilers.

The programming of the routines in this collection was
done for use by the IBM 370/168 operating under OS/VS Release
1.7 on the FORTRAN IV H-Extended Compiler, Release 2.1. Thus,
to use the programs on other types of equipment, certain adjustments must be made as required by such equipment.

Each program is written for quadruple precision and for
all variables and parameters real. The small changes to get
double or single precision or to get any of the three precision types with complex arithmetic are detailed in 3.2. In
this connection, quadruple precision is available only on the
FORTRAN (H-Extended) compiler. See Chapter 4 for an example
of complex arithmetic and double precision.

3.2. Precision and Complex Arithmetic

All computer programs are written for quadruple precision
and real arithmetic. However, double or single precision are
easily obtained by modifying the IMPLICIT, DATA, and FORMAT
statements and the imput data. Complex arithmetic is also
easily introduced. Now the general form of the IMPLICIT
statement is

IMPLICIT TYPE*SIZE(A-H,O-Z)

This will make all variables starting with A   through H   or
O   through   Z   of type TYPE with a length of SIZE bytes.
TYPE*SIZE configurations are delineated in the following
table.

Precision	Real	Complex
Single	REAL*4	COMPLEX*8
Double	REAL*8	COMPLEX*16
Quadruple	REAL*16	COMPLEX*32

The DATA statement gives a fixed value to the specified
variables for the duration of the program. If it is desired
to use a real type for the program, then the exponential terms
in the variable values must be altered according to the precision desired, namely E   for single precision, D   for double precision and Q   for quadruple precision. An example of
a data statement in double precision is

DATA ONE, TWO, THREE, FOUR/1.D0, 2.D0, 3.D0, 4.D0/

If it is desired to use a complex type for the program,
then the variable values must be represented as a pair of numbers in parantheses, separated by a comma, with each number in
the appropriate precision form. In illustration, we have

DATA ONE, TWO, THREE, FOUR/(1.D0,0.D0),

(2.D0,0.D0), (3.D0,0.D0), (4.D0,0.D0)/

In the FORMAT statements, if a real type is used, then
the format field character is the same as that used for the
precision indicator in DATA statements. We have, therefore,
the following table.

Precision	Real	Complex
Single	E15.8	2E15.8
Double	D23.16	2D23.16
Quadruple	Q39.32	2Q39.32

Thus, an example of a FORMAT statement in double precision is

FORMAT(4X, I2, 8X, D23.16)

Note that if complex type is used, each variable must have two real format specifications. An easy way to accomplish this is to precede each real number format specification with a 2 .

All data must be entered according to the specifications given in the FORMAT statements.

When complex type is used, many underflow errors might occur. The printed recordings can be suppressed by inserting a CALL ERRSET(208, 256, -1, 0, 0, 208) statement in the main line program after any IMPLICIT, DIMENSION, or DATA statements. This can be done on both the G1 and H compilers.

For an example illustrating double precision and complex arithmetic, see Chapter 4.

4 CHEBYSHEV COEFFICIENTS FOR $_2F_1(a,b;c;z)$

We now consider evaluation of $C_n(w)$ in

$$_2F_1(a,b;c;z) = \sum_{n=0}^{\infty} C_n(w)T_n^*(z/w) ,$$

$$0 \leq z/w \leq 1, \quad w \neq 1, \quad |\arg(1-w)| < \pi , \tag{1}$$

$$C_n(w) = \frac{\varepsilon_n (a)_n (b)_n w^n}{2^{2n}(c)_n n!} \; _3F_2\left(\begin{matrix} a+n,b+n,\frac{1}{2}+n \\ c+n,1+2n \end{matrix}\middle| w\right) . \tag{2}$$

This is the special case of 1.3(1,2) with $p = 2$, $q = 1$. Except for the recurrence formula for $C_n(w)$, all necessary information will be found in 1.3. Thus none of this material will be repeated, save for asymptotic forms for $C_n(w)$ for large n. We have

$$C_n(w) = \frac{2(2n)!\,(a)_n (b)_n (n\pi)^{\frac{1}{2}} e^{-n\delta} A(w) [1 + O(n^{-1})]}{2^{2n}(n!)^3 (c)_n} \tag{3}$$

$$= \frac{2\Gamma(n+\frac{1}{2})(a)_n (b)_n n^{\frac{1}{2}} e^{-n\delta} A(w) [1 + O(n^{-1})]}{(n!)^2 (c)_n} \tag{4}$$

$$= \frac{2\Gamma(c) n^{u-1} e^{-n\delta} A(w) [1 + O(n^{-1})]}{\Gamma(a)\Gamma(b)} . \tag{5}$$

Here

$$u = a+b-c, \quad A(w) = \{(1+e^{-\delta})/(1-e^{-\delta})\}^u ,$$

$$e^{-\delta} = [2-w\mp 2(1-w)^{\frac{1}{2}}]/w \tag{6}$$

where the sign in the latter is chosen so that $|e^{-\delta}| < 1$.

52

This is possible for all w except $w \geq 1$. The function $e^{-\delta}$ is the same as $-e^{-\zeta}$ of 1.8(2) if there we put $z = we^{i\pi}$. The recurrence formula for $C_n(w)$ is

$$\frac{2C_n(w)}{\varepsilon_n} = (n+1) \{2 - \frac{(2n+3)(n+a+1)(n+b+1)}{(n+2)(n+a)(n+b)}$$

$$+ \frac{4(n+c)}{w(n+a)(n+b)} \} C_{n+1}(w)$$

$$+ \frac{2}{(n+a)(n+b)} \{ [(n+2-a)(n+2-b)(n+3/2) - (n+3-a)(n+3-b)(n+1)]$$

$$+ \frac{2(n+1)(n+3-c)}{w} \} C_{n+2}(w) - \frac{(n+1)(n+3-a)(n+3-b)}{(n+2)(n+a)(n+b)} C_{n+3}(w) .$$

$$(7)$$

If $c = b$, then the $_2F_1$ in (1) reduces to the binomial function. For convenience we introduce a slight change in notation by replacing z and w by $-z$ and $-w$ respectively. We have

$$(1+z)^{-a} = \sum_{n=0}^{\infty} B_n(w) T_n^*(z/w) ,$$

$$0 \leq z/w \leq 1, \ w \neq -1, \ |arg(1+w)| < \pi , \qquad (8)$$

$$B_n(w) = C_n(-w) \ \text{with} \ c = b . \qquad (9)$$

Asymptotic forms for $B_n(w)$ follow from (3)-(5). $B_n(w)$ satisfies (7) with $c = a$ and w replaced by $-w$. However, $B_n(w)$ also satisfies a three-term recurrence relation. Thus,

$$\frac{2}{\varepsilon_n} B_n(w) = - \frac{2(n+1)}{n+a} (1+2/w) B_{n+1}(w) - \frac{(n+2-a)}{n+a} B_{n+2}(w) .$$

$$(10)$$

If $a = 1$, again with a slight change of notation, we have

$$(g+x)^{-1} = (g^2+g)^{-\frac{1}{2}} \sum_{n=0}^{\infty} \varepsilon_n(-)^n e^{-n\xi} T_n^*(x) ,$$

$$0 \leq x \leq 1, \ g \neq -1, \ |\arg(1+1/g)| < \pi \ ,$$

$$e^{-\xi} = 2g+1 \mp 2(g^2+g)^{\frac{1}{2}} \ , \tag{11}$$

where $e^{-\xi}$ is the same function described in 1.8(1) with z replaced by g .

By integration of (11), we can get the Chebyshev coefficients for $\ln(1+x/g)$. Again in (11), replace x and g by x^2 and g^2 , respectively, and integrate. We then get the Chebyshev coefficients for $\arc \tan x/g$.

We now turn to a discussion of numerical examples used to illustrate the FORTRAN programs given at the end of this chapter for computation of the coefficients $C_n(w)$, see (1), (2), (9) and (11). There are three such numerical examples and programs. The first is for a $_2F_1$ with real arithmetic and quadruple precision. The second illustrates changes required in the first program to get complex arithmetic and double precision. In this connection, see the discussion in 3.2. The third is for the binomial function.

Numerical Examples

1. Real arithmetic and quadruple precision

Let $a = b = 1$, $c = 2$, $w = -1$, $N = 40$.

In (1), replace z by $-z$. Then

$$z^{-1} \ln(1+z) = \ _2F_1(1,1;2;-z) = \sum_{n=0}^{\infty} C_n T_n^*(z) \ ,$$

$$0 \leq z \leq 1 \ . \tag{12}$$

For the error analysis,

$$u = 0, \ e^{-\delta} = -(3-2^{3/2}) = -0.17157 = -e^{-\delta}/w, \ A(w) = 1 \ . \tag{13}$$

Let t_n denote the right hand side of (3) with $O(n^{-1})$

omitted. Then

$$t_n = \frac{2(2n)!(n\pi)^{\frac{1}{2}}e^{-n\delta}}{2^{2n}n!(n+1)!} \quad , \quad \frac{t_{n+1}}{t_n} = \left(\frac{2n+1}{2n+4}\right)\left(\frac{n+1}{n}\right)^{\frac{1}{2}} e^{-\delta} \ . \quad (14)$$

Computed values of C_n determined by the backward recursion process with $N = 40$ are given in the printout accompanying the FORTRAN program at the end of this chapter. Some trun-cated values from this list are given below to enable compari-son with values from (14).

n	t_n	C_n	t_n/t_{n-1}	C_n/C_{n-1}
4	0.33598(-3)	0.32462(-3)	—	—
5	-0.48337(-4)	-0.46835(-4)	-0.14387	-0.14428
10	0.39692(-8)	0.38866(-8)	—	—
11	-0.62496(-9)	-0.61280(-9)	-0.15745	-0.15767
20	0.46246(-16)	0.45682(-16)	—	—
21	-0.75760(-17)	-0.74875(-17)	-0.16382	-0.16390

Notice that the asymptotic results give excellent accuracy even for n small.

In the present situation, it can be readily shown that

$$C_0 = 2(2^{\frac{1}{2}}-1) \ , \quad \sum_{n=0}^{\infty} C_n = \ln 2 \ , \quad (15)$$

and since these numbers are known to many decimals, they may be used as checks on the coefficients. We find that the error in the computed value of C_0 is $-0.1(-31)$. The computed value of $\ln 2$ is correct to 32 places. Suppose the magni-tude of the error in each computed C_n does not exceed $0.1(-31)$. Then from 1.3(46),

$$|ET\ 1| \le 2.382(-28) \ , \quad |ET\ 2| \le 2.458(-25)$$

and the computed $|TEST\ m| < |ET\ m|$, $m = 1,2$. See the printout noted above. From these data, together with numerics based on the use of the backward recursion process with

N = 45 and 50 , we conclude that the magnitude of the error
in each computed value of C_n with N = 40 does not exceed
0.1(-31) . In practice, we will rarely require such accuracy.
But, as pointed out in the discussion following 1.3(45), the
actual cost in taking N much larger than usually needed is
trivial. Further, it is easy to see how many coefficients
are required for accuracy less than, say, 31 places. Since
$|T_n^*(x)| \le 1$ if $0 \le x \le 1$, a bound for the error in a truncated Chebyshev expansion is given by,

$$|R_m(z)| \le \sum_{n=m}^{\infty} |C_n(w)| = R_m^*(z) , \qquad (16)$$

see 1.3(41). Thus, in the present example, for 20 decimal
accuracy we can discard all coefficients C_n , $n \ge 26$.

2. Complex arithmetic and double precision

Let $a = b = 1$, $c = 2$, $w = i$, $N = 22$.
Then from (1) with z replaced by iz , we have

$$- \frac{1}{iz} \ln(1-iz) = \sum_{n=0}^{\infty} C_n(i)T_n^*(z) . \qquad (17)$$

Let

$$C_n(i) = A_n + iB_n . \qquad (18)$$

Then

$$z^{-1} \arctan z = \sum_{n=0}^{\infty} A_n T_n^*(z) , \qquad (19)$$

$$(2z^{-1})\ln(1+z^2) = \sum_{n=0}^{\infty} B_n T_n^*(z) . \qquad (20)$$

For the error analysis,

$u = 0$, $e^{-\delta} = -0.08982 + 0.19737i = 0.21685e^{i\theta}$, $\theta = 1.14369$.

$$(21)$$

Also $A(w) = 1$.   Then from (14), we get

$$t_5 = (-1.31699 - 0.83443i)(-4) , |t_5| = 1.55909(-4) ,$$

$$t_{10} = (1.76349 + 3.73304i)(-8) , |t_{10}| = 4.12862(-8) .$$

The corresponding values deduced by use of the backward recurrence formula with $N = 22$ are

$$C_5 = (-1.27135 - 0.92287i)(-4) , |C_5| = 1.57099(-4) ,$$

$$C_{10} = (1.62876 + 3.80678i)(-8) , |C_{10}| = 4.14058(-8) .$$

For a check on the coefficients, notice from (19) and (20) that

$$\sum_{n=0}^{\infty} A_n = \pi/4 \text{ and } \sum_{n=0}^{\infty} B_n = \tfrac{1}{2} \ln 2 . \tag{22}$$

The corresponding values deduced by use of the backward recursion formula with $N = 22$ are off by two units and one unit in the 16th decimal place respectively.

3.   We consider the binomial $(1+z)^{-a}$.   Let $a = w = 1$.
Then

$$B_n = \varepsilon_n(-)^n 2^{-\frac{1}{2}} (2^{\frac{1}{2}}-1)^{2n} , S = \sum_{n=0}^{\infty} B_n = \tfrac{1}{2} . \tag{23}$$

Notice that the computed values of B_0 and S are in error by 3 units and 1 unit in the 32nd decimal place, respectively.

```
C          ---------- IBM S/370 ----------- MULTIPLE PRECISION -------------
C          ******************************************************************
C          *THIS MAINLINE PROGRAM UTILIZES THE SUBROUTINES CCOEF2 AND CHECK2 *
C          *TO COMPUTE AND TEST THE COEFFICIENTS IN THE CHEBYSHEV EXPANSION  *
C          *OF 2F1(A,B;C;Z).                                                 *
C          *                                                                 *
C          *WHERE NO AMBIGUITY CAN OCCUR, WE SHALL DENOTE 2F1(A,B;C;Z) BY    *
C          *2F1(Z).                                                          *
C          *                                                                 *
C          *DESCRIPTION OF VARIABLES.                                        *
C          *                                                                 *
C          *AP     -INPUT  - PARAMETER A IN 2F1(Z).                          *
C          *BP     -INPUT  - PARAMETER B IN 2F1(Z).                          *
C          *CP     -INPUT  - PARAMETER C IN 2F1(Z).                          *
C          *W      -INPUT  - THIS IS A PRESELECTED SCALE FACTOR SUCH THAT    *
C          *                 0.LE.(Z/W).LE.1.  SEE TEXT.                     *
C          *N      -INPUT  - TWO LESS THAN THE NUMBER OF COEFFICIENTS TO BE  *
C          *                 GENERATED.  MAXIMUM VALUE IS 100.               *
C          *C      -OUTPUT - A VECTOR CONTAINING THE N+2 CHEBYSHEV COEFFI-   *
C          *                 CIENTS FOR THE APPROXIMATION TO 2F1(Z).         *
C          *                 DIMENSION OF THE VECTOR IS SET AT 102.          *
C          *TEST1  -OUTPUT - THE RESULT OF THE FIRST TEST PERFORMED ON THE   *
C          *                 COEFFICIENTS.  SEE TEXT.                        *
C          *TEST2  -OUTPUT - THE RESULT OF THE SECOND TEST PERFORMED ON THE  *
C          *                 COEFFICIENTS.  SEE TEXT.                        *
C          *SUM    -OUTPUT - THE SUM OF THE COEFFICIENTS.                    *
C          *                                                                 *
C          *ALL OTHER VARIABLES ARE FOR INTERNAL USE.                       *
C          *                                                                 *
C          *SUBROUTINES CCOEF2 AND CHECK2 ARE REQUIRED.                      *
C          ******************************************************************
           IMPLICIT REAL*16 (A-H,O-Z)
           DIMENSION C(102)
C          ---------- THE NEXT PROGRAM CARD IS USED TO INITIALIZE ----------
C          ---------- DEVICE NUMBERS FOR THE PARTICULAR INSTALL- ----------
C          ---------- ATION OF THE OPERATING SYSTEM.  IF A SYSTEM ----------
C          ---------- HAS DIFFERENT DEVICE NUMBERS THIS CARD CAN ----------
C          ---------- BE CHANGED TO INITIALIZE THOSE VALUES AS ----------
C          ---------- DEVICE NUMBERS.                            ----------
           DATA NREAD,NWRITE,NPUNCH/5,6,7/
C
C          ---------- READ IN THE INPUT VARIABLES               ----------
           READ(NREAD,2) AP,BP,CP,W
           READ(NREAD,4) N
           N2=N+2
C
C          ---------- CALL SUBROUTINE TO COMPUTE THE COEFFICIENTS ----------
           CALL CCOEF2(AP,BP,CP,W,N,C,SUM)
C
C          ---------- CALL SUBROUTINE TO TEST THE COEFFICIENT   ----------
C          ---------- VALUES                                    ----------
           CALL CHECK2(AP,BP,CP,W,N,C,TEST1,TEST2)
C
C          ---------- WRITE OUT VALUES OF PARAMETERS AND COEFFI- ----------
C          ---------- CIENTS                                     ----------
           WRITE(NWRITE,5) AP,BP,CP,W,N
           DO 1 K=1,N2
           K1=K-1
         1 WRITE(NWRITE,6) K1,C(K)
           WRITE(NWRITE,7) TEST1,TEST2
           WRITE(NWRITE,8) SUM
C
C          ---------- PUNCH THE VALUES OF COEFFICIENTS ON CARDS  ----------
           WRITE(NPUNCH,2) (C(K),K=1,N2)
         2 FORMAT(Q39.32)
         4 FORMAT(I5)
         5 FORMAT('1',4X,'A=',Q39.32/5X,'B=',Q39.32/5X,'C=',Q39.32/5X,'W='
          :,Q39.32/5X,'N=',I5//////5X,'K',25X,'C(K)'/)
         6 FORMAT(1X,I5,8X,Q39.32)
         7 FORMAT(/1X,'TEST1=',Q39.32/1X,'TEST2=',Q39.32)
         8 FORMAT(/1X,'THE SUM OF THE COEFFICIENTS IS',Q39.32)
           STOP
           END
```

```
C
C
C      ------------ IBM S/370 ----------- MULTIPLE PRECISION ------------
C
C      ****************************************************************
C      *THIS SUBROUTINE, CCOEF2, COMPUTES MULTIPLE PRECISION VALUES OF *
C      *THE COEFFICIENTS IN THE CHEBYSHEV EXPANSION OF 2F1(A,B;C;Z).    *
C      *                                                               *
C      *WHERE NO AMBIGUITY CAN OCCUR, WE SHALL DENOTE 2F1(A,B;C;Z) BY   *
C      *2F1(Z).                                                        *
C      *                                                               *
C      *DESCRIPTION OF VARIABLES.                                      *
C      *                                                               *
C      *AP     -INPUT  - PARAMETER A IN 2F1(Z).                        *
C      *                                                               *
C      *BP     -INPUT  - PARAMETER B IN 2F1(Z).                        *
C      *                                                               *
C      *CP     -INPUT  - PARAMETER C IN 2F1(Z).                        *
C      *                                                               *
C      *W      -INPUT  - THIS IS A PRESELECTED SCALE FACTOR SUCH THAT   *
C      *                 0.LE.(Z/W).LE.1.  SEE TEXT.                    *
C      *                                                               *
C      *N      -INPUT  - TWO LESS THAN THE NUMBER OF COEFFICIENTS TO BE *
C      *                 GENERATED.  MAXIMUM VALUE IS 100.              *
C      *                                                               *
C      *C      -OUTPUT - A VECTOR CONTAINING THE N+2 CHEBYSHEV COEFFI-  *
C      *                 CIENTS FOR THE APPROXIMATION TO 2F1(Z).        *
C      *                                                               *
C      *SUM    -OUTPUT - THE SUM OF THE COEFFICIENTS.                   *
C      *                                                               *
C      *ALL OTHER VARIABLES ARE FOR INTERNAL USE.                      *
C      ****************************************************************
       SUBROUTINE CCOEF2(AP,BP,CP,W,N,C,SUM)
       IMPLICIT REAL*16 (A-H,O-Z)
       DIMENSION C(1)
       DATA ZERO,ONE,TWO,THREE,FOUR,START/0.Q0,1.Q0,2.Q0,3.Q0,4.Q0,1.Q-20
      :/
       N1=N+1
       N2=N+2
C
C      ---------- START COMPUTING COEFFICIENTS BY MEANS OF   ----------
C      ---------- BACKWARD RECURRENCE SCHEME                 ----------
       A3=ZERO
       A2=ZERO
       A1=START
       Z1=FOUR/W
       NCOUNT=N2
       C(NCOUNT)=START
       X=N1
       X1=N2
       X3=TWO*X+THREE
       X3A=X+THREE-AP
       X3B=X+THREE-BP
       DO 1 K=1,N1
       X=X-ONE
       X1=X1-ONE
       X2=ONE/(X+TWO)
       X3=X3-TWO
       X3A=X3A-ONE
       X3B=X3B-ONE
       XAB=ONE/((X+AP)*(X+BP))
       NCOUNT=NCOUNT-1
       C(NCOUNT)=X1*(TWO-(X3*X2*XAB*(X1+AP)*(X1+BP))+Z1*(X+CP)*XAB)*A1+
      :((X3A-ONE)*(X3B-ONE)*X3-X3A*X3B*(X3-ONE)+Z1*X1*(X+THREE-CP))*XAB*
      :A2-X1*X2*XAB*X3A*X3B*A3
       A3=A2
       A2=A1
     1 A1=C(NCOUNT)
       C(1)=C(1)/TWO
C
C      ---------- COMPUTE SCALE FACTOR                       ----------
       RHO=C(1)
       SUM=RHO
       P=ONE
       DO 3 I=2,N2
       RHO=RHO-P*C(I)
       SUM=SUM+C(I)
     3 P=-P
C
C      ---------- SCALE THE COEFFICIENTS                     ----------
       DO 4 L=1,N2
     4 C(L)=C(L)/RHO
       SUM=SUM/RHO
       RETURN
       END
```

```
C
C
C           ----------- IBM S/370 ----------- MULTIPLE PRECISION ------------
C     ***********************************************************************
C     *THIS SUBROUTINE, CHECK2, RUNS CHECKS ON THE COEFFICIENTS            *
C     *GENERATED BY SUBROUTINE CCOEF2.  EXCEPT FOR ROUND-OFF ERROR AND     *
C     *TRUNCATION ERROR, BOTH TEST RESULTS SHOULD BE ZERO.  SEE TEXT.      *
C     *                                                                    *
C     *WHERE NO AMBIGUITY CAN OCCUR, WE SHALL DENOTE 2F1(A,B;C;Z) BY       *
C     *2F1(Z).                                                             *
C     *                                                                    *
C     *DESCRIPTION OF VARIABLES.                                           *
C     *AP      -INPUT  - PARAMETER A IN 2F1(Z).                            *
C     *                                                                    *
C     *BP      -INPUT  - PARAMETER B IN 2F1(Z).                            *
C     *                                                                    *
C     *CP      -INPUT  - PARAMETER C IN 2F1(Z).                            *
C     *                                                                    *
C     *W       -INPUT  - THIS IS A PRESELECTED SCALE FACTOR SUCH THAT      *
C     *                  0.LE.(Z/W).LE.1.  SEE TEXT.                       *
C     *                                                                    *
C     *N       -INPUT  - TWO LESS THAN THE NUMBER OF COEFFICIENTS          *
C     *                  GENERATED.  MAXIMUM VALUE IS 100.                 *
C     *                                                                    *
C     *C       -INPUT  - A VECTOR CONTAINING THE N+2 CHEBYSHEV COEFFI-     *
C     *                  CIENTS FOR THE APPROXIMATION TO 2F1(Z).           *
C     *                                                                    *
C     *TEST1 -OUTPUT - THE RESULT OF THE FIRST TEST PERFORMED ON THE       *
C     *                  COEFFICIENTS.  SEE TEXT.                          *
C     *                                                                    *
C     *TEST2 -OUTPUT - THE RESULT OF THE SECOND TEST PERFORMED ON THE      *
C     *                  COEFFICIENTS.  SEE TEXT.                          *
C     *                                                                    *
C     *ALL OTHER VARIABLES ARE FOR INTERNAL USE.                          *
C     ***********************************************************************
      SUBROUTINE CHECK2(AP,BP,CP,W,N,C,TEST1,TEST2)
      IMPLICIT REAL*16 (A-H,O-Z)
      DIMENSION C(1)
      DATA ONE,TWO,THREE,FOUR,TWELVE/1.Q0,2.Q0,3.Q0,4.Q0,12.Q0/
C
C           ---------- COMPUTATION OF LEFT SIDES OF TEST EQUATIONS ----------
      SUM1=-C(2)+FOUR*C(3)
      P=-ONE
      SUM2=TWELVE*C(3)
      COUNT=TWO
      DO 1 I=2,N
      COUNT=COUNT+ONE
      CSQ=COUNT*COUNT
      SUM1=SUM1+P*CSQ*C(I+2)
      SUM2=SUM2+P*CSQ*(CSQ-ONE)*C(I+2)
    1 P=-P
C
C           ---------- COMPUTATION OF TEST VALUES            ----------
      TEST1=SUM1+AP*BP*W/(TWO*CP)
      TEST2=SUM2-THREE*AP*BP*(AP+ONE)*(BP+ONE)*W*W/(FOUR*CP*(CP+ONE))
      RETURN
      END
```

```
A= 0.10000000000000000000000000000000000000Q+01
B= 0.10000000000000000000000000000000000000Q+01
C= 0.20000000000000000000000000000000000000Q+01
W=-0.10000000000000000000000000000000000000Q+01
N=   40
```

K	C(K)
0	0.82842712474619009760337744841941Q+00
1	-0.15104299781559846868373726866587Q+00
2	0.17814748169295961333700053137772Q-01
3	-0.23355046144311115026040755441683Q-02
4	0.32461808182381868961291214025727Q-03
5	-0.46835103656703698622793622714178Q-04
6	0.69349734479080423748020459217567Q-05
7	-0.10467134039238451700819153040358Q-05
8	0.16032280874191251495283590572643Q-06
9	-0.24844196163274462723190778301349Q-07
10	0.38865861782240873863774804947108Q-08
11	-0.61280411591148495130864592023858Q-09
12	0.97263146505908158892659287941965Q-10
13	-0.15524919140828316280283644139281Q-10
14	0.24901844633252672585744470774101Q-11
15	-0.40113015332411126748317743383979Q-12
16	0.64858905433668469274714157379912Q-13
17	-0.10522060351830979241350204888308Q-13
18	0.17120822091353534178356403228304Q-14
19	-0.27932533615001751671265268538017Q-15
20	0.45682103338345828792334180390634Q-16
21	-0.74874767772992259748104952025751Q-17
22	0.12296876817473927517946604632074Q-17
23	-0.20232566073370802205576938317413Q-18
24	0.33345685965477189181734981819634Q-19
25	-0.55043254486368111100947951443952Q-20
26	0.90990047398786758457714063718555Q-21
27	-0.15061354915985739217174278006111Q-21
28	0.24961706453318915042223930808373Q-22
29	-0.41417926158572669760524520198475Q-23
30	0.68797645857392822690252583303071Q-24
31	-0.11439323518640972402005896952870Q-24
32	0.19038894898138773863968001093970Q-25
33	-0.31715620969127430541938071300123Q-26
34	0.52877661192763284972773290249230Q-27
35	-0.88230216067959726699038814671486Q-28
36	0.14732755797966164757704646741440Q-28
37	-0.24616134048572433573332129026495Q-29
38	0.41133360008069273507943764585051Q-30
39	-0.68528822723596146179919702653140Q-31
40	0.11169955195673937137255604775304Q-31
41	-0.15683841199405185706420540951625Q-32

```
TEST1=-0.74360155152791176286491308363204Q-29
TEST2=-0.89570710741534244764152925390769Q-26
```

THE SUM OF THE COEFFICIENTS IS 0.69314718055994530941723212145818Q+00

```
C           ----------- IBM S/370 ----------- MULTIPLE PRECISION -------------
C
C      ************************************************************************
C      *THIS MAINLINE PROGRAM UTILIZES THE SUBROUTINES CCOEF2 AND CHECK2  *
C      *TO COMPUTE AND TEST THE COEFFICIENTS IN THE CHEBYSHEV EXPANSION   *
C      *OF 2F1(A,B;C;Z).                                                  *
C      *WHERE NO AMBIGUITY CAN OCCUR, WE SHALL DENOTE 2F1(A,B;C;Z) BY     *
C      *2F1(Z).                                                           *
C      *                                                                  *
C      *DESCRIPTION OF VARIABLES.                                         *
C      *                                                                  *
C      *AP     -INPUT  - PARAMETER A IN 2F1(Z).                           *
C      *                                                                  *
C      *BP     -INPUT  - PARAMETER B IN 2F1(Z).                           *
C      *                                                                  *
C      *CP     -INPUT  - PARAMETER C IN 2F1(Z).                           *
C      *                                                                  *
C      *W      -INPUT  - THIS IS A PRESELECTED SCALE FACTOR SUCH THAT     *
C      *                 0.LE.(Z/W).LE.1.  SEE TEXT.                      *
C      *                                                                  *
C      *N      -INPUT  - TWO LESS THAN THE NUMBER OF COEFFICIENTS TO BE   *
C      *                 GENERATED.  MAXIMUM VALUE IS 100.                *
C      *                                                                  *
C      *C      -OUTPUT - A VECTOR CONTAINING THE N+2 CHEBYSHEV COEFFI-    *
C      *                 CIENTS FOR THE APPROXIMATION TO 2F1(Z).          *
C      *                 DIMENSION OF THE VECTOR IS SET AT 102.           *
C      *                                                                  *
C      *TEST1  -OUTPUT - THE RESULT OF THE FIRST TEST PERFORMED ON THE    *
C      *                 COEFFICIENTS.  SEE TEXT.                         *
C      *                                                                  *
C      *TEST2  -OUTPUT - THE RESULT OF THE SECOND TEST PERFORMED ON THE   *
C      *                 COEFFICIENTS.  SEE TEXT.                         *
C      *                                                                  *
C      *SUM    -OUTPUT - THE SUM OF THE COEFFICIENTS.                     *
C      *                                                                  *
C      *ALL OTHER VARIABLES ARE FOR INTERNAL USE.                        *
C      *SUBROUTINES CCOEF2 AND CHECK2 ARE REQUIRED.                       *
C      ************************************************************************
       IMPLICIT COMPLEX*16 (A-H,O-Z)
       DIMENSION C(102)
       CALL ERRSET(209,256,-1,1,0,209)
C           ----------- THE NEXT PROGRAM CARD IS USED TO INITIALIZE -----------
C           ----------- DEVICE NUMBERS FOR THE PARTICULAR INSTALL-  -----------
C           ----------- ATION OF THE OPERATING SYSTEM.  IF A SYSTEM -----------
C           ----------- HAS DIFFERENT DEVICE NUMBERS THIS CARD CAN  -----------
C           ----------- BE CHANGED TO INITIALIZE THOSE VALUES AS    -----------
C           ----------- DEVICE NUMBERS.                             -----------
       DATA NREAD,NWRITE,NPUNCH/5,6,7/
C           ----------- READ IN THE INPUT VARIABLES               -----------
       READ(NREAD,2) AP,BP,CP,W
       READ(NREAD,4) N
       N2=N+2
C           ----------- CALL SUBROUTINE TO COMPUTE THE COEFFICIENTS -----------
       CALL CCOEF2(AP,BP,CP,W,N,C,SUM)
C           ----------- CALL SUBROUTINE TO TEST THE COEFFICIENT    -----------
C           ----------- VALUES                                     -----------
       CALL CHECK2(AP,BP,CP,W,N,C,TEST1,TEST2)
C           ----------- WRITE OUT VALUES OF PARAMETERS AND COEFFI- -----------
C           ----------- CIENTS                                     -----------
       WRITE(NWRITE,5) AP,BP,CP,W,N
       DO 1 K=1,N2
       K1=K-1
     1 WRITE(NWRITE,6) K1,C(K)
       WRITE(NWRITE,7) TEST1,TEST2
       WRITE(NWRITE,8) SUM
C           ----------- PUNCH THE VALUES OF COEFFICIENTS ON CARDS  -----------
       WRITE(NPUNCH,2) (C(K),K=1,N2)
     2 FORMAT(2D23.16)
     4 FORMAT(I5)
     5 FORMAT('1',4X,'A=',D23.16,5X,D23.16/5X,'B=',D23.16,5X,D23.16/5X,
      :'C=',D23.16,5X,D23.16/5X,'W=',D23.16,5X,D23.16/5X,'N=',I5/////5X,
      :'K',30X,'C(K)'//24X,'REAL',21X,'IMAGINARY'/)
     6 FORMAT(1X,I5,8X,D23.16,5X,D23.16)
     7 FORMAT(/1X,'TEST1=',D23.16,5X,D23.16/1X,'TEST2=',D23.16,5X,D23.16)
     8 FORMAT(/1X,'THE SUM OF THE COEFFICIENTS IS',D23.16,5X,D23.16)
       STOP
       END
```

```
C
C
C             ----------- IBM S/370 ----------- MULTIPLE PRECISION ------------
C
C   ****************************************************************************
C   *THIS SUBROUTINE, CCOEF2, COMPUTES MULTIPLE PRECISION VALUES OF      *
C   *THE COEFFICIENTS IN THE CHEBYSHEV EXPANSION OF 2F1(A,B;C;Z).        *
C   *                                                                    *
C   *WHERE NO AMBIGUITY CAN OCCUR, WE SHALL DENOTE 2F1(A,B;C;Z) BY       *
C   *2F1(Z).                                                             *
C   *                                                                    *
C   *DESCRIPTION OF VARIABLES.                                           *
C   *                                                                    *
C   *AP      -INPUT  - PARAMETER A IN 2F1(Z).                            *
C   *                                                                    *
C   *BP      -INPUT  - PARAMETER B IN 2F1(Z).                            *
C   *                                                                    *
C   *CP      -INPUT  - PARAMETER C IN 2F1(Z).                            *
C   *                                                                    *
C   *W       -INPUT  - THIS IS A PRESELECTED SCALE FACTOR SUCH THAT      *
C   *                  0.LE.(Z/W).LE.1.  SEE TEXT.                       *
C   *                                                                    *
C   *N       -INPUT  - TWO LESS THAN THE NUMBER OF COEFFICIENTS TO BE    *
C   *                  GENERATED.  MAXIMUM VALUE IS 100.                 *
C   *                                                                    *
C   *C       -OUTPUT - A VECTOR CONTAINING THE N+2 CHEBYSHEV COEFFI-     *
C   *                  CIENTS FOR THE APPROXIMATION TO 2F1(Z).           *
C   *                                                                    *
C   *SUM     -OUTPUT - THE SUM OF THE COEFFICIENTS.                      *
C   *                                                                    *
C   *ALL OTHER VARIABLES ARE FOR INTERNAL USE.                          *
C   ****************************************************************************
      SUBROUTINE CCOEF2(AP,BP,CP,W,N,C,SUM)
      IMPLICIT COMPLEX*16 (A-H,O-Z)
      DIMENSION C(1)
      DATA ZERO,ONE,TWO,THREE,FOUR,START/(0.D0,0.D0),(1.D0,0.D0),(2.D0,
     :0.D0),(3.D0,0.D0),(4.D0,0.D0),(1.D-20,0.D0)/
      N1=N+1
      N2=N+2
C
C             ---------- START COMPUTING COEFFICIENTS BY MEANS OF    ----------
C             ---------- BACKWARD RECURRENCE SCHEME                  ----------
      A3=ZERO
      A2=ZERO
      A1=START
      Z1=FOUR/W
      NCOUNT=N2
      C(NCOUNT)=START
      X=N1
      X1=N2
      X3=TWO*X+THREE
      X3A=X+THREE-AP
      X3B=X+THREE-BP
      DO 1 K=1,N1
      X=X-ONE
      X1=X1-ONE
      X2=ONE/(X+TWO)
      X3=X3-TWO
      X3A=X3A-ONE
      X3B=X3B-ONE
      XAB=ONE/((X+AP)*(X+BP))
      NCOUNT=NCOUNT-1
      C(NCOUNT)=X1*(TWO-(X3*X2*XAB*(X1+AP)*(X1+BP))+Z1*(X+CP)*XAB)*A1+
     :((X3A-ONE)*(X3B-ONE)*X3-X3A*X3B*(X3-ONE)+Z1*X1*(X+THREE-CP))*XAB*
     :A2-X1*X2*XAB*X3A*X3B*A3
      A3=A2
      A2=A1
    1 A1=C(NCOUNT)
      C(1)=C(1)/TWO
C
C             ---------- COMPUTE SCALE FACTOR                        ----------
      RHO=C(1)
      SUM=RHO
      P=ONE
      DO 3 I=2,N2
      RHO=RHO-P*C(I)
      SUM=SUM+C(I)
    3 P=-P
C
C             ---------- SCALE THE COEFFICIENTS                      ----------
      DO 4 L=1,N2
    4 C(L)=C(L)/RHO
      SUM=SUM/RHO
      RETURN
      END
```

```
C
C
C          ----------- IBM S/370 ----------- MULTIPLE PRECISION ------------
C
C     **********************************************************************
C     *THIS SUBROUTINE, CHECK2, RUNS CHECKS ON THE COEFFICIENTS           *
C     *GENERATED BY SUBROUTINE CCOEF2.  EXCEPT FOR ROUND-OFF ERROR AND    *
C     *TRUNCATION ERROR, BOTH TEST RESULTS SHOULD BE ZERO.  SEE TEXT.     *
C     *                                                                   *
C     *WHERE NO AMBIGUITY CAN OCCUR, WE SHALL DENOTE 2F1(A,B;C;Z) BY      *
C     *2F1(Z).                                                            *
C     *                                                                   *
C     *DESCRIPTION OF VARIABLES.                                          *
C     *                                                                   *
C     *AP     -INPUT  - PARAMETER A IN 2F1(Z).                            *
C     *                                                                   *
C     *BP     -INPUT  - PARAMETER B IN 2F1(Z).                            *
C     *                                                                   *
C     *CP     -INPUT  - PARAMETER C IN 2F1(Z).                            *
C     *                                                                   *
C     *W      -INPUT  - THIS IS A PRESELECTED SCALE FACTOR SUCH THAT      *
C     *                 0.LE.(Z/W).LE.1.  SEE TEXT.                       *
C     *                                                                   *
C     *N      -INPUT  - TWO LESS THAN THE NUMBER OF COEFFICIENTS          *
C     *                 GENERATED.  MAXIMUM VALUE IS 100.                 *
C     *                                                                   *
C     *C      -INPUT  - A VECTOR CONTAINING THE N+2 CHEBYSHEV COEFFI-     *
C     *                 CIENTS FOR THE APPROXIMATION TO 2F1(Z).           *
C     *                                                                   *
C     *TEST1 -OUTPUT - THE RESULT OF THE FIRST TEST PERFORMED ON THE      *
C     *                 COEFFICIENTS.  SEE TEXT.                          *
C     *                                                                   *
C     *TEST2 -OUTPUT - THE RESULT OF THE SECOND TEST PERFORMED ON THE     *
C     *                 COEFFICIENTS.  SEE TEXT.                          *
C     *                                                                   *
C     *ALL OTHER VARIABLES ARE FOR INTERNAL USE.                         *
C     **********************************************************************
      SUBROUTINE CHECK2(AP,BP,CP,W,N,C,TEST1,TEST2)
      IMPLICIT COMPLEX*16 (A-H,O-Z)
      DIMENSION C(1)
      DATA ONE,TWO,THREE,FOUR,TWELVE/(1.D0,0.D0),(2.D0,0.D0),(3.D0,0.D0)
     :,(4.D0,0.D0),(12.D0,0.D0)/
C
C          ---------- COMPUTATION OF LEFT SIDES OF TEST EQUATIONS ----------
      SUM1=-C(2)+FOUR*C(3)
      P=-ONE
      SUM2=TWELVE*C(3)
      COUNT=TWO
      DO 1 I=2,N
      COUNT=COUNT+ONE
      CSQ=COUNT*COUNT
      SUM1=SUM1+P*CSQ*C(I+2)
      SUM2=SUM2+P*CSQ*(CSQ-ONE)*C(I+2)
    1 P=-P
C
C          ---------- COMPUTATION OF TEST VALUES            ----------
      TEST1=SUM1+AP*BP*W/(TWO*CP)
      TEST2=SUM2-THREE*AP*BP*(AP+ONE)*(BP+ONE)*W*W/(FOUR*CP*(CP+ONE))
      RETURN
      END
```

```
A=  0.1000000000000000D+01        0.0
B=  0.1000000000000000D+01        0.0
C=  0.2000000000000000D+01        0.0
W=  0.0                           0.1000000000000000D+01
N=    22
```

K	C(K)	
	REAL	IMAGINARY
0	0.9101797211244548D+00	0.1973682269356198D+00
1	-0.1120450966790049D+00	0.1743697795731904D+00
2	-0.1732343340594020D-01	-0.2491378201325795D-01
3	0.4870610014650850D-02	-0.9960775209558887D-03
4	-0.1815524147860348D-03	0.8472852893824749D-03
5	-0.1271350571865098D-03	-0.9228704771392222D-04
6	0.2534903169058566D-04	-0.1463415988977011D-04
7	0.5695178741396069D-06	0.5534628883498776D-05
8	-0.1019571456350395D-05	-0.3375099709370514D-06
9	0.1419216857564363D-06	-0.1545697589607988D-06
10	0.1628758290847193D-07	0.3806781117612367D-07
11	-0.8235326142383239D-08	-0.1644102534206923D-09
12	0.7096138401714257D-09	-0.1489457365437794D-08
13	0.2144480165318662D-09	0.2539732780098760D-09
14	-0.6476066648045335D-10	0.1834127041749770D-10
15	0.2038312318454317D-11	-0.1353655183014563D-10
16	0.2345333850871514D-11	0.1520150628633570D-11
17	-0.4820837402728298D-12	0.3089518090680222D-12
18	-0.1688295569971633D-13	-0.1164402621128035D-12
19	0.2328156170462764D-13	0.6741634494177207D-14
20	-0.3250516002953789D-14	0.3809605881636246D-14
21	-0.4456584930492396D-15	-0.9451143840636146D-15
22	0.2228005709064968D-15	0.3546040030793479D-17
23	-0.2596617848876818D-16	0.3444563163162144D-16

```
TEST1=  0.3094862196604678D-13     0.2341182803178299D-13
TEST2=  0.1240577085503958D-10     0.9254711976764309D-11
```

THE SUM OF THE COEFFICIENTS IS

```
0.7853981633974481D+00        0.3465735902799726D+00
```

```
C     ----------- IBM S/370 ----------- MULTIPLE PRECISION -------------
C
C     **********************************************************************
C     *THIS MAINLINE PROGRAM UTILIZES THE SUBROUTINES CCOEF1 AND CHECK1 *
C     *TO COMPUTE AND TEST THE COEFFICIENTS IN THE CHEBYSHEV EXPANSION  *
C     *OF 1F0(A;Z).                                                     *
C     *                                                                 *
C     *WHERE NO AMBIGUITY CAN OCCUR, WE SHALL DENOTE 1F0(A;Z) BY        *
C     *1F0(Z).                                                          *
C     *                                                                 *
C     *                                                                 *
C     *DESCRIPTION OF VARIABLES.                                        *
C     *                                                                 *
C     *AP     -INPUT  - PARAMETER A IN 1F0(Z).                          *
C     *                                                                 *
C     *W      -INPUT  - THIS IS A PRESELECTED SCALE FACTOR SUCH THAT    *
C     *                 0.LE.(Z/W).LE.1.  SEE TEXT.                     *
C     *                                                                 *
C     *N      -INPUT  - TWO LESS THAN THE NUMBER OF COEFFICIENTS TO BE  *
C     *                 GENERATED.  MAXIMUM VALUE IS 100.               *
C     *                                                                 *
C     *C      -OUTPUT - A VECTOR CONTAINING THE N+2 CHEBYSHEV COEFFI-   *
C     *                 CIENTS FOR THE APPROXIMATION TO 1F0(Z).         *
C     *                 DIMENSION OF THE VECTOR IS SET AT 102.          *
C     *                                                                 *
C     *TEST1  -OUTPUT - THE RESULT OF THE FIRST TEST PERFORMED ON THE   *
C     *                 COEFFICIENTS.  SEE TEXT.                        *
C     *                                                                 *
C     *TEST2  -OUTPUT - THE RESULT OF THE SECOND TEST PERFORMED ON THE  *
C     *                 COEFFICIENTS.  SEE TEXT.                        *
C     *                                                                 *
C     *SUM    -OUTPUT - THE SUM OF THE COEFFICIENTS.                    *
C     *                                                                 *
C     *ALL OTHER VARIABLES ARE FOR INTERNAL USE.                       *
C     *                                                                 *
C     *SUBROUTINES CCOEF1 AND CHECK1 ARE REQUIRED.                     *
C     **********************************************************************
      IMPLICIT REAL*16 (A-H,O-Z)
      DIMENSION C(102)
C     ----------- THE NEXT PROGRAM CARD IS USED TO INITIALIZE -----------
C     ----------- DEVICE NUMBERS FOR THE PARTICULAR INSTALL-  -----------
C     ----------- ATION OF THE OPERATING SYSTEM.  IF A SYSTEM -----------
C     ----------- HAS DIFFERENT DEVICE NUMBERS THIS CARD CAN  -----------
C     ----------- BE CHANGED TO INITIALIZE THOSE VALUES AS    -----------
C     ----------- DEVICE NUMBERS.                             -----------
      DATA NREAD,NWRITE,NPUNCH/5,6,7/
C     ----------- READ IN THE INPUT VARIABLES -----------
      READ(NREAD,2) AP,W
      READ(NREAD,4) N
      N2=N+2
C     ----------- CALL SUBROUTINE TO COMPUTE THE COEFFICIENTS -----------
      CALL CCOEF1(AP,W,N,C,SUM)
C     ----------- CALL SUBROUTINE TO TEST THE COEFFICIENT -----------
C     ----------- VALUES                                  -----------
      CALL CHECK1(AP,W,N,C,TEST1,TEST2)
C     ----------- WRITE OUT VALUES OF PARAMETERS AND COEFFI- -----------
C     ----------- CIENTS                                     -----------
      WRITE(NWRITE,5) AP,W,N
      DO 1 K=1,N2
      K1=K-1
    1 WRITE(NWRITE,6) K1,C(K)
      WRITE(NWRITE,7) TEST1,TEST2
      WRITE(NWRITE,8) SUM
C     ----------- PUNCH THE VALUES OF COEFFICIENTS ON CARDS  -----------
      WRITE(NPUNCH,2) (C(K),K=1,N2)
    2 FORMAT(Q39.32)
    4 FORMAT(I5)
    5 FORMAT('1',3X,'AP=',Q39.32/5X,'W=',Q39.32/5X,'N=',I5/////5X,'K',
     125X,'C(K)'/)
    6 FORMAT(1X,I5,8X,Q39.32)
    7 FORMAT(/1X,'TEST1=',Q12.5/1X,'TEST2=',Q12.5)
    8 FORMAT(/1X,'THE SUM OF THE COEFFICIENTS IS',Q39.32)
      STOP
      END
```

```
C
C          ---------- IBM S/370 ---------- MULTIPLE PRECISION -------------
C
C     *********************************************************************
C     *THIS SUBROUTINE, CCOEF1, COMPUTES MULTIPLE PRECISION VALUES OF     *
C     *THE COEFFICIENTS IN THE CHEBYSHEV EXPANSION OF 1F0(A;Z).           *
C     *                                                                   *
C     *WHERE NO AMBIGUITY CAN OCCUR, WE SHALL DENOTE 1F0(A;Z) BY          *
C     *1F0(Z).                                                            *
C     *                                                                   *
C     *DESCRIPTION OF VARIABLES.                                          *
C     *                                                                   *
C     *AP      -INPUT  - PARAMETER A IN 1F0(Z).                           *
C     *                                                                   *
C     *W       -INPUT  - THIS IS A PRESELECTED SCALE FACTOR SUCH THAT     *
C     *                  0.LE.(Z/W).LE.1.  SEE TEXT.                      *
C     *                                                                   *
C     *N       -INPUT  - TWO LESS THAN THE NUMBER OF COEFFICIENTS TO BE   *
C     *                  GENERATED.  MAXIMUM VALUE IS 100.                *
C     *                                                                   *
C     *C       -OUTPUT - A VECTOR CONTAINING THE N+2 CHEBYSHEV COEFFI-    *
C     *                  CIENTS FOR THE APPROXIMATION TO 1F0(Z).          *
C     *                                                                   *
C     *SUM     -OUTPUT - THE SUM OF THE COEFFICIENTS.                     *
C     *                                                                   *
C     *ALL OTHER VARIABLES ARE FOR INTERNAL USE.                         *
C     *********************************************************************
      SUBROUTINE CCOEF1(AP,W,N,C,SUM)
      IMPLICIT REAL*16 (A-H,O-Z)
      DIMENSION C(1)
      DATA ZERO,ONE,TWO,FOUR,START/0.Q0,1.Q0,2.Q0,4.Q0,1.Q-20/
      N1=N+1
      N2=N+2
C
C          ---------- START COMPUTING COEFFICIENTS BY MEANS OF   ----------
C          ---------- BACKWARD RECURRENCE SCHEME
      A2=ZERO
      A1=START
      NCOUNT=N2
      C(NCOUNT)=START
      X1=N2
      V1=ONE-AP
      AFAC=TWO+FOUR/W
      DO 2 K=1,N1
      X1=X1-ONE
      NCOUNT=NCOUNT-1
      C(NCOUNT)=-(X1*AFAC*A1+(X1+V1)*A2)/(X1-V1)
      A2=A1
    2 A1=C(NCOUNT)
      C(1)=C(1)/TWO
C
C          ---------- COMPUTE SCALE FACTOR                       ----------
      RHO=C(1)
      SUM=RHO
      P=ONE
      DO 3 I=2,N2
      RHO=RHO-P*C(I)
      SUM=SUM+C(I)
    3 P=-P
C
C          ---------- SCALE THE COEFFICIENTS                     ----------
      DO 4 L=1,N2
    4 C(L)=C(L)/RHO
      SUM=SUM/RHO
      RETURN
      END
```

```
C
C
C          ----------- IBM S/370 ----------- MULTIPLE PRECISION -------------
C     ******************************************************************
C     *THIS SUBROUTINE, CHECK1, RUNS CHECKS ON THE COEFFICIENTS        *
C     *GENERATED BY SUBROUTINE CCOEF1.   EXCEPT FOR ROUND-OFF ERROR AND *
C     *TRUNCATION ERROR, BOTH TEST RESULTS SHOULD BE ZERO.  SEE TEXT.   *
C     *                                                                *
C     *WHERE NO AMBIGUITY CAN OCCUR, WE SHALL DENOTE 1F0(A;Z) BY        *
C     *1F0(Z).                                                         *
C     *                                                                *
C     *DESCRIPTION OF VARIABLES.                                       *
C     *                                                                *
C     *AP    -INPUT  - PARAMETER A IN 1F0(Z).                          *
C     *                                                                *
C     *W     -INPUT  - THIS IS A PRESELECTED SCALE FACTOR SUCH THAT     *
C     *                0.LE.(Z/W).LE.1.  SEE TEXT.                     *
C     *                                                                *
C     *N     -INPUT  - TWO LESS THAN THE NUMBER OF COEFFICIENTS        *
C     *                GENERATED.  MAXIMUM VALUE IS 100.               *
C     *                                                                *
C     *C     -INPUT  - A VECTOR CONTAINING THE N+2 CHEBYSHEV COEFFI-    *
C     *                CIENTS FOR THE APPROXIMATION TO 1F0(Z).          *
C     *                                                                *
C     *TEST1 -OUTPUT - THE RESULT OF THE FIRST TEST PERFORMED ON THE    *
C     *                COEFFICIENTS.  SEE TEXT.                         *
C     *                                                                *
C     *TEST2 -OUTPUT - THE RESULT OF THE SECOND TEST PERFORMED ON THE   *
C     *                COEFFICIENTS.  SEE TEXT.                         *
C     *                                                                *
C     *ALL OTHER VARIABLES ARE FOR INTERNAL SUBROUTINE USE.            *
C     ******************************************************************
      SUBROUTINE CHECK1(AP,W,N,C,TEST1,TEST2)
      IMPLICIT REAL*16 (A-H,O-Z)
      DIMENSION C(1)
      DATA ONE,TWO,THREE,FOUR,TWELVE/1.Q0,2.Q0,3.Q0,4.Q0,12.Q0/
C
C          ---------- COMPUTATION OF LEFT SIDES OF TEST EQUATIONS -----------
      SUM1=-C(2)+FOUR*C(3)
      P=-ONE
      SUM2=TWELVE*C(3)
      COUNT=TWO
      DO 1 I=2,N
      COUNT=COUNT+ONE
      CSQ=COUNT*COUNT
      SUM1=SUM1+P*CSQ*C(I+2)
      SUM2=SUM2+P*CSQ*(CSQ-ONE)*C(I+2)
    1 P=-P
C
C          ---------- COMPUTATION OF TEST VALUES                -----------
      TEST1=SUM1-AP*W/TWO
      TEST2=SUM2-THREE*AP*(AP+ONE)*W*W/FOUR
      RETURN
      END
```

```
AP= 0.10000000000000000000000000000000000000Q+01
 W= 0.10000000000000000000000000000000000000Q+01
 N=    40
```

K	C(K)
0	0.70710678118654752440084436210488Q+00
1	-0.24264068711928514640506617262910Q+00
2	0.41630560342615829628708311564869Q-01
3	-0.71426749364098313671836967601101Q-02
4	0.12254892758431585743938689957914Q-02
5	-0.21026071864912007917951721463824Q-03
6	0.36075036051561900683234292038046Q-04
7	-0.61894976602513249198885375900353Q-05
8	0.10619499099460488360969335021661Q-05
9	-0.18220179942496809669306342296102Q-06
10	0.31260886603759744061447035600065Q-07
11	-0.53635201975903676756187906393665Q-08
12	0.92023458178246199226570823613437Q-09
13	-0.15788729310440427797545877743971Q-09
14	0.27089176843963675587044428503871Q-10
15	-0.46477679593777755468077935835197Q-11
16	0.79743091230297769380233299724731Q-12
17	-0.13681751444009061600620439996410Q-12
18	0.23474174337566002234893402537290Q-13
19	-0.40275315853053974031560152596386Q-14
20	0.69101517426638218404268902054144Q-15
21	-0.11855946029289570110011886361010Q-15
22	0.20341587490992022558024161119143Q-16
23	-0.34900646530564342480261031047583Q-17
24	0.59880042734658293013245750940715Q-18
25	-0.10273791102306333276864195168460Q-18
26	0.17627038791797066479394200700468Q-19
27	-0.30243212727190661077232525182024Q-20
28	0.51889157451733016694531440874637Q-21
29	-0.89027719384914893486633934275895Q-22
30	0.15274741792159196746489196908994Q-22
31	-0.26207313680402865303012471780689Q-23
32	0.44964641608252243531828615941924Q-24
33	-0.77147128454848081608469778446587Q-25
34	0.13236354646566054332532511260277Q-25
35	-0.22709994245482443867252891150722Q-26
36	0.38964190072341198781922343015648Q-27
37	-0.66851979792227540190051465866674Q-28
38	0.11469978029953253321085365043567Q-28
39	-0.19678883749197973646072439472297Q-29
40	0.33735229499862509767898132481081Q-30
41	-0.56225382499770849613163554135134Q-31

```
TEST1=-0.24785Q-28
TEST2=-0.43789Q-25

THE SUM OF THE COEFFICIENTS IS 0.49999999999999999999999999999999Q+00
```

V COEFFICIENTS FOR THE EXPANSION OF THE CONFLUENT HYPERGEOMETRIC FUNCTION $_1F_1(a;c;z)$ IN ASCENDING SERIES OF CHEBYSHEV POLYNOMIALS

We are concerned with evaluation of $C_n(w)$ in

$$_1F_1(a;c;z) = \sum_{n=0}^{\infty} C_n(w)T_n^*(z/w) \quad , \quad 0 \leq z/w \leq 1 \quad , \tag{1}$$

$$C_n(w) = \frac{\varepsilon_n(a)_n w^n}{2^{2n}(c)_n n!} \; {}_2F_2 \left(\begin{array}{c} a+n,\frac{1}{2}+n \\ c+n,1+2n \end{array} \middle| \; w \right) \quad . \tag{2}$$

This is the special case of 1.3(1,2) with $p = q = 1$. It is also a confluent form of 4(1,2), for if in the latter we replace z by z/b and w by w/b and then let $b \to \infty$, we get (1) and (2) above. Except for the recurrence formula for $C_n(w)$, all necessary information will be found in 1.3. Thus none of this material will be repeated, save for asymptotic forms for $C_n(w)$ for large n . We have

$$C_n(w) = \frac{2(a)_n w^n e^{w/2}}{2^{2n}(c)_n n!} \; [1 + \frac{w^2+8w(a-c)}{16n} + O(n^{-2})] \quad , \tag{3}$$

or

$$C_n(w) = \frac{2\Gamma(c)n^{a-c}w^n e^{w/2}}{\Gamma(a)} \; [1 + \frac{(a-c)(a+c-1)}{2n} + O(n^{-2})]$$

$$\times \; [1 + \frac{w^2+8w(a-c)}{16n} + O(n^{-2})] \quad . \tag{4}$$

The recurrence formula for $C_n(w)$ follows from 4(7) by confluence. Thus

$$\frac{2C_n(w)}{\varepsilon_n} = \frac{(n+1)}{(n+2)(n+a)} \{ \frac{4(n+c)(n+2)}{w} - (n+3-a) \} \; C_{n+1}(w)$$

$$+ \; \{ 1 + \frac{4(n+1)(n+3-c)}{(n+a)w} \} C_{n+2}(w) \; + \; \frac{(n+1)(n+3-a)}{(n+2)(n+a)} \; C_{n+3}(w) \quad . \tag{5}$$

If $a = c$, the analysis simplifies since (1) and (2) become

$$e^z = \sum_{n=0}^{\infty} C_n(w) T_n^*(z/w) \ , \ 0 \le z/w \le 1 \ , \tag{6}$$

$$C_n(w) = \varepsilon_n e^{w/2} I_n(w/2) \tag{7}$$

where $I_n(w/2)$ is the modified Bessel function of the first kind. The previous developments are valid for arbitrary $a = c$, say $a = c = 1$, but in place of (5) , we can do with the simpler recursion formula

$$\frac{2C_n(w)}{\varepsilon_n} = \frac{4(n+1)}{w} C_{n+1}(w) + C_{n+2}(w) \ , \tag{8}$$

which follows from 4(10) by confluence. That is, replace w by $-w/a$ and let $a \to \infty$. For this situation, we have the relations

$$e^w = \sum_{n=0}^{\infty} C_n(w) \ , \tag{9}$$

$$e^w = C_0^2 + \tfrac{1}{2} \sum_{n=1}^{\infty} (-)^n C_n^2(w) \ , \tag{10}$$

which are useful for further checks on the computations. We do not give a separate program for the exponential function, as the one given will do with $a = c = 1$. For details on the expansion

$$z^a [\Gamma(b)\Gamma(c)]^{-1} {}_1F_1(a;c;-z) = z^a e^{-z} [\Gamma(b)\Gamma(c)]^{-1} {}_1F_1(c-a;c;z)$$

$$= \sum_{n=0}^{\infty} D_n(w) T_n^*(w/z) \ , \ z/w \ge 1, \tag{11}$$

see 8(7).

Numerical Example

Let

$$a = 0.5 \ , \ c = 1.5 \ , \ w = -8 \ , \ N = 35 \ .$$

Then we can write

$$f(z) = {}_1F_1(0.5;1.5;-z) = \sum_{n=0}^{\infty} C_n T_n^*(z/8), \quad 0 \le z \le 8 ,$$

$$f(z) = z^{-\frac{1}{2}} \mathrm{Erf}(z^{\frac{1}{2}}) . \tag{12}$$

Let t_n denote the right hand side of (3) with $O(n^{-2})$ omitted. Then

$$t_n = \frac{(-)^n 2^{n+1}(1.83156) \cdot 10^{-2}(n+8)}{(n!)(2n+1)n} ,$$

$$\frac{t_{n+1}}{t_n} = - \frac{2n(2n+1)(n+9)}{(n+1)^2(n+8)(2n+3)} . \tag{13}$$

Computed values of C_n determined by the backward recursion process with $N = 35$ are given in the printout accompanying the FORTRAN program at the end of this chapter. Some truncated values from this list are given below to enable comparison with values from (13).

n	t_n	C_n	t_n/t_{n-1}	C_n/C_{n-1}
20	0.53910(-15)	0.55941(-15)	—	—
21	-0.48289(-16)	-0.49966(-16)	-0.08957	-0.08932
25	-0.20510(-20)	-0.21040(-20)	—	—
26	0.15040(-21)	0.15403(-21)	-0.07333	-0.07321

```
C    ---------- IBM S/370 ---------- MULTIPLE PRECISION -----------
C    ***************************************************************
C    *THIS MAINLINE PROGRAM UTILIZES THE SUBROUTINES CCOEF3 AND CHECK3 *
C    *TO COMPUTE AND TEST THE COEFFICIENTS IN THE CHEBYSHEV EXPANSION *
C    *OF 1F1(A;C;Z).                                                *
C    *                                                              *
C    *WHERE NO AMBIGUITY CAN OCCUR, WE SHALL DENOTE 1F1(A;C;Z) BY   *
C    *1F1(Z).                                                       *
C    *                                                              *
C    *DESCRIPTION OF VARIABLES.                                     *
C    *                                                              *
C    *AP      -INPUT  - PARAMETER A IN 1F1(Z).                      *
C    *                                                              *
C    *CP      -INPUT  - PARAMETER C IN 1F1(Z).                      *
C    *                                                              *
C    *W       -INPUT  - THIS IS A PRESELECTED SCALE FACTOR SUCH THAT*
C    *                  0.LE.(Z/W).LE.1.  SEE TEXT.                 *
C    *                                                              *
C    *N       -INPUT  - TWO LESS THAN THE NUMBER OF COEFFICIENTS TO BE*
C    *                  GENERATED.  MAXIMUM VALUE IS 100.           *
C    *                                                              *
C    *C       -OUTPUT - A VECTOR CONTAINING THE N+2 CHEBYSHEV COEFFI-*
C    *                  CIENTS FOR THE APPROXIMATION TO 1F1(Z).     *
C    *                  DIMENSION OF THE VECTOR IS SET AT 102.      *
C    *TEST1  -OUTPUT - THE RESULT OF THE FIRST TEST PERFORMED ON THE*
C    *                  COEFFICIENTS.  SEE TEXT.                    *
C    *TEST2  -OUTPUT - THE RESULT OF THE SECOND TEST PERFORMED ON THE*
C    *                  COEFFICIENTS.  SEE TEXT.                    *
C    *SUM    -OUTPUT - THE SUM OF THE COEFFICIENTS.                 *
C    *                                                              *
C    *ALL OTHER VARIABLES ARE FOR INTERNAL USE.                    *
C    *SUBROUTINES CCOEF3 AND CHECK3 ARE REQUIRED.                  *
C    ***************************************************************
      IMPLICIT REAL*16 (A-H,O-Z)
      DIMENSION C(102)
C    ---------- THE NEXT PROGRAM CARD IS USED TO INITIALIZE -----------
C    ---------- DEVICE NUMBERS FOR THE PARTICULAR INSTALL- -----------
C    ---------- ATION OF THE OPERATING SYSTEM.  IF A SYSTEM -----------
C    ---------- HAS DIFFERENT DEVICE NUMBERS THIS CARD CAN -----------
C    ---------- BE CHANGED TO INITIALIZE THOSE VALUES AS  -----------
C    ---------- DEVICE NUMBERS.                           -----------
      DATA NREAD,NWRITE,NPUNCH/5,6,7/
C    ---------- READ IN THE INPUT VARIABLES               -----------
      READ(NREAD,2) AP,CP,W
      READ(NREAD,4) N
      N2=N+2
C    ---------- CALL SUBROUTINE TO COMPUTE THE COEFFICIENTS -----------
      CALL CCOEF3(AP,CP,W,N,C,SUM)
C    ---------- CALL SUBROUTINE TO TEST THE COEFFICIENT   -----------
C    ---------- VALUES                                    -----------
      CALL CHECK3(AP,CP,W,N,C,TEST1,TEST2)
C    ---------- WRITE OUT VALUES OF PARAMETERS AND COEFFI- -----------
C    ---------- CIENTS                                    -----------
      WRITE(NWRITE,5) AP,CP,W,N
      DO 1 K=1,N2
      K1=K-1
    1 WRITE(NWRITE,6) K1,C(K)
      WRITE(NWRITE,7) TEST1,TEST2
      WRITE(NWRITE,8) SUM
C    ---------- PUNCH THE VALUES OF COEFFICIENTS ON CARDS -----------
      WRITE(NPUNCH,2) (C(K),K=1,N2)
    2 FORMAT(Q39.32)
    4 FORMAT(I5)
    5 FORMAT('1',4X,'A=',Q39.32/5X,'C=',Q39.32/5X,'W=',Q39.32/5X,'N=',
     :I5/////5X,'K',25X,'C(K)'/)
    6 FORMAT(1X,I5,8X,Q39.32)
    7 FORMAT(/1X,'TEST1=',Q39.32/1X,'TEST2=',Q39.32)
    8 FORMAT(/1X,'THE SUM OF THE COEFFICIENTS IS',Q39.32)
      STOP
      END
```

```
C
C
C          ------------ IBM S/370 ----------- MULTIPLE PRECISION -----------
C
C          *****************************************************************
C          *THIS SUBROUTINE, CCOEFF3, COMPUTES MULTIPLE PRECISION VALUES OF*
C          *THE COEFFICIENTS IN THE CHEBYSHEV EXPANSION OF 1F1(A;C;Z).     *
C          *                                                               *
C          *WHERE NO AMBIGUITY CAN OCCUR, WE SHALL DENOTE 1F1(A;C;Z) BY    *
C          *1F1(Z).                                                        *
C          *                                                               *
C          *                                                               *
C          *DESCRIPTION OF VARIABLES.                                      *
C          *                                                               *
C          *AP      -INPUT  - PARAMETER A IN 1F1(Z).                       *
C          *                                                               *
C          *CP      -INPUT  - PARAMETER C IN 1F1(Z).                       *
C          *                                                               *
C          *W       -INPUT  - THIS IS A PRESELECTED SCALE FACTOR SUCH THAT *
C          *                  0.LE.(Z/W).LE.1.  SEE TEXT.                  *
C          *                                                               *
C          *N       -INPUT  - TWO LESS THAN THE NUMBER OF COEFFICIENTS TO BE*
C          *                  GENERATED.  MAXIMUM VALUE IS 100.            *
C          *                                                               *
C          *C       -OUTPUT - A VECTOR CONTAINING THE N+2 CHEBYSHEV COEFFI-*
C          *                  CIENTS FOR THE APPROXIMATION TO 1F1(Z).      *
C          *                                                               *
C          *SUM     -OUTPUT - THE SUM OF THE COEFFICIENTS.                 *
C          *                                                               *
C          *ALL OTHER VARIABLES ARE FOR INTERNAL USE.                     *
C          *****************************************************************
           SUBROUTINE CCOEFF3(AP,CP,W,N,C,SUM)
           IMPLICIT REAL*16 (A-H,O-Z)
           DIMENSION C(1)
           DATA ZERO,ONE,TWO,THREE,FOUR,START/0.Q0,1.Q0,2.Q0,3.Q0,4.Q0,1.Q-20
          :/
           N1=N+1
           N2=N+2
C
C          ---------- START COMPUTING COEFFICIENTS BY MEANS OF  -----------
C          ---------- BACKWARD RECURRENCE SCHEME
           A3=ZERO
           A2=ZERO
           A1=START
           Z1=FOUR/W
           NCOUNT=N2
           C(NCOUNT)=START
           X=N1
           X1=N2
           XA=X+AP
           X3A=X+THREE-AP
           DO 1 K=1,N1
           X=X-ONE
           X1=X1-ONE
           XA=XA-ONE
           X2=ONE/(X+TWO)
           X3A=X3A-ONE
           NCOUNT=NCOUNT-1
           C(NCOUNT)=X1*(((X+CP)*Z1-X3A*X2)*A1+(Z1*(X+THREE-CP)+XA/X1)*
          :A2+X3A*X2*A3)/XA
           A3=A2
           A2=A1
         1 A1=C(NCOUNT)
           C(1)=C(1)/TWO
C
C          ---------- COMPUTE SCALE FACTOR                      -----------
           RHO=C(1)
           SUM=RHO
           P=ONE
           DO 3 I=2,N2
           RHO=RHO-P*C(I)
           SUM=SUM+C(I)
         3 P=-P
C
C          ---------- SCALE THE COEFFICIENTS                    -----------
           DO 4 L=1,N2
         4 C(L)=C(L)/RHO
           SUM=SUM/RHO
           RETURN
           END
```

```
C       ----------- IBM S/370 ----------- MULTIPLE PRECISION -------------
C       ****************************************************************
C       *THIS SUBROUTINE, CHECK3, RUNS CHECKS ON THE COEFFICIENTS     *
C       *GENERATED BY SUBROUTINE CCOEF3.  EXCEPT FOR ROUND-OFF ERROR AND *
C       *TRUNCATION ERROR, BOTH TEST RESULTS SHOULD BE ZERO.  SEE TEXT. *
C       *                                                             *
C       *WHERE NO AMBIGUITY CAN OCCUR, WE SHALL DENOTE 1F1(A;C;Z) BY  *
C       *1F1(Z).                                                      *
C       *                                                             *
C       *DESCRIPTION OF VARIABLES.                                    *
C       *                                                             *
C       *AP      -INPUT  - PARAMETER A IN 1F1(Z).                     *
C       *                                                             *
C       *CP      -INPUT  - PARAMETER C IN 1F1(Z).                     *
C       *                                                             *
C       *W       -INPUT  - THIS IS A PRESELECTED SCALE FACTOR SUCH THAT *
C       *                  0.LE.(Z/W).LE.1.  SEE TEXT.                *
C       *N       -INPUT  - TWO LESS THAN THE NUMBER OF COEFFICIENTS   *
C       *                  GENERATED.  MAXIMUM VALUE IS 100.          *
C       *C       -INPUT  - A VECTOR CONTAINING THE N+2 CHEBYSHEV COEFFI- *
C       *                  CIENTS FOR THE APPROXIMATION TO 1F1(Z).    *
C       *TEST1  -OUTPUT - THE RESULT OF THE FIRST TEST PERFORMED ON THE *
C       *                  COEFFICIENTS.  SEE TEXT.                   *
C       *TEST2  -OUTPUT - THE RESULT OF THE SECOND TEST PERFORMED ON THE *
C       *                  COEFFICIENTS.  SEE TEXT.                   *
C       *                                                             *
C       *ALL OTHER VARIABLES ARE FOR INTERNAL USE.                    *
C       ****************************************************************
        SUBROUTINE CHECK3(AP,CP,W,N,C,TEST1,TEST2)
        IMPLICIT REAL*16 (A-H,O-Z)
        DIMENSION C(1)
        DATA ONE,TWO,THREE,FOUR,TWELVE/1.Q0,2.Q0,3.Q0,4.Q0,12.Q0/
C
C       ---------- COMPUTATION OF LEFT SIDES OF TEST EQUATIONS -----------
        SUM1=-C(2)+FOUR*C(3)
        P=-ONE
        SUM2=TWELVE*C(3)
        COUNT=TWO
        DO 1 I=2,N
        COUNT=COUNT+ONE
        CSQ=COUNT*COUNT
        SUM1=SUM1+P*CSQ*C(I+2)
        SUM2=SUM2+P*CSQ*(CSQ-ONE)*C(I+2)
      1 P=-P
C
C       ---------- COMPUTATION OF TEST VALUES           -----------
        TEST1=SUM1+AP*W/(TWO*CP)
        TEST2=SUM2-THREE*AP*(AP+ONE)*W*W/(FOUR*CP*(CP+ONE))
        RETURN
        END
```

```
A=  0.5000000000000000000000000000000000000Q+00
C=  0.1500000000000000000000000000000000000Q+01
W=-0.8000000000000000000000000000000000000Q+01
N=   35
```

K	C(K)
0	0.5380941125171047870461324762229600Q+00
1	-0.3046827035813655242746843698928000Q+00
2	0.1070978095978959500067273234783100Q+00
3	-0.3559747379692181920124354614914500Q-01
4	0.1072195023751156546539084106721300Q-01
5	-0.2903207214978966093061470786889300Q-02
6	0.7080307297827946636469930569004100Q-03
7	-0.1563220863638070555110251389090800Q-03
8	0.3143411589408767619960335412073700Q-04
9	-0.5791832186131531277988336494432400Q-05
10	0.9834140209492875191324492610083900Q-06
11	-0.1546798728896917972567479146632100Q-06
12	0.2264492330271624649563639951673600Q-07
13	-0.3098930584575162169076783377573500Q-08
14	0.3979604976089642289005452480956900Q-09
15	-0.4812560284677900105415231683894700Q-10
16	0.5497909938132053193404455651589900Q-11
17	-0.5950512122345899710588136529017400Q-12
18	0.6117645674439301679189334992302100Q-13
19	-0.5988617171216476468599364536734580Q-14
20	0.5594129509106647065687426771321000Q-15
21	-0.4996617385403336943179629484974000Q-16
22	0.4275256539858735039619376058726300Q-17
23	-0.3510218532181917579593813925432300Q-18
24	0.2770003946838159887888463254344500Q-19
25	-0.2103971710610274013039982108167000Q-20
26	0.1540307002690432342483641438557700Q-21
27	-0.1088271123706968664018716093840900Q-22
28	0.7429295172261383096003479554363500Q-24
29	-0.4905961903901353975211835459462400Q-25
30	0.3137055414708653878278482454924500Q-26
31	-0.1944329000876972803046291379462100Q-27
32	0.1169148832138828070231640001819200Q-28
33	-0.6826522294745374284022606050444100Q-30
34	0.3873150106859167295280842627147700Q-31
35	-0.2132179271538023565114019798465700Q-32
36	0.1091475690075682588088830450465Q-33

```
TEST1=-0.1448299318179201361499095412282390Q-30
TEST2=-0.8556983780110264384561832238022250Q-28
```

THE SUM OF THE COEFFICIENTS IS 0.3133086873213071691635316143060040Q+00

VI CHEBYSHEV COEFFICIENTS FOR $_0F_1(c;z)$

We are concerned with the evaluation of $C_n(w)$ in

$$_0F_1(c;z) = \sum_{n=0}^{\infty} C_n(w)T_n^*(z/w) \;,\; 0 \le z/w \le 1 \;, \tag{1}$$

$$C_n(w) = \frac{\varepsilon_n w^n}{2^{2n}(c)_n n!} \; _1F_2(\tfrac{1}{2}+n;c+n,1+2n;w) \;. \tag{2}$$

This is the special case of 1.3(1,2) with $p = 0$, $q = 1$. It is also a confluent form of 5(1,2) for if in the latter we replace z by z/a and w by w/a and then let $a \to \infty$, we get (1) and (2) above. Except for the recurrence formula for $C_n(w)$, all necessary information will be found in 1.3. Thus, none of this material will be repeated, save for the asymptotic form of $C_n(w)$ for large n . We have

$$C_n(w) = \frac{2w^n \Gamma(c)n^{1-c}}{2^{2n}(n!)^2} \left[1 - \frac{c(c-1)}{2n} + \frac{c(c^2-1)(3c-2)}{24n^2} + O(n^{-3})\right] \tag{3}$$

$$\times \; [1 + \frac{w}{2(n+c)} + \frac{w^2(2n+3)}{8(n+c)_2(2n+2)} + \frac{w^3(2n+5)}{48(n+c)_3(2n+2)} + O(n^{-4})].$$

The recurrence formula for $C_n(w)$ follows from 5(5) by confluence. Thus,

$$\frac{2C_n(w)}{\varepsilon_n} = \frac{(n+1)}{(n+2)} \{C_{n+1}(w) - C_{n+3}(w)\} + \frac{4(n+1)(n+c)}{w} C_{n+1}(w)$$

$$+ \{1 + \frac{4(n+1)(n+3-c)}{w}\} C_{n+2}(w) \;. \tag{4}$$

Numerical Example

Let

$$c = -0.5 \text{ , } w = -16 \text{ .}$$

Then we can write

$$f(z) = {}_0F_1(-0.5;-\frac{z^2}{4}) = \sum_{n=0}^{\infty} C_n T_{2n}(z/8), \ 0 \le z \le 8 \text{ ,}$$

$$f(z) = \cos z + z \sin z \text{ .} \tag{5}$$

Let t_n be the right hand side of (3) with the order terms omitted. Then

$$t_n = \frac{(-)^{n+1} 2^{2n+2} (n\pi)^{\frac{1}{2}} n}{(n!)^2} [1 - \frac{3}{8n} - \frac{21}{384n^2}]$$

$$\times [1 - \frac{16}{2n-1} + \frac{128(2n+3)}{(2n-1)(2n+1)(2n+2)}$$

$$- \frac{2048(2n+5)}{3(2n-1)(2n+1)(2n+2)(2n+3)}] \text{ .} \tag{6}$$

We find $t_{20} = -0.76399(-22)$ and $t_{25} = 0.29419(-32)$, which should be compared with $C_{20} = -0.76508(-22)$ and $C_{25} = 0.29433(-32)$, respectively, as found by use of the backward recurrence formula with $N = 25$. For a further check, from (5),

$$\sum_{n=0}^{\infty} C_n = f(8) = \cos 8 + 8 \sin 8 \text{ .} \tag{7}$$

The sum of the coefficients found by use of the backward re-currence formula with $N = 25$ differs by a unit in the 31st decimal place from the value of $f(8)$ determined by use of the quadruple precision routines for cos and sin already built into the computer system.

```
C
C           ------------ IBM S/370 ----------- MULTIPLE PRECISION ------------
C
C           *************************************************************************
C           *THIS MAINLINE PROGRAM UTILIZES THE SUBROUTINES CCOEF4 AND CHECK4 *
C           *TO COMPUTE AND TEST THE COEFFICIENTS IN THE CHEBYSHEV EXPANSION  *
C           *OF 0F1(;C;Z).                                                    *
C           *                                                                 *
C           *WHERE NO AMBIGUITY CAN OCCUR, WE SHALL DENOTE 0F1(;C;Z) BY       *
C           *0F1(Z).                                                          *
C           *                                                                 *
C           *                                                                 *
C           *DESCRIPTION OF VARIABLES.                                        *
C           *                                                                 *
C           *CP     -INPUT  - PARAMETER C IN 0F1(Z).                          *
C           *                                                                 *
C           *W      -INPUT  - THIS IS A PRESELECTED SCALE FACTOR SUCH THAT    *
C           *                 0.LE.(Z/W).LE.1.  SEE TEXT.                      *
C           *                                                                 *
C           *N      -INPUT  - TWO LESS THAN THE NUMBER OF COEFFICIENTS TO BE  *
C           *                 GENERATED.  MAXIMUM VALUE IS 100.               *
C           *                                                                 *
C           *C      -OUTPUT - A VECTOR CONTAINING THE N+2 CHEBYSHEV COEFFI-   *
C           *                 CIENTS FOR THE APPROXIMATION TO 0F1(Z).         *
C           *                 DIMENSION OF THE VECTOR IS SET AT 102.          *
C           *                                                                 *
C           *TEST1  -OUTPUT - THE RESULT OF THE FIRST TEST PERFORMED ON THE   *
C           *                 COEFFICIENTS.  SEE TEXT.                        *
C           *                                                                 *
C           *TEST2  -OUTPUT - THE RESULT OF THE SECOND TEST PERFORMED ON THE  *
C           *                 COEFFICIENTS.  SEE TEXT.                        *
C           *                                                                 *
C           *SUM    -OUTPUT - THE SUM OF THE COEFFICIENTS.                    *
C           *                                                                 *
C           *ALL OTHER VARIABLES ARE FOR INTERNAL USE.                        *
C           *                                                                 *
C           *SUBROUTINES CCOEF4 AND CHECK4 ARE REQUIRED.                      *
C           *************************************************************************
            IMPLICIT REAL*16 (A-H,O-Z)
            DIMENSION C(102)
C
C           ---------- THE NEXT PROGRAM CARD IS USED TO INITIALIZE -----------
C           ---------- DEVICE NUMBERS FOR THE PARTICULAR INSTALL- -----------
C           ---------- ATION OF THE OPERATING SYSTEM.  IF A SYSTEM -----------
C           ---------- HAS DIFFERENT DEVICE NUMBERS THIS CARD CAN -----------
C           ---------- BE CHANGED TO INITIALIZE THOSE VALUES AS -----------
C           ---------- DEVICE NUMBERS.                            -----------
            DATA NREAD,NWRITE,NPUNCH/5,6,7/
C
C           ---------- READ IN THE INPUT VARIABLES                -----------
            READ(NREAD,2) CP,W
            READ(NREAD,4) N
            N2=N+2
C
C           ---------- CALL SUBROUTINE TO COMPUTE THE COEFFICIENTS -----------
            CALL CCOEF4(CP,W,N,C,SUM)
C
C           ---------- CALL SUBROUTINE TO TEST THE COEFFICIENT    -----------
C           ---------- VALUES                                     -----------
            CALL CHECK4(CP,W,N,C,TEST1,TEST2)
C
C           ---------- WRITE OUT VALUES OF PARAMETERS AND COEFFI- -----------
C           ---------- CIENTS                                     -----------
            WRITE(NWRITE,5) CP,W,N
            DO 1 K=1,N2
            K1=K-1
          1 WRITE(NWRITE,6) K1,C(K)
            WRITE(NWRITE,7) TEST1,TEST2
            WRITE(NWRITE,8) SUM
C
C           ---------- PUNCH THE VALUES OF COEFFICIENTS ON CARDS  -----------
            WRITE(NPUNCH,2) (C(K),K=1,N2)
          2 FORMAT(Q39.32)
          4 FORMAT(I5)
          5 FORMAT('1',4X,'C=',Q39.32/5X,'W=',Q39.32/5X,'N=',I5/////5X,'K',25X
           :,'C(K)'/)
          6 FORMAT(1X,I5,8X,Q39.32)
          7 FORMAT(/1X,'TEST1=',Q12.5/1X,'TEST2=',Q12.5)
          8 FORMAT(/1X,'THE SUM OF THE COEFFICIENTS IS',Q39.32)
            STOP
            END
```

```
C
C        ------------ IBM S/370 ----------- MULTIPLE PRECISION ------------
C
C        ****************************************************************
C        *THIS SUBROUTINE, CCOEF4, COMPUTES MULTIPLE PRECISION VALUES OF *
C        *THE COEFFICIENTS IN THE CHEBYSHEV EXPANSION OF 0F1(#C#Z).      *
C        *                                                              *
C        *WHERE NO AMBIGUITY CAN OCCUR, WE SHALL DENOTE 0F1(#C#Z) BY    *
C        *0F1(Z).                                                       *
C        *                                                              *
C        *DESCRIPTION OF VARIABLES.                                     *
C        *                                                              *
C        *CP      -INPUT  - PARAMETER C IN 0F1(Z).                      *
C        *                                                              *
C        *W       -INPUT  - THIS IS A PRESELECTED SCALE FACTOR SUCH THAT *
C        *                  0.LE.(Z/W).LE.1.  SEE TEXT.                 *
C        *                                                              *
C        *N       -INPUT  - TWO LESS THAN THE NUMBER OF COEFFICIENTS TO BE *
C        *                  GENERATED.  MAXIMUM VALUE IS 100.           *
C        *                                                              *
C        *C       -OUTPUT - A VECTOR CONTAINING THE N+2 CHEBYSHEV COEFFI- *
C        *                  CIENTS FOR THE APPROXIMATION TO 0F1(Z).     *
C        *                                                              *
C        *SUM     -OUTPUT - THE SUM OF THE COEFFICIENTS.                *
C        *                                                              *
C        *ALL OTHER VARIABLES ARE FOR INTERNAL USE.                    *
C        ****************************************************************
         SUBROUTINE CCOEF4(CP,W,N,C,SUM)
         IMPLICIT REAL*16 (A-H,O-Z)
         DIMENSION C(1)
         DATA ZERO,ONE,TWO,FOUR,START/0.Q0,1.Q0,2.Q0,4.Q0,1.Q-20/
         N1=N+1
         N2=N+2
C
C        ---------- START COMPUTING COEFFICIENTS BY MEANS OF  -----------
C        ---------- BACKWARD RECURRENCE SCHEME
         A3=ZERO
         A2=ZERO
         A1=START
         Z1=FOUR/W
         NCOUNT=N2
         C(NCOUNT)=START
         X1=N2
         C1=ONE-CP
         DO 1 K=1,N1
         DIVFAC=ONE/X1
         X1=X1-ONE
         NCOUNT=NCOUNT-1
         C(NCOUNT)=X1*((DIVFAC+Z1*(X1-C)))*A1+(ONE/X1+Z1*(X1+C1+ONE))*
        :A2-DIVFAC*A3)
         A3=A2
         A2=A1
       1 A1=C(NCOUNT)
         C(1)=C(1)/TWO
C
C        ---------- COMPUTE SCALE FACTOR                      -----------
         RHO=C(1)
         SUM=RHO
         P=ONE
         DO 3 I=2,N2
         RHO=RHO-P*C(I)
         SUM=SUM+C(I)
       3 P=-P
C
C        ---------- SCALE THE COEFFICIENTS                    -----------
         DO 4 L=1,N2
       4 C(L)=C(L)/RHO
         SUM=SUM/RHO
         RETURN
         END
```

```
C
C        ----------- IBM S/370 ----------- MULTIPLE PRECISION -----------
C
C    ****************************************************************
C    *THIS SUBROUTINE, CHECK4, RUNS CHECKS ON THE COEFFICIENTS      *
C    *GENERATED BY SUBROUTINE CCOEF4.  EXCEPT FOR ROUND-OFF ERROR AND*
C    *TRUNCATION ERROR, BOTH TEST RESULTS SHOULD BE ZERO.  SEE TEXT. *
C    *                                                              *
C    *WHERE NO AMBIGUITY CAN OCCUR, WE SHALL DENOTE 0F1(#C#Z) BY    *
C    *0F1(Z).                                                       *
C    *                                                              *
C    *DESCRIPTION OF VARIABLES.                                     *
C    *CP     -INPUT  - PARAMETER C IN 0F1(Z).                       *
C    *                                                              *
C    *W      -INPUT  - THIS IS A PRESELECTED SCALE FACTOR SUCH THAT *
C    *                 0.LE.(Z/W).LE.1.  SEE TEXT.                  *
C    *                                                              *
C    *N      -INPUT  - TWO LESS THAN THE NUMBER OF COEFFICIENTS     *
C    *                 GENERATED.  MAXIMUM VALUE IS 100.            *
C    *                                                              *
C    *C      -INPUT  - A VECTOR CONTAINING THE N+2 CHEBYSHEV COEFFI-*
C    *                 CIENTS FOR THE APPROXIMATION TO 0F1(Z).      *
C    *                                                              *
C    *TEST1  -OUTPUT - THE RESULT OF THE FIRST TEST PERFORMED ON THE*
C    *                 COEFFICIENTS.  SEE TEXT.                     *
C    *                                                              *
C    *TEST2  -OUTPUT - THE RESULT OF THE SECOND TEST PERFORMED ON THE*
C    *                 COEFFICIENTS.  SEE TEXT.                     *
C    *                                                              *
C    *ALL OTHER VARIABLES ARE FOR INTERNAL USE.                    *
C    ****************************************************************
      SUBROUTINE CHECK4(CP,W,N,C,TEST1,TEST2)
      IMPLICIT REAL*16 (A-H,O-Z)
      DIMENSION C(1)
      DATA ONE,TWO,THREE,FOUR,TWELVE/1.Q0,2.Q0,3.Q0,4.Q0,1.2Q1/
C
C        ---------- COMPUTATION OF LEFT SIDES OF TEST EQUATIONS -----------
      SUM1=-C(2)+FOUR*C(3)
      P=-ONE
      SUM2=TWELVE*C(3)
      COUNT=TWO
      DO 1 I=2,N
      COUNT=COUNT+ONE
      CSQ=COUNT*COUNT
      SUM1=SUM1+P*CSQ*C(I+2)
      SUM2=SUM2+P*CSQ*(CSQ-1)*C(I+2)
    1 P=-P
C
C        ---------- COMPUTATION OF TEST VALUES                -----------
      TEST1=SUM1+W/(TWO*CP)
      TEST2=SUM2-THREE*W*W/(FOUR*CP*(CP+ONE))
      RETURN
      END
```

```
        C=-0.50000000000000000000000000000000000Q+00
        W=-0.16000000000000000000000000000000000Q+02
        N=   25
        K                          C(K)

        0          0.20487415819688709011410826205756Q+01
        1          0.44321318722070854900737278951735Q+01
        2          0.36045409643013466196774425291185Q+01
        3         -0.17536662465412900870528099810561Q+01
        4         -0.11072354933573676845881715475810Q+01
        5          0.68425972652454823640719639995766Q+00
        6         -0.15932738829525018673359823477908Q+00
        7         -0.21819006666775593447883416806971Q-01
        8         -0.20294361391766688629764412399868Q-02
        9          0.13830815497050866270002652489999Q-03
        10        -0.72485906768734783208001155212868Q-05
        11         0.30228512224591097456306756067927Q-06
        12        -0.10291515248099950975262192760120Q-07
        13         0.29186392227261745905542831590280Q-09
        14        -0.70079277277752317734009349294516Q-11
        15         0.14439812468339492321513248137691Q-12
        16        -0.25824687399434108254146338437258Q-14
        17         0.40479880415414124224504733016234Q-16
        18        -0.56084175838263547923304165905856Q-18
        19         0.69191107380122113645658458649593Q-20
        20        -0.76508193414889838177329059368464Q-22
        21         0.76267953973182686913251278460522Q-24
        22        -0.68900203986870207159053386710760Q-26
        23         0.56674696392919121890226873174032Q-28
        24        -0.42628856069259933837539642166141Q-30
        25         0.29432919821548246482034756968142Q-32
        26        -0.18594649550905005089556909930151Q-34

        TEST1=-0.29582Q-30
        TEST2=-0.78886Q-29

        THE SUM OF THE COEFFICIENTS IS 0.77693659391784406965961474041312Q+01
```

VII COEFFICIENTS FOR THE EXPANSION OF $_1F_2(a;b,c;z)$ IN ASCENDING SERIES OF CHEBYSHEV POLYNOMIALS

We are concerned with the evaluation of $C_n(w)$ in

$$_1F_2(a;b,c;z) = \sum_{n=0}^{\infty} C_n(w)T_n^*(z/w) \ , \ 0 \le z/w \le 1 \ , \tag{1}$$

$$C_n(w) = \frac{\varepsilon_n (a)_n w^n}{2^{2n}(b)_n(c)_n n!} \ _2F_3\left(\begin{matrix} \frac{1}{2}+n,a+n \\ 1+2n,b+n,c+n \end{matrix}\middle| w\right) \ . \tag{2}$$

This is the special case of 1.3(1,2) with $p = 1$, $q = 2$. Except for the recurrence formula for $C_n(w)$, all necessary information will be found in 1.3 and will not be repeated. The recurrence formula is

$$\frac{2C_n(w)}{\varepsilon_n} = \frac{2(n+1)}{(n+a)}\left\{\frac{2a-3}{2n+5} + \frac{2(n+b)(n+c)}{w}\right\} C_{n+1}(w) + \frac{4(n+2)}{(2n+5)(n+a)}$$

$$\times \left\{\frac{2n^2+8n+3(a+1)}{2n+6} + \frac{(2n+2)[2n^2+8n-2bc+3(b+c+1)]}{w}\right\} C_{n+2}(w)$$

$$+ \frac{(2n+2)}{(2n+5)(n+a)}\left\{(3-2a) + \frac{2(2n+3)(n+4-b)(n+4-c)}{w}\right\} C_{n+3}(w)$$

$$- \frac{(2n+2)(2n+3)(n+4-a)}{(2n+5)(2n+6)(n+a)} C_{n+4}(w) \ . \tag{3}$$

For an expansion of certain combinations of two and three $_1F_2$'s in descending series of Chebyshev polynomials, see Chapter 9.

Numerical Example

Let

$$a = 1.0 \ , \ b = c = 1.5 \ \text{and} \ w = 16 \ .$$

In (1), replace z by $z^2/4$. Then we can write

$$_1F_2(1;\ 3/2,\ 3/2;\ z^2/4) = (\pi/2z)L_0(z)$$

$$= \sum_{n=0}^{\infty} C_n T_{2n}(z/8)\ ,\ 0 \le z \le 8\ . \qquad (4)$$

That is,

$$L_0(z) = (2z/\pi) \sum_{n=0}^{\infty} C_n T_{2n}(z/8)\ . \qquad (5)$$

Using (2) and 1.3(14), we have

$$C_n = \frac{2^{2n-1}\pi}{n(n!)^2} \{1 - \frac{3}{4n} + \frac{17}{32n^2} + O(n^{-3})\}$$

$$\times\ _2F_3 \left(\begin{matrix} n+\tfrac{1}{2},n+1 \\ 2n+1,n+3/2,n+3/2 \end{matrix} \middle| 16 \right)\ . \qquad (6)$$

Let $n = 20$.   Use the first five terms of the $_2F_3$ series in
(6) and so deduce $C_{20} = 0.20196(-25)$, which agrees with the
value deduced by the backward recursion process with $N = 23$.

```
C     ---------- IBM S/370 ----------- MULTIPLE PRECISION -------------
C
C     *********************************************************************
C     *THIS MAINLINE PROGRAM UTILIZES THE SUBROUTINES CCOEF6 AND CHECK6 *
C     *TO COMPUTE AND TEST THE COEFFICIENTS IN THE CHEBYSHEV EXPANSION  *
C     *OF 1F2(A;B,C;Z).                                                 *
C     *                                                                 *
C     *WHERE NO AMBIGUITY CAN OCCUR, WE SHALL DENOTE 1F2(A;B,C;Z) BY    *
C     *1F2(Z).                                                          *
C     *                                                                 *
C     *DESCRIPTION OF VARIABLES.                                        *
C     *                                                                 *
C     *AP     -INPUT  - PARAMETER A IN 1F2(Z).                          *
C     *                                                                 *
C     *BP     -INPUT  - PARAMETER B IN 1F2(Z).                          *
C     *                                                                 *
C     *CP     -INPUT  - PARAMETER C IN 1F2(Z).                          *
C     *                                                                 *
C     *W      -INPUT  - THIS IS A PRESELECTED SCALE FACTOR SUCH THAT    *
C     *                 0.LE.(Z/W).LE.1.  SEE TEXT.                     *
C     *                                                                 *
C     *N      -INPUT  - TWO LESS THAN THE NUMBER OF COEFFICIENTS TO BE  *
C     *                 GENERATED.  MAXIMUM VALUE IS 100.               *
C     *                                                                 *
C     *C      -OUTPUT - A VECTOR CONTAINING THE N+2 CHEBYSHEV COEFFI-   *
C     *                 CIENTS FOR THE APPROXIMATION TO 1F2(Z).         *
C     *                 DIMENSION OF THE VECTOR IS SET AT 102.          *
C     *                                                                 *
C     *TEST1  -OUTPUT - THE RESULT OF THE FIRST TEST PERFORMED ON THE   *
C     *                 COEFFICIENTS.  SEE TEXT.                        *
C     *                                                                 *
C     *TEST2  -OUTPUT - THE RESULT OF THE SECOND TEST PERFORMED ON THE  *
C     *                 COEFFICIENTS.  SEE TEXT.                        *
C     *                                                                 *
C     *SUM    -OUTPUT - THE SUM OF THE COEFFICIENTS.                    *
C     *                                                                 *
C     *ALL OTHER VARIABLES ARE FOR INTERNAL USE.                       *
C     *SUBROUTINES CCOEF6 AND CHECK6 ARE REQUIRED.                     *
C     *********************************************************************
      IMPLICIT REAL*16 (A-H,O-Z)
      DIMENSION C(102)
C     ---------- THE NEXT PROGRAM CARD IS USED TO INITIALIZE -----------
C     ---------- DEVICE NUMBERS FOR THE PARTICULAR INSTALL-  -----------
C     ---------- ATION OF THE OPERATING SYSTEM.  IF A SYSTEM -----------
C     ---------- HAS DIFFERENT DEVICE NUMBERS THIS CARD CAN  -----------
C     ---------- BE CHANGED TO INITIALIZE THOSE VALUES AS    -----------
C     ---------- DEVICE NUMBERS.                             -----------
      DATA NREAD,NWRITE,NPUNCH/5,6,7/
C
C     ---------- READ IN THE INPUT VARIABLES                 -----------
      READ(NREAD,2) AP,BP,CP,W
      READ(NREAD,4) N
      N2=N+2
C
C     ---------- CALL SUBROUTINE TO COMPUTE THE COEFFICIENTS -----------
      CALL CCOEF6(AP,BP,CP,W,N,C,SUM)
C
C     ---------- CALL SUBROUTINE TO TEST THE COEFFFICIENT    -----------
C     ---------- VALUES                                      -----------
      CALL CHECK6(AP,BP,CP,W,N,C,TEST1,TEST2)
C
C     ---------- WRITE OUT VALUES OF PARAMETERS AND          -----------
C     ---------- COEFFICIENTS                                -----------
      WRITE(NWRITE,5) AP,BP,CP,W,N
      DO 1 K=1,N2
      K1=K-1
    1 WRITE(NWRITE,6) K1,C(K)
      WRITE(NWRITE,7) TEST1,TEST2
      WRITE(NWRITE,8) SUM
C
C     ---------- PUNCH THE VALUES OF COEFFICIENTS ON CARDS   -----------
      WRITE(NPUNCH,2) (C(K),K=1,N2)
    2 FORMAT(Q39.32)
    4 FORMAT(I5)
    5 FORMAT('1',4X,'A=',Q39.32/5X,'B=',Q39.32/5X,'C=',Q39.32/5X,'W=',
     :Q39.32/5X,'N=',I5/////5X,'K',25X,'C(K)'/)
    6 FORMAT(1X,I5,8X,Q39.32)
    7 FORMAT(/1X,'TEST1=',Q12.5/1X,'TEST2=',Q12.5)
    8 FORMAT(/1X,'THE SUM OF THE COEFFICIENTS IS',Q39.32)
      STOP
      END
```

```
C
C
C           ----------- IBM S/370 ----------- MULTIPLE PRECISION -------------
C
C      ****************************************************************
C      *THIS SUBROUTINE, CCOEF6, COMPUTES MULTIPLE PRECISION VALUES OF   *
C      *THE COEFFICIENTS IN THE CHEBYSHEV EXPANSION OF 1F2(A;B,C;Z).     *
C      *                                                                *
C      *WHERE NO AMBIGUITY CAN OCCUR, WE SHALL DENOTE 1F2(A;B,C;Z) BY    *
C      *1F2(Z).                                                         *
C      *                                                                *
C      *DESCRIPTION OF VARIABLES.                                        *
C      *                                                                *
C      *AP     -INPUT  - PARAMETER A IN 1F2(Z).                          *
C      *                                                                *
C      *BP     -INPUT  - PARAMETER B IN 1F2(Z).                          *
C      *                                                                *
C      *CP     -INPUT  - PARAMETER C IN 1F2(Z).                          *
C      *                                                                *
C      *W      -INPUT  - THIS IS A PRESELECTED SCALE FACTOR SUCH THAT    *
C      *                 0.LE.(Z/W).LE.1.  SEE TEXT.                     *
C      *                                                                *
C      *N      -INPUT  - TWO LESS THAN THE NUMBER OF COEFFICIENTS TO BE  *
C      *                 GENERATED.  MAXIMUM VALUE IS 100.               *
C      *                                                                *
C      *C      -OUTPUT - A VECTOR CONTAINING THE N+2 CHEBYSHEV COEFFI-   *
C      *                 CIENTS FOR THE APPROXIMATION TO 1F2(Z).         *
C      *                                                                *
C      *SUM    -OUTPUT - THE SUM OF THE COEFFICIENTS.                    *
C      *                                                                *
C      *ALL OTHER VARIABLES ARE FOR INTERNAL USE.                        *
C      ****************************************************************
       SUBROUTINE CCOEF6(AP,BP,CP,W,N,C,SUM)
       IMPLICIT REAL*16 (A-H,O-Z)
       DIMENSION C(1)
       DATA ZERO,ONE,TWO,THREE,FOUR,FIVE,START/0.Q0,1.Q0,2.Q0,3.Q0,4.Q0,
      :5.Q0,1.Q-20/
       N1=N+1
       N2=N+2
C
C           ----------- START COMPUTING COEFFICIENTS BY MEANS OF  -----------
C           ----------- BACKWARD RECURRENCE SCHEME
       A4=ZERO
       A3=ZERO
       A2=ZERO
       A1=START
       Z1=FOUR/W
       NCOUNT=N2
       C(NCOUNT)=START
       X=N1
       P=X+AP
       DO 1 K=1,N1
       X=X-ONE
       P=P-ONE
       Q=(TWO*X+THREE)/(TWO*X+FIVE)
       S=(X+BP)*(X+CP)
       NCOUNT=NCOUNT-1
       C(NCOUNT)=(X+ONE)*((TWO*(P-Q*(P+ONE))+S*Z1)*A1-(X+TWO)*((TWO*X+
      :ONE)*P/(X+ONE)-FOUR*Q*(P+ONE)+(TWO*X+THREE)*(P+TWO)/(X+THREE)+TWO
      :*Z1*(S-Q*(X+BP+ONE)*(X+CP+ONE)))*A2+(TWO*(X+THREE-AP-Q*(X+FOUR
      :-AP))+Q*Z1*(X+FOUR-BP)*(X+FOUR-CP))*A3-Q*(X+FOUR-AP)/(X+THREE
      :)*A4)/P
       A4=A3
       A3=A2
       A2=A1
     1 A1=C(NCOUNT)
       C(1)=C(1)/TWO
C
C           ----------- COMPUTE SCALE FACTOR                  -----------
       RHO=C(1)
       SUM=RHO
       P=ONE
       DO 3 I=2,N2
       RHO=RHO-P*C(I)
       SUM=SUM+C(I)
     3 P=-P
C
C           ----------- SCALE THE COEFFICIENTS                -----------
       DO 4 L=1,N2
     4 C(L)=C(L)/RHO
       SUM=SUM/RHO
       RETURN
       END
```

```
C
C
C      ----------- IBM S/370 ----------- MULTIPLE PRECISION -------------
C
C      ****************************************************************
C      *THIS SUBROUTINE, CHECK6, RUNS CHECKS ON THE COEFFICIENTS      *
C      *GENERATED BY SUBROUTINE CCOEF6, EXCEPT FOR ROUND-OFF ERROR AND *
C      *TRUNCATION ERROR, BOTH TEST RESULTS SHOULD BE ZERO.  SEE TEXT. *
C      *                                                              *
C      *WHERE NO AMBIGUITY CAN OCCUR, WE SHALL DENOTE 1F2(A;B,C;Z) BY  *
C      *1F2(Z).                                                       *
C      *                                                              *
C      *DESCRIPTION OF VARIABLES.                                     *
C      *                                                              *
C      *AP     -INPUT  - PARAMETER A IN 1F2(Z).                       *
C      *                                                              *
C      *BP     -INPUT  - PARAMETER B IN 1F2(Z).                       *
C      *                                                              *
C      *CP     -INPUT  - PARAMETER C IN 1F2(Z).                       *
C      *                                                              *
C      *W      -INPUT  - THIS IS A PRESFLECTED SCALE FACTOR SUCH THAT  *
C      *                 0.LE.(Z/W).LE.1.  SEE TEXT.                  *
C      *                                                              *
C      *N      -INPUT  - TWO LESS THAN THE NUMBER OF COEFFICIENTS      *
C      *                 GENERATED.  MAXIMUM VALUE IS 100.             *
C      *                                                              *
C      *C      -INPUT  - A VECTOR CONTAINING THE N+2 CHEBYSHEV COEFFI- *
C      *                 CIENTS FOR THE APPROXIMATION TO 1F2(Z).       *
C      *                                                              *
C      *TEST1  -OUTPUT - THE RESULT OF THE FIRST TEST PERFORMED ON THE *
C      *                 COEFFICIENTS.  SEE TEXT.                      *
C      *                                                              *
C      *TEST2  -OUTPUT - THE RESULT OF THE SECOND TEST PERFORMED ON THE*
C      *                 COEFFICIENTS.  SEE TEXT.                      *
C      *                                                              *
C      *ALL OTHER VARIABLES ARE FOR INTERNAL USE.                     *
C      ****************************************************************
       SUBROUTINE CHECK6(AP,BP,CP,W,N,C,TEST1,TEST2)
       IMPLICIT REAL*16 (A-H,O-Z)
       DIMENSION C(1)
       DATA ONE,TWO,THREE,FOUR,TWELVE/1.Q0,2.Q0,3.Q0,4.Q0,1.2Q1/
C
C      ---------- COMPUTATION OF LEFT SIDES OF TEST EQUATIONS ----------
       SUM1=-C(2)+FOUR*C(3)
       P=-ONE
       SUM2=TWELVE*C(3)
       COUNT=TWO
       DO 1 I=2,N
       COUNT=COUNT+ONE
       CSQ=COUNT*COUNT
       SUM1=SUM1+P*CSQ*C(I+2)
       SUM2=SUM2+P*CSQ*(CSQ-ONE)*C(I+2)
     1 P=-P
C
C      ---------- COMPUTATION OF TEST VALUES           ----------
       TEST1=SUM1+AP*W/(TWO*BP*CP)
       TEST2=SUM2-THREE*AP*(AP+ONE)*W*W/(FOUR*BP*(BP+ONE)*CP*(CP+ONE))
       RETURN
       END
```

```
A=  0.10000000000000000000000000000000000Q+01
B=  0.15000000000000000000000000000000000Q+01
C=  0.15000000000000000000000000000000000Q+01
W=  0.16000000000000000000000000000000000Q+02
N=    23
```

K	C(K)
0	0.27523746075028819557645398076086Q+02
1	0.38044955428203491423941420124138Q+02
2	0.14413306817775516457871664699697Q+02
3	0.33641911979264000187290579844023Q+01
4	0.52611661559348214258660040199601Q+00
5	0.58589782918239736911009949502576Q-01
6	0.48633952715202532662755958859605Q-02
7	0.31172757694691661384282572124888Q-03
8	0.15867259608265356465374477306790Q-04
9	0.65613221962057931228515613170437Q-06
10	0.22459300441954180635932280652265Q-07
11	0.64647916873383724766187048762183Q-09
12	0.15859089648551715062877150644511Q-10
13	0.33540095291296474219038510729471Q-12
14	0.61766374312227570135054701322524Q-14
15	0.99917973322013230711740949543591Q-16
16	0.14308559658862007602694167157715Q-17
17	0.18263924353423351878541712182967Q-19
18	0.20907837105862874180438933491513Q-21
19	0.21584616286765506359089210371741Q-23
20	0.20196377888323139759941714080170Q-25
21	0.17205675165407083192505478804592Q-27
22	0.13401145972622904270542948557561Q-29
23	0.95780999522225318270891720394241Q-32
24	0.63844089242793896134625870038357Q-34

```
TEST1= 0.46222Q-30
TEST2= 0.38950Q-29

THE SUM OF THE COEFFICIENTS IS 0.83936097586808224772795265572648Q+02
```

VIII COEFFICIENTS FOR THE EXPANSION OF THE CONFLUENT HYPERGEOMETRIC FUNCTIONS $U(a;c;z)$ AND $_1F_1(a;c;-z)$ IN DESCENDING SERIES OF CHEBYSHEV POLYNOMIALS

We first consider the evaluation of $G_n(w)$ in

$$z^a \, U(a;c;z) = [\Gamma(a)\Gamma(b)]^{-1} \, G_{1,2}^{2,1}(z \, |_{a,b}^{1})$$

$$= z^a [\frac{\Gamma(1-c)}{\Gamma(b)} \, _1F_1(_c^a|z) + \frac{\Gamma(c-1)z^{1-c}}{\Gamma(a)} \, _1F_1(_{2-c}^b|z)]$$

$$= \sum_{n=0}^{\infty} G_n(w) T_n^*(w/z) \; , \; z/w \geq 1 \; ,$$

$$b = a + 1-c \; , \; z \neq 0 \; , \; |\arg z| < 3\pi/2 \; . \tag{1}$$

This is the special case of 1.3(15) with $q = 2$ and $p = 0$. Except for the recurrence formula for $G_n(w)$, all necessary data will be found in 1.3. Thus, virtually none of this material will be repeated, save for the asymptotic expansion of $G_n(w)$ for large n. From 1.3(23), we have

$$G_n(w) = \frac{4(-)^n(\pi/3)^{\frac{1}{2}}(n^2w)^u}{n\Gamma(a)\Gamma(b)} \exp\{-3s+2sP(y)+S/s\}$$

$$\times \left[1 + O\left(\frac{(1+|w|^{8/3})}{s^2}\right)\right] \; ,$$

$$u = (2a+2b-1)/6 \; , \; s = (n^2w)^{1/3} \; ,$$

$$P(y) = \frac{1}{2}(1 - \{\frac{y}{\sinh y}\}^2) \; , \; y = (w/n)^{1/3} \; ,$$

$$S = -w(a+b-1)/3 - \frac{1}{2}(a+b+2ab) + (2a+2b+1)^2/12 - 1/9 \; ,$$

$$n^2 w \to \infty \ , \quad |\arg w| \leq 4\pi - \varepsilon \ , \quad \varepsilon > 0 \ ,$$

$$w = o(n^{2/3}) \ , \quad (w/n^2) = o(1) \ . \tag{2}$$

An alternative form is

$$G_n(w) = \frac{4(-)^n (\pi/3)^{\frac{1}{2}} (n^2 w)^u e^{w/3}}{n \Gamma(a) \Gamma(b)} \exp\{-3s + (S - w^2/15)/s\}$$

$$\times \left[1 + O\left(\frac{(1 + |w|^{8/3})}{s^2} \right) \right] \ , \tag{3}$$

under the same conditions as for (2).

The recurrence formula for $G_n(w)$ is

$$\frac{2 G_n(w)}{\varepsilon_n} = - \frac{(n+1)}{(n+2)(n+a)(n+b)}$$

$$\times \ [3n^2 + n(a+b+8) + 3(a+b+1) - ab + 4w(n+2)] \ G_{n+1}(w)$$

$$+ \ [1 - \frac{2(n+1)(2n+3-2w)}{(n+a)(n+b)}] \ G_{n+2}(w)$$

$$- \ \frac{(n+1)(n+3-a)(n+3-b)}{(n+2)(n+a)(n+b)} \ G_{n+3}(w) \ . \tag{4}$$

In Luke [1969, v.2, p.238-240], it was shown how the recurrence formula for $G_n(w)$ could be derived by considering an integral equation satisfied by (1). This procedure produced two further relations,

$$\frac{6G_0 + 4G_1 + G_2}{16} + \frac{(a-1)(b-1)}{4} \left(\frac{3G_0}{4} - \frac{G_1}{3} - \frac{3G_2}{16} - \sum_{n=3}^{\infty} \frac{(-)^n G_n}{n^2 - 4} \right)$$

$$+ \frac{a+b-3}{4} \left(\frac{3G_0}{4} + \frac{G_1}{12} - \frac{19G_2}{48} + 3 \sum_{n=3}^{\infty} \frac{(-)^n G_n}{(n^2-1)(n^2-4)} \right)$$

$$+ \frac{w}{2} \left(G_0 - \frac{G_1}{4} - \sum_{n=2}^{\infty} \frac{(-)^n G_n}{n^2 - 1} \right) = \frac{w}{2} \ , \tag{5}$$

$$\frac{8G_0 + 7G_1 + 4G_2 + G_3}{16}$$

$$+ \frac{(a-1)(b-1)}{4}\left(G_0 - \frac{3G_1}{8} - \frac{G_2}{3} + \frac{G_3}{4} - \sum_{n=4}^{\infty} \frac{(-)^n G_n}{n^2-1}\right)$$

$$+ \frac{a+b-3}{4}\left(G_0 + \frac{G_1}{4} - \frac{G_2}{2} - \frac{G_3}{4}\right) + \frac{w}{4}(2G_0 - G_2) = \frac{w}{2} , \qquad (6)$$

which can be used for check purposes, as in the manner of the remarks surrounding 1.3(19-22).

Next we turn our attention to

$$Ez^a {}_1F_1(a;c;-z) = Ez^a e^{-z} {}_1F_1(c-a;c;z)$$

$$= \frac{E\Gamma(c)}{\Gamma(a)} G_{1,2}^{1,1}\left(z \Big| \begin{matrix} 1 \\ a,b \end{matrix}\right) = \sum_{n=0}^{\infty} D_n(v) T_n^*(v/z) ,$$

$$b = a+1-c , \quad z/v \geq 1 , \quad z \neq 0 , \quad |\arg z| < \pi ,$$

$$E = \frac{\pi}{(\sin \pi b)\Gamma(b)\Gamma(c)} = \frac{\Gamma(1-b)}{\Gamma(c)} , \qquad (7)$$

which is the special case of 1.3(24) with $q = 2$, $p = 0$ and w   replaced by v .   From (3), (7) and 1.3(25), we get

$$D_n(v) = \frac{4\pi(-)^n (n^2 v)^u e^{-v/3} \sin \theta}{(3\pi)^{\frac{1}{2}} n\Gamma(a)\Gamma(b)\sin \pi b} \exp\{-3s/2 + (S-v^2/15)/2s\}$$

$$\times \left[1 + O\left(\frac{(1+|v|^{8/3})}{s^2}\right)\right] , \qquad (8)$$

$$\theta = \pi(b-u) + 3s(3)^{\frac{1}{2}}/2 + (S-v^2/15)3^{\frac{1}{2}}/2s , \quad s = (n^2 v)^{1/3} ,$$

where S is the same as in (2) with $w = -v$, and all the conditions given for (2) apply with $w = -v$, except that we now have $|\arg v| \leq 3\pi - \epsilon$, $\epsilon > 0$.

As previously noted by the comments in the paragraph

after 1.3(27), $D_n(v)$ satisfies (4) and 1.3(19-22) (where
$q = 2$, $p = 0$, $b_1 = a$, $b_2 = b$) with w replaced by -v .
$D_n(v)$ also satisfies (5) and (6) with w replaced by -v .
The coefficients $G_n(w)$ and $D_n(v)$ are readily found by the
backward recursion process described by 1.3(28-35). The
reader is reminded that convergence of this scheme for $G_n(w)$
and $D_n(v)$ weakens as $|\arg w| \to \pi$ and $\arg v \to 0$, respec-
tively, and fails when $|\arg w| = \pi$ and $\arg v = 0$, respec-
tively. In these situations, the process must be modified as
outlined in 1.3(36-40). In connection with this and the discussion following 1.3(39), we were not successful when W was
based on either of the formulas (5) or (6).

An important special case of (7) is

$$I_\nu(z) = (2\pi z)^{-\frac{1}{2}} e^z \ G_{1,2}^{1,1}(2z\,|_{\frac{1}{2}+\nu,\frac{1}{2}-\nu}^{\quad 1}) . \tag{9}$$

For further special cases of the $_1F_1$ in (7), see the listings given in Chapter 2.

In the computer programs given at the end of this chap-
ter, for a given w , $|\arg w| < \pi$, subroutine CCOEF produces
values of $G_n(w)$ for the expansion (1). If $w < 0$, say
$w = -v$, $v > 0$, then this subroutine produces values of
$G_n(w)$ for the expansion (1) provided the value of
$(ve^{i\pi})^a U(a;c;ve^{i\pi})$ is specified. Again, if $w < 0$, say
$w = -v$, $v > 0$, then this subroutine produces values of
$D_n(v)$ for the expansion (7) provided the value of
$Ev^a {}_1F_1(a;c;-v)$ is specified.

Numerical Examples
1. Let $a = b = \frac{1}{2}$ (whence $c = 1$) and $w = 10$. Then,
from (1) and 2.2(50), we have

$$K_0(z) = (\pi/2z)^{\frac{1}{2}} e^{-z} \sum_{n=0}^{\infty} G_n T_n^*(5/z) \quad , \quad z \geq 5 \; .$$

In (2) and (3), $u = 1/6$, $\Gamma(a)\Gamma(b) = \pi$ and $S = -1/9$. Let $n = 25,30$ and use (2) without the order term. Then G_{25} and G_{30} are $-0.44744(-23)$ and $0.32498(-26)$, respectively. The corresponding values deduced by use of the backward recursion formula with $N = 40$ are $-0.44136(-23)$ and $0.32139(-26)$, respectively.

2.   Again, let $a = b = \frac{1}{2}$ (whence $c = 1$) and $v = 16 = -w$. Then, from (7) and (9),

$$I_0(z) = (2\pi z)^{-\frac{1}{2}} e^z \sum_{n=0}^{\infty} D_n(v) T_n^*(8/z) \quad , \quad z \geq 8 \; .$$

In (8), $u = 1/6$, $\Gamma(a)\Gamma(b) = \pi$ and $S = -1/9$. Let $n = 30,35$ and use (8) without the order term. Then D_{30} and D_{35} are $0.74684(-19)$ and $-0.18329(-20)$, respectively. The computed values deduced by the modified backward recursion process with $N = 76$, $M = 77$ and $W = 4\pi^{\frac{1}{2}} e^{-8} I_0(8)$ are $0.72343(-19)$ and $-0.18925(-20)$, respectively.

```
C
C     ----------- IBM S/370 ----------- MULTIPLE PRECISION -----------
C
C     ***********************************************************************
C     *THIS MAINLINE PROGRAM UTILIZES THE SUBROUTINES CCOEF, LINCOM AND *
C     *CHECK TO COMPUTE AND TEST THE COEFFICIENTS IN THE CHEBYSHEV      *
C     *EXPANSION OF (Z**A)*U(A;C;Z) OR OF E*(Z**A)*1F1(A;C;-Z) UNDER    *
C     *THE CONDITIONS NOTED IN THE TEXT.                               *
C     *                                                               *
C     *WHERE NO AMBIGUITY CAN OCCUR, WE SHALL DENOTE U(A;C;Z) BY U(Z)  *
C     *AND E*1F1(A;C;-Z) BY F(Z).                                      *
C     *                                                               *
C     *DESCRIPTION OF VARIABLES.                                       *
C     *                                                               *
C     *AP    -INPUT  - PARAMETER A IN U(Z) OR F(Z).                    *
C     *                                                               *
C     *CP    -INPUT  - PARAMETER C IN U(Z) OR F(Z).                    *
C     *                                                               *
C     *W     -INPUT  - THIS IS A PRESELECTED SCALE FACTOR SUCH THAT    *
C     *                0.LE.(W/Z).LE.1.  SEE TEXT.                     *
C     *                                                               *
C     *N     -INPUT  - TWO LESS THAN THE NUMBER OF COEFFICIENTS TO BE  *
C     *                COMPUTED IN THE FIRST OR ONLY SET OF COEFFI-    *
C     *                CIENTS.  MAXIMUM VALUE IS 100.                  *
C     *                                                               *
C     *NN    -INPUT  - TWO LESS THAN THE NUMBER OF COEFFICIENTS TO BE  *
C     *                COMPUTED IN THE SECOND SET OF COEFFICIENTS.     *
C     *                NN IS TO BE ENTERED ONLY IF W IS NEGATIVE.  NN  *
C     *                MUST BE GREATER THAN N.  MAXIMUM VALUE IS 100.  *
C     *                                                               *
C     *T     -INPUT  - THE VALUE OF (Z**A)*U(Z) AT Z=W OR THE VALUE OF *
C     *                (Z**A)*F(Z) AT Z=V=-W ACCORDING TO WHETHER ONE  *
C     *                DESIRES THE COEFFICIENTS C(W) IN EXPANSION (1)  *
C     *                OR THE COEFFICIENTS D(V) IN EXPANSION (7).      *
C     *                T IS ENTERED ONLY IF W IS NEGATIVE.             *
C     *                                                               *
C     *S1    -OUTPUT - THE SUM OF THE FIRST SET OF COEFFICIENTS.       *
C     *                                                               *
C     *S2    -OUTPUT - THE SUM OF THE SECOND SET OF COEFFICIENTS.      *
C     *                                                               *
C     *C1    -OUTPUT - THE FIRST VECTOR CONTAINING THE N+2 CHEBYSHEV   *
C     *                COEFFICIENTS FOR THE APPROXIMATION TO (Z**A)*   *
C     *                U(Z) OR (Z**A)*F(Z).  DIMENSION OF THE VECTOR IS*
C     *                SET AT 102.                                     *
C     *                                                               *
C     *C2    -OUTPUT - THE SECOND VECTOR CONTAINING THE NN+2 CHEBYSHEV *
C     *                COEFFICIENTS FOR THE APPROXIMATION TO (Z**A)*   *
C     *                U(Z) OR (Z**A)*F(Z) IF W IS NEGATIVE.  DIMENSION*
C     *                OF THE VECTOR IS SET AT 102.                    *
C     *                                                               *
C     *C     -OUTPUT - THE VECTOR CONTAINING THE FINAL N+2 OR NN+2     *
C     *                CHEBYSHEV COEFFICIENTS FOR THE APPROXIMATION TO *
C     *                (Z**A)*U(Z) OR (Z**A)*F(Z).  DIMENSION OF THE   *
C     *                VECTOR IS SET AT 102.                           *
C     *                                                               *
C     *TEST1 -OUTPUT - THE RESULT OF THE FIRST TEST PERFORMED ON THE   *
C     *                COEFFICIENTS.  SEE TEXT.                        *
C     *                                                               *
C     *TEST2 -OUTPUT - THE RESULT OF THE SECOND TEST PERFORMED ON THE  *
C     *                COEFFICIENTS.  SEE TEXT.                        *
C     *                                                               *
C     *SUMM  -OUTPUT - THE SUM OF THE COEFFICIENTS.                    *
C     *                                                               *
C     *ALL OTHER VARIABLES ARE FOR INTERNAL USE.                      *
C     *                                                               *
C     *SUBROUTINES CCOEF, LINCOM AND CHECK ARE REQUIRED.              *
C     ***********************************************************************
      IMPLICIT REAL*16 (A-H,O-Z)
      DIMENSION C1(102),C2(102),C(102)
C     ----------- THE NEXT PROGRAM CARD IS USED TO INITIALIZE -----------
C     ----------- DEVICE NUMBERS FOR THE PARTICULAR INSTALL- -----------
C     ----------- ATION OF THE OPERATING SYSTEM.  IF A SYSTEM -----------
C     ----------- HAS DIFFERENT DEVICE NUMBERS THIS CARD CAN -----------
C     ----------- BE CHANGED TO INITIALIZE THOSE VALUES AS -----------
C     ----------- DEVICE NUMBERS.                           -----------
      DATA NREAD,NWRITE,NPUNCH/5,6,7/
C
C     ----------- READ IN THE INPUT VARIABLES               -----------
      READ(NREAD,2) AP,CP,W
      BP=1.Q0+AP-CP
      READ(NREAD,4) N
      N2=N+2
      NUM=1
      IF (W) 7,9,9
    7 READ(NREAD,4) NN
      NN2=NN+2
      READ(NREAD,2) T
      NUM=2
```

```
C
C
C         ---------- CALL SUBROUTINE TO COMPUTE THE FIRST OR    ----------
C         ---------- ONLY SET OF COEFFICIENTS                   ----------
    9 CALL CCOEF(AP,BP,W,N,C1,S1)
      WRITE(NWRITE,5) AP,CP,W,N
      DO 14 K=1,N2
      K1=K-1
   14 WRITE(NWRITE,6) K1,C1(K)
      WRITE(NWRITE,18) S1
      IF (NUM.EQ.2) GO TO 10
      DO 13 K=1,N2
   13 C(K)=C1(K)
      NN=N
      NN2=N2
      GO TO 12
C
C         ---------- CALL SUBROUTINE TO COMPUTE THE SECOND SET  ----------
C         ---------- OF COEFFICIENTS                            ----------
   10 CALL CCOEF(AP,BP,W,NN,C2,S2)
      WRITE(NWRITE,5) AP,CP,W,NN
      DO 15 K=1,NN2
      K1=K-1
   15 WRITE(NWRITE,6) K1,C2(K)
      WRITE(NWRITE,18) S2
C
C         ---------- CALL SUBROUTINE TO COMPUTE THE FINAL VALUES ----------
C         ---------- OF COEFFICIENTS                             ----------
      CALL LINCOM(N,NN,C1,C2,S1,S2,T,C,SUMM)
C
C         ---------- WRITE OUT VALUES OF PARAMETERS AND COEFFI-  ----------
C         ---------- CIENTS                                      ----------
      WRITE(NWRITE,5) AP,CP,W,NN
      DO 1 K=1,NN2
      K1=K-1
    1 WRITE(NWRITE,6) K1,C(K)
      WRITE(NWRITE,18) SUMM
C
C         ---------- CALL SUBROUTINE TO TEST THE COEFFICIENT    ----------
C         ---------- VALUES                                     ----------
   12 WRITE(NWRITE,19)
      CALL CHECK(AP,BP,W,NN,C,TEST1,TEST2)
      WRITE(NWRITE,17) TEST1,TEST2
C
C         ---------- PUNCH THE VALUES OF COEFFICIENTS ON CARDS  ----------
      WRITE (NPUNCH,2) (C(K),K=1,NN2)
    2 FORMAT(Q39.32)
    4 FORMAT(I5)
    5 FORMAT('1',4X,'A=',Q39.32/5X,'C=',Q39.32/5X,'W=',Q39.32/5X,'N=',
     :I5/////5X,'K',25X,'C(K)'/)
    6 FORMAT(1X,I5,8X,Q39.32)
   17 FORMAT(/1X,'TEST1=',Q12.5/1X,'TEST2=',Q12.5)
   18 FORMAT(/1X,'THE SUM OF THE COEFFICIENTS IS',Q39.32)
   19 FORMAT(/1X,'THE FINAL COEFFICIENTS ARE DIRECTLY ABOVE.')
      STOP
      END
C
C
C         ----------- IBM S/370 ----------- MULTIPLE PRECISION -----------
C
C     *********************************************************************
C     *THIS SUBROUTINE, CCOEF, COMPUTES MULTIPLE PRECISION VALUES OF    *
C     *THE COEFFICIENTS IN THE CHEBYSHEV EXPANSION OF (Z**A)*U(A;C;Z)   *
C     *OR OF E*(Z**A)*1F1(A;C;-Z) UNDER THE CONDITIONS NOTED IN THE     *
C     *TEXT.                                                            *
C     *                                                                 *
C     *WHERE NO AMBIGUITY CAN OCCUR, WE SHALL DENOTE U(A;C;Z) BY U(Z)   *
C     *AND E*1F1(A;C;-Z) BY F(Z).                                       *
C     *                                                                 *
C     *DESCRIPTION OF VARIABLES.                                        *
C     *                                                                 *
C     *AP    -INPUT  - PARAMETER A IN U(Z) OR F(Z).                     *
C     *                                                                 *
C     *CP    -INPUT  - PARAMETER C IN U(Z) OR F(Z).                     *
C     *                                                                 *
C     *W     -INPUT  - THIS IS A PRESELECTED SCALE FACTOR SUCH THAT     *
C     *                0.LE.(W/Z).LE.1.  SEE TEXT.                      *
C     *                                                                 *
C     *N     -INPUT  - TWO LESS THAN THE NUMBER OF COEFFICIENTS TO BE   *
C     *                GENERATED.  MAXIMUM VALUE IS 100.                *
C     *                                                                 *
C     *CC    -OUTPUT - A VECTOR CONTAINING THE N+2 CHEBYSHEV COEFFI-    *
C     *                CIENTS FOR THE APPROXIMATION TO (Z**A)*U(Z) OR   *
C     *                (Z**A)*F(Z).                                     *
C     *                                                                 *
C     *SUM   -OUTPUT - THE SUM OF THE COEFFICIENTS.                     *
C     *                                                                 *
C     *ALL OTHER VARIABLES ARE FOR INTERNAL USE.                       *
C     *********************************************************************
```

```
      SUBROUTINE CCOEF(AP,BP,W,N,CC,SUM)
      IMPLICIT REAL*16 (A-H,O-Z)
      DIMENSION CC(1)
      DATA ZERO,ONE,TWO,THREE,FOUR,START/0.Q0,1.Q0,2.Q0,3.Q0,4.Q0,
     :1.Q-20/
      N1=N+1
      N2=N1+1
C
C    ----------- START COMPUTING COEFFICIENTS BY MEANS OF    -----------
C    ----------- BACKWARD RECURRENCE SCHEME                  -----------
      A3=ZERO
      A2=ZERO
      A1=START
      NCOUNT=N2
      CC(NCOUNT)=START
      X=N1
      DO 1 K=1,N1
      X=X-ONE
      D=(X+AP)*(X+BP)
      NCOUNT=NCOUNT-1
      CC(NCOUNT)=(X+ONE)*((TWO*D-(TWO*X+THREE)*(X+AP+ONE)*(X+BP+ONE)/
     :(X+TWO)-FOUR*W)*A1+(D/(X+ONE)-TWO*(TWO*X+THREE-TWO*W))*A2-
     :(X+THREE-AP)*(X+THREE-BP)*A3/(X+TWO))/D
      A3=A2
      A2=A1
    1 A1=CC(NCOUNT)
      CC(1)=CC(1)/TWO
C
C    ----------- COMPUTE SCALE FACTOR                        -----------
      RHO=CC(1)
      SUM=RHO
      P=ONE
      DO 3 I=2,N2
      RHO=RHO-P*CC(I)
      SUM=SUM+CC(I)
    3 P=-P
C
C    ----------- SCALE THE COEFFICIENTS                      -----------
      DO 4 L=1,N2
    4 CC(L)=CC(L)/RHO
      SUM=SUM/RHO
      RETURN
      END
C
C
C    ------------ IBM S/370 ------------ MULTIPLE PRECISION ------------
C
C    *****************************************************************
C    *THIS SUBROUTINE, LINCOM, FORMS A LINEAR COMBINATION OF TWO SETS *
C    *OF CHEBYSHEV COEFFICIENTS IN ORDER TO SATISFY A GIVEN VALUE OF  *
C    *THE FUNCTION THAT IS TO BE APPROXIMATED.                        *
C    *                                                               *
C    *DESCRIPTION OF VARIABLES.                                       *
C    *                                                               *
C    *N      -INPUT  - TWO LESS THAN THE NUMBER OF COEFFICIENTS       *
C    *                 COMPUTED IN THE FIRST SET OF COEFFICIENTS.     *
C    *                                                               *
C    *NN     -INPUT  - TWO LESS THAN THE NUMBER OF COEFFICIENTS       *
C    *                 COMPUTED IN THE SECOND SET OF COEFFICIENTS. NN *
C    *                 MUST BE GREATER THAN N.                        *
C    *                                                               *
C    *C1     -INPUT  - THE FIRST VECTOR CONTAINING THE N+2 CHEBYSHEV  *
C    *                 COEFFICIENTS FOR THE APPROXIMATION TO THE      *
C    *                 FUNCTION.                                      *
C    *                                                               *
C    *C2     -INPUT  - THE SECOND VECTOR CONTAINING THE NN+2 CHEBYSHEV*
C    *                 COEFFICIENTS FOR THE APPROXIMATION TO THE      *
C    *                 FUNCTION.                                      *
C    *                                                               *
C    *S1     -INPUT  - THE SUM OF THE FIRST SET OF COEFFICIENTS.      *
C    *                                                               *
C    *S2     -INPUT  - THE SUM OF THE SECOND SET OF COEFFICIENTS.     *
C    *                                                               *
C    *T      -INPUT  - THE VALUE OF THE FUNCTION AT Z=W.              *
C    *                                                               *
C    *C      -OUTPUT - THE FINAL VECTOR CONTAINING THE NN+2 CHEBYSHEV *
C    *                 COEFFICIENTS FOR THE APPROXIMATION TO THE      *
C    *                 FUNCTION.                                      *
C    *                                                               *
C    *SUM    -OUTPUT - THE SUM OF THE COEFFICIENTS.                   *
C    *                                                               *
C    *ALL OTHER VARIABLES ARE FOR INTERNAL USE.                      *
C    *****************************************************************
```

```
      SUBROUTINE LINCOM(N,NN,C1,C2,S1,S2,T,C,SUM)
      IMPLICIT REAL*16 (A-H,O-Z)
      DIMENSION C(1),C1(1),C2(1)
      N2=N+2
      N3=N2+1
      NN2=NN+2
      FAC1=(T-S2)/(S1-S2)
      FAC2=(S1-T)/(S1-S2)
      C(1)=C2(1)+FAC1*(C1(1)-C2(1))
      SUM=C(1)
      DO 1 K=2,N2
      C(K)=C2(K)+FAC1*(C1(K)-C2(K))
    1 SUM=SUM+C(K)
      DO 2 K=N3,NN2
      C(K)=FAC2*C2(K)
    2 SUM=SUM+C(K)
      RETURN
      END
```

```
C
C
C          ----------- IBM S/370 ----------- MULTIPLE PRECISION -----------
C
C     **********************************************************************
C     *THIS SUBROUTINE, CHECK, RUNS CHECKS ON THE COEFFICIENTS             *
C     *GENERATED BY SUBROUTINE CCOEF.  EXCEPT FOR ROUND-OFF ERROR AND      *
C     *TRUNCATION ERROR, BOTH TEST RESULTS SHOULD BE ZERO.  SEE TEXT.      *
C     *                                                                    *
C     *WHERE NO AMBIGUITY CAN OCCUR, WE SHALL DENOTE U(A;C;Z) BY U(Z)      *
C     *AND E*1F1(A;C;-Z) BY F(Z).                                          *
C     *                                                                    *
C     *DESCRIPTION OF VARIABLES.                                           *
C     *                                                                    *
C     *AP      -INPUT   - PARAMETER A IN U(Z) OR F(Z).                     *
C     *                                                                    *
C     *CP      -INPUT   - PARAMETER C IN U(Z) OR F(Z).                     *
C     *                                                                    *
C     *W       -INPUT   - THIS IS A PRESELECTED SCALE FACTOR SUCH THAT     *
C     *                   0.LE.(W/Z).LE.1.  SEE TEXT.                      *
C     *                                                                    *
C     *N       -INPUT   - TWO LESS THAN THE NUMBER OF COEFFICIENTS         *
C     *                   GENERATED.  MAXIMUM VALUE IS 100.                *
C     *                                                                    *
C     *C       -INPUT   - THE FINAL VECTOR CONTAINING THE N+2 CHEBYSHEV    *
C     *                   COEFFICIENTS FOR THE APPROXIMATION TO (Z**A)*    *
C     *                   U(Z) OR (Z**A)*F(Z).                             *
C     *                                                                    *
C     *TEST1 -OUTPUT - THE RESULT OF THE FIRST TEST PERFORMED ON THE       *
C     *                   COEFFICIENTS.  SEE TEXT.                         *
C     *                                                                    *
C     *TEST2 -OUTPUT - THE RESULT OF THE SECOND TEST PERFORMED ON THE      *
C     *                   COEFFICIENTS.  SEE TEXT.                         *
C     *                                                                    *
C     *ALL OTHER VARIABLES ARE FOR INTERNAL USE.                          *
C     **********************************************************************
      SUBROUTINE CHECK(AP,BP,W,N,C,TEST1,TEST2)
      IMPLICIT REAL*16 (A-H,O-Z)
      DIMENSION C(1)
      DATA ONE,TWO,THREE,FOUR,TWELVE/1.Q0,2.Q0,3.Q0,4.Q0,1.2Q1/
C
C          ---------- COMPUTATION OF LEFT SIDES OF TEST EQUATIONS ----------
      SUM1=-C(2)+FOUR*C(3)
      P=-ONE
      SUM2=TWELVE*C(3)
      COUNT=TWO
      DO 1 I=2,N
      COUNT=COUNT+ONE
      CSQ=COUNT*COUNT
      SUM1=SUM1+P*CSQ*C(I+2)
      SUM2=SUM2+P*CSQ*(CSQ-ONE)*C(I+2)
    1 P=-P
C
C          ---------- COMPUTATION OF TEST VALUES             ----------
      TEST1=SUM1-AP*BP/(TWO*W)
      TEST2=SUM2-THREE*AP*BP*(AP+ONE)*(BP+ONE)/(W*W*FOUR)
      RETURN
      END
```

```
A= 0.50000000000000000000000000000000000Q+00
C= 0.10000000000000000000000000000000000Q+01
W= 0.10000000000000000000000000000000000Q+02
N=   40
```

K	C(K)
0	0.99840817423082580035073782202167Q+00
1	-0.11310504664928280686311510337326Q-01
2	0.26953261276272369397276497904651Q-03
3	-0.11106685196665353982735125007893Q-04
4	0.63257510850049334011891650668190Q-06
5	-0.45047337641102343005719383143431Q-07
6	0.37929964556843109698786404155762Q-08
7	-0.36454717920606431807716908504321Q-09
8	0.39043755762714479731298705593994Q-10
9	-0.45799362176458863574882337766902Q-11
10	0.58081062853312258904664265403099Q-12
11	-0.78832361194333059107098663596282Q-13
12	0.11360423305347206642647833732852Q-13
13	-0.17269721011624011275659813713276Q-14
14	0.27545461269618617055771450344120Q-15
15	-0.45892310304496126149858831652677Q-16
16	0.79561384643686484497992055858337Q-17
17	-0.14306077403419639306785852176887Q-17
18	0.26605324654581862466352817988274Q-18
19	-0.51047713413056892625454588294765Q-19
20	0.10083305999573732668324405070054Q-19
21	-0.20465189149724090927496443519562Q-20
22	0.42606236601876853643984380396629Q-21
23	-0.90847130214211487634156962554950Q-22
24	0.19812187435939917839722710917569Q-22
25	-0.44136387477623247588769986665136Q-23
26	0.10032614001182761267093247361934Q-23
27	-0.23245439372056778094504887315081Q-24
28	0.54847818771123245354494809930772Q-25
29	-0.13167528085991803884134128617987Q-25
30	0.32138653869540954229550643403484Q-26
31	-0.79691408890215660842210316579819Q-27
32	0.20061405517143584158546709914509Q-27
33	-0.51239420350014749220636202467951Q-28
34	0.13270467830928832065555562749344850Q-28
35	-0.34831360314842795288214477564974Q-29
36	0.92604458906109239981547365648546Q-30
37	-0.24926609795501680168958440123068Q-30
38	0.67929220595321378466061524051837Q-31
39	-0.18782080976320368336895826519511Q-31
40	0.52471009627236541885414464385080Q-32
41	-0.12878754536952832543449675145503Q-32

THE SUM OF THE COEFFICIENTS IS 0.97735668650265938144849852902896Q+00

THE FINAL COEFFICIENTS ARE DIRECTLY ABOVE.

TEST1=-0.11923Q-29
TEST2=-0.21511Q-26

```
A= 0.5000000000000000000000000000000000000Q+00
C= 0.1000000000000000000000000000000000000Q+01
W=-0.1600000000000000000000000000000000000Q+02
N=   76
```

K	C(K)
0	0.10082791932872055749812451528669Q+01
1	0.84450996092461476430289268035629Q-02
2	0.17268120195942544944114740521564Q-03
3	0.72330031673798697824829748554425Q-05
4	0.50391369434385579848042844650871Q-06
5	0.51216762359836890092247194276772Q-07
6	0.57375887135898496521367351673838Q-08
7	0.10076168885632976600456766806794Q-09
8	-0.32908014290692107310448747719505Q-09
9	-0.14917591994529969882353778702164Q-09
10	-0.30962043184953981104541095139276Q-10
11	0.25501882078849575257392024298311Q-11
12	0.37924028320946605345032759072763Q-11
13	0.78942610779799956712071905192002Q-12
14	-0.22393752289362852113208787202288Q-12
15	-0.13976323407740772354421480087763Q-12
16	0.15423090865466712927938165902762Q-14
17	0.18128011872665464581099943188120Q-13
18	0.21590040505325213570684715785952Q-14
19	-0.22885598141519173037337922820089Q-14
20	-0.49595268326173242164182799269127Q-15
21	0.31184297332040360955266059342220Q-15
22	0.87287719130360422330560276273754Q-16
23	-0.48064072064331503068445675669872Q-16
24	-0.13906092228725235903312975829246Q-16
25	0.83812471101384907975377079902060Q-17
26	0.20067297780173191157068764214 3Q-17
27	-0.15946231144006034143288772749 96Q-17
28	-0.23110910838339177662351058282303Q-18
29	0.31438221459736451286845955964761Q-18
30	0.73426974567901498469487039432878Q-20
31	-0.60817065473128367171469314657458Q-19
32	0.74111784108758129385471844496639Q-20
33	0.10820191377984773222266252081070Q-19
34	-0.33787784095967809752911482595108Q-20
35	-0.15707949510844608566639695941 6123Q-20
36	0.10185858579469382991569199409808 0Q-20
37	0.11091952772645162232326321172105Q-21
38	-0.24291235801248285388161722653729Q-21
39	0.34565890434762874105831890921230Q-22
40	0.44570601000284296306267857401032Q-22
41	-0.19515422559349515281748429982185Q-22
42	-0.46351746990916049208622337037171Q-23
43	0.57864753968493441476247645843408Q-23
44	-0.66247922377675388671126383526937Q-24
45	-0.11452256409202650442376606215850Q-23
46	0.52356682360293951007633309971149Q-24
47	0.99987865185590979584069585358168Q-25
48	-0.16033121350058630020984492924664Q-24
49	0.31887139568550507250564104291393Q-25
50	0.27951313018908239675394355142524Q-25
51	-0.18256940719042577907098523179137Q-25
52	-0.89228638821248492553996658306732Q-28
53	0.46544296063321324375479166785296Q-26
54	-0.18241936324690251332117999393886Q-26
55	-0.45904859512356187892136450684360Q-27
56	0.66380986629736812700295945263666Q-27
57	-0.16006228725741575178967375190499Q-27
58	-0.10636557129119211154698324572961Q-27
59	0.89218063465740179425050603434369 10Q-28
60	-0.11285307962348784591637739721439Q-28
61	-0.18983101201123771226880995360573Q-28
62	0.11830812894274450387507138047703Q-28
63	-0.34247823009567008900192233379753Q-30
64	-0.31032590879065680784493540532029Q-29
65	0.15948750129925072578162896016544Q-29
66	0.86605491805945331235699624302277Q-31
67	-0.49239565550874048935837364476204Q-30
68	0.22359440251568374974713635246730Q-30
69	0.26341252832300122196449173753947Q-31
70	-0.77831253369668969574840979296647Q-31
71	0.33146879851429422607823388411005Q-31
72	0.50536567393439157380190657605799Q-32
73	-0.12400976001662664533656704100650Q-31
74	0.52396368475382046356569245951765Q-32
75	0.79610546014104157687875160285319Q-33
76	-0.20078566894103499724402858296895Q-32
77	0.91381171433599850929107732144994Q-33

THE SUM OF THE COEFFICIENTS IS 0.10169047675679552352457315190721Q+01

```
A= 0.500000000000000000000000000000000000Q+00
C= 0.100000000000000000000000000000000000Q+01
W=-0.160000000000000000000000000000000000Q+02
N=  77
```

K	C(K)
0	0.10082793254489916399375295865917Q+01
1	0.84453495196168771635431941622596Q-02
2	0.17289216527845057494440682112515Q-03
3	0.73914028524380961409371161327759Q-05
4	0.60895693459375414547300669222169Q-06
5	0.11202527361806790154580779801157Q-06
6	0.35874739468925431708293960155885Q-07
7	0.12453819044204012399804538112886Q-07
8	0.35677548742532646223971188478432Q-08
9	0.61328322432474826119114135514194Q-09
10	-0.54471110874504929901979211919170Q-10
11	-0.73973748723321594035523710847628Q-10
12	-0.17843918584027419644780700480380Q-10
13	0.27039800177796550606712358489096Q-11
14	0.26268113265482787573477192369781Q-11
15	0.24813633820130616184308220156473Q-12
16	-0.27526489533783509261009399296825Q-12
17	-0.74965843144791624026231771307955Q-13
18	0.26960866397024955147740469334149Q-13
19	0.13237872345589130474028078034529Q-13
20	-0.28068911879844587186653315271803Q-14
21	-0.21321684251653977927549512190908Q-14
22	0.35493111029970609718262204200424Q-15
23	0.34235357256748558744284973261369Q-15
24	-0.59780131023401640009297615686771Q-16
25	-0.55620665446119099001893549756407Q-16
26	0.12711957318571971226508321395803Q-16
27	0.89465410219043805998510231139537Q-17
28	-0.29976041805951517197428377503572Q-17
29	-0.13451806020100427742106880937714Q-17
30	0.71312891579806470462143865057123Q-18
31	0.16353659048837516769799555313677Q-18
32	-0.16208525740769836057895376861189Q-18
33	-0.63187453599237693139922942038569Q-20
34	0.33654104612075033073193780099370Q-19
35	-0.50635256160723892800549667838964Q-20
36	-0.59530392093772271762908489308730Q-20
37	0.23500904282809250355054549452928Q-20
38	0.73414727837466796880449173569487Q-21
39	-0.68609988849630267917645210804824Q-21
40	0.12700478397077452273462752668963Q-22
41	0.14924301823467280449588391247326Q-21
42	-0.44195425408914328682672745301639Q-22
43	-0.21292205164130576369754093608751Q-22
44	0.16571595640169698990018668551118Q-22
45	-0.16907204243371778359986885572000Q-24
46	-0.38572051357369322163967484860191Q-23
47	0.13167611860841237910005388441857Q-23
48	0.48104954509633933863024293682236Q-24
49	-0.48948973642772979749447850051900Q-24
50	0.58066911007313178877099075448298Q-25
51	0.10011450601835911153329748736000Q-24
52	-0.52502629584422114574166897674450Q-25
53	-0.49623359881674530610223284700868Q-25
54	0.15373566515735916479198845665396Q-25
55	-0.49554043026413542270253062677083Q-26
56	-0.19417747799364192717759485697440Q-26
57	0.21176264830735908280238940857921Q-26
58	-0.39871726774904262981675927978021Q-27
59	-0.39871805564132050409663032278463Q-27
60	0.28035934457111852930447630963576Q-27
61	-0.22222752113629300255595150973238Q-28
62	-0.66054066652396857717417251108466U-28
63	0.37060391844713513823944060315052Q-28
64	0.59186617330401082789209964908538Q-30
65	-0.10588676912419041957652145660751Q-28
66	0.50257663830225802626332457116840Q-29
67	0.48450086105473880239308805237386Q-30
68	-0.16693308622933879421256266055448Q-29
69	0.71416877937903251732120799813472Q-30
70	0.10999173378622131422096955853079Q-30
71	-0.26393565945826643564061244633537Q-30
72	0.10792017811207864836890872355163Q-30
73	0.19292483528794886899515111629862Q-31
74	-0.42198791257249570351816484947260Q-31
75	0.17436782556547524588110934600618Q-31
76	0.28904078369844115323871203190199Q-32
77	-0.68620818072978732006239611877478Q-32
78	0.31083066643690071193822522503215Q-32

THE SUM OF THE COEFFICIENTS IS 0.10169057318875200076759918727730Q+01

```
A= 0.50000000000000000000000000000000Q+00
C= 0.10000000000000000000000000000000Q+01
W=-0.16000000000000000000000000000000Q+02
N=    77
```

K   C(K)

```
 0    0.10082792054587400318800127115733Q+01
 1    0.84451226249209431977092419274831Q-02
 2    0.17270063077756652703661790745472Q-03
 3    0.72475910999589629334265554516387Q-05
 4    0.51358772687802224829372929233877660Q-06
 5    0.56816965808120397540106933986955Q-07
 6    0.85130912228479602062131324116510Q-08
 7    0.12384253640034156034406813318550Q-08
 8    0.29801672301775267397624773185895Q-10
 9    0.78956598317201713487514457554173Q-10
10   -0.33127127631469398433264408224314Q-10
11   -0.44973863681088759306648249389680Q-11
12    0.17997902965207698339329794634642Q-11
13    0.96574832300685536412293987464517Q-12
14    0.38604236582086752495018249304270Q-13
15   -0.10403934481726874247059761750720Q-12
16   -0.23950448895047711691273534795764Q-13
17    0.95544665396637010532983514714542Q-14
18    0.44431493480031774078919180258549Q-14
19   -0.85864191086248113835647231256981Q-15
20   -0.70878022195875254415860403969684Q-15
21    0.86759991049698506224620789432281Q-16
22    0.11193652915913705443917567094487Q-15
23   -0.12108279125199573123164972875204Q-16
24   -0.18130894726461129552116749570900Q-16
25    0.24869450794210852676900796327785Q-17
26    0.29920390730330207866558758251030Q-17
27   -0.62382704451545261765890076884186Q-18
28   -0.48589145582924398135171917756183Q-18
29    0.16154358697926317552205018327052Q-18
30    0.72342585333025031855671304989845Q-19
31   -0.40155054650086641480121070603133Q-19
32   -0.81987173866381914816560803267700Q-20
33    0.92417687092891262847300151845270Q-20
34    0.31791507452023765191700623725669Q-22
35   -0.18924604862087616031010840013290Q-20
36    0.37652904822881262633063871509870Q-21
37    0.31713757745520346079911384433608Q-21
38   -0.15292935017929402842881748071411Q-21
39   -0.31804340918275147174231651703608Q-22
40    0.41635499206034428621074551349173Q-22
41   -0.39734929428156764077070152370994Q-23
42   -0.82785043556801141924586340106603Q-23
43    0.32926590320721552381668936434556Q-23
44    0.92470525885576429080079814249433Q-24
45   -0.10553260752972248647762599409419Q-23
46    0.12011648868756131790444559311472Q-24
47    0.21204747668533764147478380203953Q-24
48   -0.10126279262272766034684995698555Q-24
49   -0.16129437690022430228481358159413Q-25
50    0.30724830610641030959605828253365Q-25
51   -0.73554282415813107918516578354230Q-26
52   -0.49162783829816681576916651641807Q-26
53    0.37687666830129750137063345600177Q-26
54   -0.24035357962229131919101923657184Q-27
55   -0.87313698209019328336357302196840Q-27
56    0.42384667991598998855299313852489Q-28
57    0.49703093308413240280578995713310Q-27
58   -0.13328991041309707676237926694811Q-27
59    0.45202193862941718485385327440969Q-28
60    0.15573915442340150994172748914955Q-28
61   -0.19281459174967317466435785654687Q-28
62    0.46579490753440065642313278245913Q-29
63    0.31021659099184220428620755535145Q-29
64   -0.27629538773120573619649309640853Q-29
65    0.47282206196264145500402667863331Q-30
66    0.54148105580644182839691393959784Q-30
67   -0.40242766993136122011645673429624Q-30
68    0.49264092893248782041603946039240Q-31
69    0.89687222110078241131528749510215Q-31
70   -0.60533606085325407554878642134268Q-31
71    0.57868250364738467291220678873009Q-32
72    0.14527222787658264778487006682141Q-31
73   -0.94821441238662648935472935544033Q-32
74    0.87076079455604119477377201880276Q-33
75    0.23286405498642529239880074578308Q-32
76   -0.15567475054536774597032068532626Q-32
77    0.19768522460147225361991420400549Q-33
78    0.28626172922395709292701944237424Q-33
```

THE SUM OF THE COEFFICIENTS IS 0.10169048563776570375542263743884Q+01

THE FINAL COEFFICIENTS ARE DIRECTLY ABOVE.

TEST1=-0.19727Q-29
TEST2=-0.11979Q-25

IX COEFFICIENTS FOR THE EXPANSION OF THE FUNCTIONS $G_{1,3}^{m,1}(z^2/4 \mid \begin{smallmatrix} 1 \\ a,b,c \end{smallmatrix})$, m = 3 OR m = 2 , IN DESCENDING SERIES OF CHEBYSHEV POLYNOMIALS

We first consider the evaluation of $G_n(w)$ in

$$B \ G_{1,3}^{3,1}(z^2/4 \mid \begin{smallmatrix} 1 \\ a,b,c \end{smallmatrix}) = \sum_{n=0}^{\infty} G_n(w) T_{2n}(z/w) \ , \ z/w \geq 1 \ ,$$

$$B = [\Gamma(a)\Gamma(b)\Gamma(c)]^{-1} \ , \ z \neq 0 \ , \ |\arg z| < 3\pi/4 \ . \tag{1}$$

This is the special case of 1.3(15) with $q = 3$, $p = 0$, and z and w replaced by $z^2/4$ and $w^2/4$, respectively. Note that the G-function on the left of (1) can be expressed as a linear combination of three $_1F_2$'s , see 1.2(5) . Except for the recurrence formula for $G_n(w)$, all necessary data will be found in 1.3. Thus, virtually none of this material will be repeated, save for the asymptotic expansion of $G_n(w)$ for large n . From 1.3(23), we have

$$G_n(w) = \frac{2^{\frac{1}{2}}(-)^n(2\pi)Bs^u}{n} \exp\{-4s+2sP(y)+S/s\}$$

$$\times \left[1 + 0 \left(\frac{(1+|w^2|^{4/3})}{s^2} \right) \right] \ ,$$

$$u = a+b+c-1 \ , \ s = (nw/2)^{\frac{1}{2}} \ , \ P(y) = 1 - \frac{y}{\sinh y} \ ,$$

$$y = (w/2n)^{\frac{1}{2}} \ , \ S = -E_2 + (3E_1^2/8) - 5/32 \ , \ E_1 = u+3/2 \ ,$$

$$E_2 = \tfrac{1}{2}(u+1)+ab+ac+bc \ , \ n^2w \to \infty \ , \ |\arg w| \leq \frac{5\pi}{2} - \varepsilon \ , \ \varepsilon > 0$$

$$w = o(n^{3/5}) \quad , \quad (wn)^{-1} = o(1) \quad , \tag{2}$$

or the same as above with

$$2sP(y) = (sy^2/3)\left(1 - \{7y^2/60\} + \{31y^4/2520\} + O(y^6)\right) . \tag{3}$$

The recursion formula for $G_n(w)$ is given by

$$\frac{2G_n(w)}{\varepsilon_n} = (P_1 + w^2 Q_1/4)G_{n+1}(w) + (P_2 + w^2 Q_2/4)G_{n+2}(w)$$

$$+ (P_3 + w^2 Q_3/4)G_{n+3}(w) + P_4 G_{n+4}(w) ,$$

$$P_1 = 2(n+1) \left[1 - \frac{(2n+3)(n+1+A)}{(2n+5)(n+A)}\right] , \quad Q_1 = \frac{-4(n+1)}{(n+A)} ,$$

$$P_2 = -(n+2)(2n+1)\left[1 - \frac{2(2n+2)(2n+3)(n+1+A)}{(2n+5)(2n+1)(n+A)}\right.$$

$$\left. + \frac{(2n+2)(2n+3)(n+2+A)}{(2n+1)(2n+6)(n+A)}\right] ,$$

$$Q_2 = \frac{16(n+1)(n+2)}{(2n+5)(n+A)} , \quad Q_3 = \frac{-2(2n+2)(2n+3)}{(2n+5)(n+A)} , \tag{4}$$

$$P_3 = \frac{(2n+2)(n+3-A)}{(n+A)}\left[1 - \frac{(2n+3)(n+4-A)}{(2n+5)(n+3-A)}\right] ,$$

$$P_4 = \frac{-(2n+2)(2n+3)(n+4-A)}{(2n+5)(2n+6)(n+A)} .$$

Here $(n+r+tA)$ is shorthand for the product $(n+r+ta)(n+r+tb)(n+r+tc)$.

In the present instance 1.3(18) holds, and likewise for 1.3(19-22) with $q = 3$, $p = 0$, etc., provided that in the right hand sides of these equations w is replaced by $w^2/4$. It is convenient to repeat 1.3(18) and the analogs of 1.3(21,22). Thus

$$\sum_{n=0}^{\infty} (-)^n G_n(w) = 1 \; , \tag{5}$$

$$\text{TEST 1} = \sum_{n=1}^{\infty} (-)^n n^2 G_{n,N}(w) - \frac{2abc}{w^2} \; , \tag{6}$$

$$\text{TEST 2} = \sum_{n=2}^{\infty} (-)^n n^2 (n^2-1) G_{n,N}(w) - \frac{12a(a+1)b(b+1)c(c+1)}{w^4} \; . \tag{7}$$

Next we consider

$$\frac{\pi B}{\sin \pi c} \; G_{1,3}^{2,1}(z^2/4 \,|\, {}_{a,b,c}^{\quad 1}) = \sum_{n=0}^{\infty} D_n(w) T_{2n}(w/z) \; ,$$

$$B \quad \text{as in (1)} \, , \; z \neq 0 \, , \; |\arg z| < \pi/4 \; , \tag{8}$$

which is the special case of 1.3(24) with $q = 3$, $p = 0$ **and** z and w replaced by $z^2/4$ and $w^2/4$, respectively. Note that the G-function on the left of (8) can be expressed as a linear combination of two ${}_1F_2$'s , see 1.2(5). From (2), (3), (8) and 1.3(25), we have

$$D_n(w) = \frac{(2\pi)2^{\frac{1}{2}}(-)^n B s^u}{n \sin \pi c} (\sin\theta) \exp\{-2^{3/2}s - E + S/2^{\frac{1}{2}}s\}$$

$$\times \left[1 + O\left(\frac{(1+|w^2|^{4/3})}{s^2} \right) \right] \; ,$$

$$E = \frac{2^{\frac{1}{2}}sy^2}{6} [1 - \frac{7y^2}{60} - \frac{31y^4}{2520}] \; , \quad F = \frac{2^{\frac{1}{2}}sy^2}{6} [1 + \frac{7y^2}{60} - \frac{31y^4}{2520}] \; ,$$

$$\theta = \pi c - \pi/4 + 2^{3/2}s - F + S/2^{\frac{1}{2}}s \; , \tag{9}$$

where all the conditions in (2) apply, save that in the present case $|\arg w| \leq 2\pi - \varepsilon$, $\varepsilon > 0$.

$D_n(w)$ satisfies the same recurrence relation as does $G_n(w)$, that is , (4), and $D_n(w)$ also satisfies (6), if w^2

is replaced by $-w^2$. Further, $D_n(w)$ satisfies (5) and (7).
The coefficients $G_n(w)$ and $D_n(w)$ are readily found by the
backward recursion process described by 1.3(28-35). The reader is reminded that convergence of this scheme for $G_n(w)$ and
$D_n(w)$ weakens as $|\arg w| \to \pi/2$ and $\arg w \to 0$, respective-
ly, and fails when $|\arg w| = \pi/2$ and $\arg w = 0$, respec-
tively. In these instances, the process must be modified as
outlined by 1.3(36-40).

For simplification, the algorithms developed for the
evaluation of the coefficients in (1) and (8) use the same
subroutines called CCOEF 5 and CHECK 5. These subroutines
differ only in the parameter VAL defined in the comments of
the subroutines. When the subroutines are used with the respective main line programs for computation of the coefficients in (1) and (8), the proper value of VAL is automati-
cally specified. But if the subroutines are used independent-
ly of the main line program, then the parameter VAL must be
specified by the user. For the subroutine LINCOM, see Chapter 8.

Numerical Examples

1. Let $a = 1$, $b = c = \frac{1}{2}$ and $w = 8$. Then from (1) and
2.3(2) we have

$$H_0(z) - Y_0(z) = (2/\pi z) \sum_{n=0}^{\infty} G_n T_{2n}(8/z) , \quad z \geq 8 .$$

In (2), $u = 1$, $\Gamma(a)\Gamma(b)\Gamma(c) = \pi$ and $S = -1/16$. Let
$n = 16$ and 25 and use (2,3) without the order terms. Then
G_{16} and G_{25} are $0.33963(-13)$ and $-0.80627(-17)$, respectively. The corresponding values deduced by the backward recurrence process with $N = 85$ are $0.33937(-13)$ and
$-0.80647(-17)$, respectively. Note that

$$G_0(w) = (\pi w/4) [J_0^2(w/2) + Y_0^2(w/2)] .$$

2.   Let   a,b,c,w , etc. be as in the above example.   Then from (8) and 2.3(4), we have

$$I_0(z) - L_0(z) = (2/\pi z) \sum_{n=0}^{\infty} D_n T_{2n}(8/z) , z \geq 8 .$$

Let   n = 16   and   36   and use (9) without the order term. Then D_{16} and D_{36} are   -0.10640(-9)   and   0.19081(-15) , respectively.   The corresponding values deduced by the back- ward recurrence process with   N = 99 , M = 100   are -0.10653(-9)   and   0.18822(-15) , respectively.   Notice that with   N = 99 , only about 25 decimal accuracy in the evalua- tion of $\sum D_n T_{2n}(8/z)$ is possible.   For quadruple accuracy, we would need about 150 terms.   If   n = 144 , then from (9) without the order term, D_{144} = -0.75821(-30) .

```
C            ------------ IBM S/370 ----------- MULTIPLE PRECISION ------------
C
C      **************************************************************************
C      *THIS MAINLINE PROGRAM UTILIZES THE SUBROUTINES CCOEF5 AND CHECK5 *
C      *TO COMPUTE AND TEST THE COEFFICIENTS IN THE CHEBYSHEV EXPANSION  *
C      *     3,1                                                         *
C      *OF G    ((Z/2)**2||A,B,C).                                       *
C      *     1,3                                                         *
C      *                                                                 *
C      *WHERE NO AMBIGUITY CAN OCCUR, WE SHALL DENOTE                    *
C      * 3,1                                                             *
C      *G    ((Z/2)**2||A,B,C) BY (3,1 G 1,3)(Z).                        *
C      * 1,3                                                             *
C      *                                                                 *
C      *NOTE: THE PARAMETER VAL, AS DEFINED IN THE COMMENTS OF THE SUB-  *
C      *ROUTINES, IS AUTOMATICALLY SPECIFIED IF THIS MAINLINE PROGRAM IS *
C      *USED.  IF THE SUBROUTINES ARE USED INDEPENDENTLY OF THE MAINLINE *
C      *PROGRAM, VAL MUST BE SPECIFIED BY THE USER.                      *
C      *                                                                 *
C      *DESCRIPTION OF VARIABLES.                                        *
C      *                                                                 *
C      *AP     -INPUT  - PARAMETER A IN (3,1 G 1,3)(Z).                  *
C      *                                                                 *
C      *BP     -INPUT  - PARAMETER B IN (3,1 G 1,3)(Z).                  *
C      *                                                                 *
C      *CP     -INPUT  - PARAMETER C IN (3,1 G 1,3)(Z).                  *
C      *                                                                 *
C      *W      -INPUT  - THIS IS A PRESELECTED SCALE FACTOR SUCH THAT    *
C      *                 0.LE.(W/Z).LE.1.  SEE TEXT.                     *
C      *                                                                 *
C      *N      -INPUT  - TWO LESS THAN THE NUMBER OF COEFFICIENTS TO BE  *
C      *                 GENERATED.  MAXIMUM VALUE IS 100.               *
C      *                                                                 *
C      *C      -OUTPUT - A VECTOR CONTAINING THE N+2 CHEBYSHEV COEFFI-   *
C      *                 CIENTS FOR THE APPROXIMATION TO (3,1 G 1,3)(Z). *
C      *                 DIMENSION OF THE VECTOR IS SET AT 102.          *
C      *                                                                 *
C      *TEST1  -OUTPUT - THE RESULT OF THE FIRST TEST PERFORMED ON THE   *
C      *                 COEFFICIENTS.  SEE TEXT.                        *
C      *                                                                 *
C      *TEST2  -OUTPUT - THE RESULT OF THE SECOND TEST PERFORMED ON THE  *
C      *                 COEFFICIENTS.  SEE TEXT.                        *
C      *                                                                 *
C      *SUM    -OUTPUT - THE SUM OF THE COEFFICIENTS.                    *
C      *                                                                 *
C      *ALL OTHER VARIABLES ARE FOR INTERNAL USE.                       *
C      *                                                                 *
C      *SUBROUTINES CCOEF5 AND CHECK5 ARE REQUIRED.                     *
C      **************************************************************************
       IMPLICIT REAL*16 (A-H,O-Z)
       DIMENSION C(102)
       DATA ONE/1.Q0/
C            ---------- THE NEXT PROGRAM CARD IS USED TO INITIALIZE ----------
C            ---------- DEVICE NUMBERS FOR THE PARTICULAR INSTALL-  ----------
C            ---------- ATION OF THE OPERATING SYSTEM.  IF A SYSTEM ----------
C            ---------- HAS DIFFERENT DEVICE NUMBERS THIS CARD CAN  ----------
C            ---------- BE CHANGED TO INITIALIZE THOSE VALUES AS    ----------
C            ---------- DEVICE NUMBERS.                             ----------
       DATA NREAD,NWRITE,NPUNCH/5,6,7/
C
C            ---------- READ IN THE INPUT VARIABLES              ----------
       READ(NREAD,2) AP,BP,CP,W
       READ(NREAD,4) N
       N2=N+2
C
C            ---------- CALL SUBROUTINE TO COMPUTE THE COEFFICIENTS ----------
       CALL CCOEF5(AP,BP,CP,W,N,ONE,C,SUM)
C
C            ---------- CALL SUBROUTINE TO TEST THE COEFFICIENT  ----------
C            ---------- VALUES                                   ----------
       CALL CHECK5(AP,BP,CP,W,N,ONE,C,TEST1,TEST2)
C
C            ---------- WRITE OUT VALUES OF PARAMETERS AND COEFFI- ----------
C            ---------- CIENTS                                     ----------
       WRITE(NWRITE,5) AP,BP,CP,W,N
       DO 1 K=1,N2
       K1=K-1
    1  WRITE(NWRITE,6) K1,C(K)
       WRITE(NWRITE,7) TEST1,TEST2
       WRITE(NWRITE,8) SUM
C
C            ---------- PUNCH THE VALUES OF COEFFICIENTS ON CARDS ----------
       WRITE(NPUNCH,2) (C(K),K=1,N2)
    2  FORMAT(Q39.32)
    4  FORMAT(I5)
    5  FORMAT('1',4X,'A=',Q39.32/5X,'B=',Q39.32/5X,'C=',Q39.32/5X,'W=',
      :Q39.32/5X,'N=',I5/////5X,'K',25X,'C(K)'/)
    6  FORMAT(1X,I5,8X,Q39.32)
    7  FORMAT(/1X,'TEST1=',Q12.5/1X,'TEST2=',Q12.5)
    8  FORMAT(/1X,'THE SUM OF THE COEFFICIENTS IS',Q39.32)
       STOP
       END
```

```
C
C
C          ----------- IBM S/370 ----------- MULTIPLE PRECISION -----------
C       **********************************************************************
C       *THIS SUBROUTINE, CCOEF5, COMPUTES MULTIPLE PRECISION VALUES OF      *
C       *THE COEFFICIENTS IN THE CHEBYSHEV EXPANSION                         *
C       *    3,1                                                             *
C       *OF G    ((Z/2)**2||$A,B,C).                                         *
C       *    1,3                                                             *
C       *                                                                   *
C       *WHERE NO AMBIGUITY CAN OCCUR, WE SHALL DENOTE                       *
C       *  3,1                                                               *
C       *G    ((Z/2)**2||$A,B,C) BY (3,1 G 1,3)(Z).                          *
C       *  1,3                                                               *
C       *                                                                   *
C       *DESCRIPTION OF VARIABLES.                                           *
C       *                                                                   *
C       *AP    -INPUT  - PARAMETER A IN (3,1 G 1,3)(Z).                      *
C       *                                                                   *
C       *BP    -INPUT  - PARAMETER B IN (3,1 G 1,3)(Z).                      *
C       *                                                                   *
C       *CP    -INPUT  - PARAMETER C IN (3,1 G 1,3)(Z).                      *
C       *                                                                   *
C       *W     -INPUT  - THIS IS A PRESELECTED SCALE FACTOR SUCH THAT        *
C       *                0.LE.(W/Z).LE.1.  SEE TEXT.                         *
C       *                                                                   *
C       *N     -INPUT  - TWO LESS THAN THE NUMBER OF COEFFICIENTS TO BE      *
C       *                GENERATED.  MAXIMUM VALUE IS 100.                   *
C       *                                                                   *
C       *VAL   -INPUT  - A PARAMETER TO ALLOW USE OF THE SUBROUTINE FOR      *
C       *                VAL*(Z/2)**2 AS THE ARGUMENT.  FOR THE              *
C       *                (3,1 G 1,3) VAL IS 1 AND FOR THE (2,1 G 1,3) VAL    *
C       *                IS -1.                                              *
C       *                                                                   *
C       *C     -OUTPUT - A VECTOR CONTAINING THE N+2 CHEBYSHEV COEFFI-       *
C       *                CIENTS FOR THE APPROXIMATION TO (3,1 G 1,3)(Z).     *
C       *                                                                   *
C       *SUM   -OUTPUT - THE SUM OF THE COEFFICIENTS.                        *
C       *                                                                   *
C       *ALL OTHER VARIABLES ARE FOR INTERNAL USE.                          *
C       **********************************************************************
        SUBROUTINE CCOEF5(AP,BP,CP,W,N,VAL,C,SUM)
        IMPLICIT REAL*16 (A-H,O-Z)
        DIMENSION C(1)
        DATA ZERO,ONE,TWO,THREE,FOUR,FIVE,START/0.Q0,1.Q0,2.Q0,3.Q0,4.Q0,
       :5.Q0,1.Q-20/
        N1=N+1
        N2=N+2
C
C          ---------- START COMPUTING COEFFICIENTS BY MEANS OF    ----------
C          ---------- BACKWARD RECURRENCE SCHEME                  ----------
        A4=ZERO
        A3=ZERO
        A2=ZERO
        A1=START
        NCOUNT=N2
        C(NCOUNT)=START
        X=N1
        AR=VAL*W*W
        DO 1 K=1,N1
        X=X-ONE
        S0=(X+AP)*(X+BP)*(X+CP)
        S1=(X+ONE+AP)*(X+ONE+BP)*(X+ONE+CP)
        S2=(X+TWO+AP)*(X+TWO+BP)*(X+TWO+CP)
        S3=(X+THREE-AP)*(X+THREE-BP)*(X+THREE-CP)
        S4=(X+FOUR-AP)*(X+FOUR-BP)*(X+FOUR-CP)
        Z1=AR/((TWO*X+FIVE)*S0)
        P=(TWO*X+THREE)/((TWO*X+FIVE)*S0)
        NCOUNT=NCOUNT-1
        C(NCOUNT)=(X+ONE)*((TWO*(ONE-P*S1)-(TWO*X+FIVE)*Z1)*A1+
       :(-(X+TWO)*(TWO*X+ONE)-(ONE/(X+FOUR)-FOUR*S1*P/(TWO*X+ONE)+
       :P*(TWO*X+FIVE)*S2/((X+THREE)*(TWO*X+ONE)))+FOUR*(X+TWO)*Z1
       :)*A2+(TWO*S3/S0*(ONE-P*S0*S4/S3)-(TWO*X+THREE)*Z1)*A3
       :-P*S4*A4/(X+THREE))
        A4=A3
        A3=A2
        A2=A1
      1 A1=C(NCOUNT)
        C(1)=C(1)/TWO
C
C          ---------- COMPUTE SCALE FACTOR                        ----------
        RHO=C(1)
        SUM=RHO
        P=ONE
        DO 3 I=2,N2
        RHO=RHO-P*C(I)
        SUM=SUM+C(I)
      3 P=-P
C
C          ---------- SCALE THE COEFFICIENTS                      ----------
        DO 4 L=1,N2
      4 C(L)=C(L)/RHO
        SUM=SUM/RHO
        RETURN
        END
```

```
C
C
C
C          ---------- IBM S/370 ----------- MULTIPLE PRECISION --------------
C    ******************************************************************************
C    *THIS SUBROUTINE, CHECK5, RUNS CHECKS ON THE COEFFICIENTS             *
C    *GENERATED BY SUBROUTINE CCOEF5.  EXCEPT FOR ROUND-OFF ERROR AND      *
C    *TRUNCATION ERROR, BOTH TEST RESULTS SHOULD BE ZERO.  SEE TEXT.       *
C    *                                                                     *
C    *WHERE NO AMBIGUITY CAN OCCUR, WE SHALL DENOTE                        *
C    *  3,1                                                                *
C    * G    ((Z/2)**2||A,B,C) BY (3,1 G 1,3)(Z).                          *
C    * 1,3                                                                 *
C    *                                                                     *
C    *DESCRIPTION OF VARIABLES.                                            *
C    *                                                                     *
C    *AP     -INPUT - PARAMETER A IN (3,1 G 1,3)(Z).                      *
C    *                                                                     *
C    *BP     -INPUT - PARAMETER B IN (3,1 G 1,3)(Z).                      *
C    *                                                                     *
C    *CP     -INPUT - PARAMETER C IN (3,1 G 1,3)(Z).                      *
C    *                                                                     *
C    *W      -INPUT - THIS IS A PRESELECTED SCALE FACTOR SUCH THAT         *
C    *                0.LE.(W/7).LE.1.  SEE TEXT.                         *
C    *                                                                     *
C    *N      -INPUT - TWO LESS THAN THE NUMBER OF COEFFICIENTS             *
C    *                GENERATED.  MAXIMUM VALUE IS 100.                    *
C    *                                                                     *
C    *VAL    -INPUT - A PARAMETER TO ALLOW USE OF THE SUBROUTINE FOR       *
C    *                VAL*(Z/2)**2 AS THE ARGUMENT.  FOR THE               *
C    *                (3,1 G 1,3) VAL IS 1 AND FOR THE (2,1 G 1,3) VAL     *
C    *                IS -1.                                               *
C    *                                                                     *
C    *C      -INPUT - A VECTOR CONTAINING THE N+2 CHEBYSHEV COEFFI-        *
C    *                CIENTS FOR THE APPROXIMATION TO (3,1 G 1,3)(Z).      *
C    *                                                                     *
C    *TEST1  -OUTPUT - THE RESULT OF THE FIRST TEST PERFORMED ON THE       *
C    *                COEFFICIENTS.  SEE TEXT.                             *
C    *                                                                     *
C    *TEST2  -OUTPUT - THE RESULT OF THE SECOND TEST PERFORMED ON THE      *
C    *                COEFFICIENTS.  SEE TEXT.                             *
C    *                                                                     *
C    *ALL OTHER VARIABLES ARE FOR INTERNAL USE.                           *
C    ******************************************************************************
C          SUBROUTINE CHECK5(AP,BP,CP,W,N,VAL,C,TEST1,TEST2)
C          IMPLICIT REAL*16 (A-H,O-Z)
C          DIMENSION C(1)
C          DATA ONE,TWO,FOUR,TWELVE/1.Q0,2.Q0,4.Q0,12.Q0/
C
C          ---------- COMPUTATION OF LEFT SIDES OF TEST EQUATIONS -----------
          SUM1=-C(2)+FOUR*C(3)
          P=-ONE
          SUM2=TWELVE*C(3)
          COUNT=TWO
          DO 1 I=2,N
          COUNT=COUNT+ONE
          CSQ=COUNT*COUNT
          SUM1=SUM1+P*CSQ*C(I+2)
          SUM2=SUM2+P*CSQ*(CSQ-ONE)*C(I+2)
        1 P=-P
C
C          ---------- COMPUTATION OF TEST VALUES           -----------
          TEST1=SUM1-TWO*VAL*AP*BP*CP/(W*W)
          TEST2=SUM2-TWELVE*AP*BP*CP*(AP+ONE)*(BP+ONE)*(CP+ONE)/W**4
          RETURN
          END
```

```
A=  0.100000000000000000000000000000000000Q+01
B=  0.500000000000000000000000000000000000Q+00
C=  0.500000000000000000000000000000000000Q+00
W=  0.800000000000000000000000000000000000Q+01
N=   85
```

K	C(K)
0	0.99283727576423943189107819779817Q+00
1	-0.69688128113862475697120993392402Q-02
2	0.18205103787037122732928350228075Q-03
3	-0.10632582252844160782718661598233Q-04
4	0.98198294286525278995335914171957Q-06
5	-0.12250645444976943118330317243676Q-06
6	0.18940833117999227126761426677707Q-07
7	-0.34435822256042778975046550130114Q-08
8	0.71119101711062246929458544621459Q-09
9	-0.16288744136658632368603709670183Q-09
10	0.40656807283543286967366273762365Q-10

```
11      -0.10915047958955359959194077295815Q-10
12       0.31200524275853974417987355363797Q-11
13      -0.94202070057994452989998191498196Q-12
14       0.29447947171326430560449885034707Q-12
15      -0.98724164743142757906732713155758Q-13
16       0.33937123558726253683413066775221Q-13
17      -0.12079796031988521513544102808464Q-13
18       0.44382094662674738089320155105358Q-14
19      -0.16785850636581731428113288624159Q-14
20       0.65199951727001322892829601129270Q-15
21      -0.25955537426842441884633685879613Q-15
22       0.10570846206147261554324322227513Q-15
23      -0.43973897939726431889322982105660Q-16
24       0.18658237198191404585883399826319Q-16
25      -0.80646740081597837118986554198861Q-17
26       0.35469057326079837093779126039703Q-17
27      -0.15856724145630677853123655472543Q-17
28       0.71990408042516210021105625371558Q-18
29      -0.33163941482142682049486781548517Q-18
30       0.15490066272244041910009841259240Q-18
31      -0.73304267311822540490822372569590Q-19
32       0.35124738616263504373284196730378Q-19
33      -0.17031218465106653227742462566198Q-19
34       0.83519537718171763511794639077933Q-20
35      -0.41401780590485852786323143046603Q-20
36       0.20736402756791536754586172337682Q-20
37      -0.10489158348272965310435945614774Q-20
38       0.53562782879704254912867400112216Q-21
39      -0.27601660210247589627331534423168Q-21
40       0.14348349874286854518437786855597Q-21
41      -0.75217400877318321725503794855723Q-22
42       0.39751013493867987489241047627617Q-22
43      -0.21172091275827465043051185457215Q-22
44       0.11361734780498511877847357455591Q-22
45      -0.61415429749305436430897662342050Q-23
46       0.33431457956839910083330820476460Q-23
47      -0.18322174801073702777129188853154Q-23
48       0.10107575129350324031577133522313Q-23
49      -0.56114436244739251021601538840171Q-24
50       0.31345413798384231722935071664564Q-24
51      -0.17614229492869106731349666152037Q-24
52       0.99555680739871226807286219309504Q-25
53      -0.56585802637911632213453634601570Q-25
54       0.32338342309746689017608350202034Q-25
55      -0.18579325742541537683697955652924Q-25
56       0.10729518222305724109072488164963Q-25
57      -0.62274123172310915747256020362692Q-26
58       0.36320678260443722658891740160496Q-26
59      -0.21284510116661804683225300055708Q-26
60       0.12530943469197493881840947133773Q-26
61      -0.74107777047821104166804837593071Q-27
62       0.44020552628205305604993220268272Q-27
63      -0.26261076860819608290953794028685Q-27
64       0.15732206208704214933751272484551Q-27
65      -0.94632728301597705488292256060273Q-28
66       0.57150579575988848210941993844931Q-28
67      -0.34647948874437678661054968188181Q-28
68       0.21084321437512671565265031132863Q-28
69      -0.12877038689816013738860891184440Q-28
70       0.78923654046551248466916571644024Q-29
71      -0.48542597946764300551445262431927Q-29
72       0.29964555393306148759896557115881Q-29
73      -0.18568908118125281663850464992874Q-29
74       0.11557769238539946840235033440374Q-29
75      -0.72300706099973709825668855635680Q-30
76       0.45477991844665905040521315696751Q-30
77      -0.28757878993812002514781143575411Q-30
78       0.18249054575024732265545629555128Q-30
79      -0.11570120124355823507163305904754Q-30
80       0.72692194161881717070001600831698Q-31
81      -0.44663843030674832619620482287030Q-31
82       0.26309422512836988449767025797140Q-31
83      -0.14418789920088708051934313865476Q-31
84       0.70013169097612556270043843850054Q-32
85      -0.27417226099095211776188978902463Q-32
86       0.67407705131765005277268120518576Q-33
```

TEST1=-0.153420-27
TEST2=-0.118750-23

THE SUM OF THE COEFFICIENTS IS 0.98604065696238259766225262177064Q+00

```
C          ----------- IBM S/370 ----------- MULTIPLE PRECISION ------------
C
C          ******************************************************************
C          *THIS PROGRAM UTILIZES THE SUBROUTINES CCOEF5, LINCOM AND CHECK5 *
C          *TO COMPUTE AND TEST THE COEFFICIENTS IN THE CHEBYSHEV EXPANSION *
C          *      2,1                                                       *
C          *OF G    ((Z/2)**2|1|A,B,C).                                     *
C          *      1,3                                                       *
C          *                                                                *
C          *WHERE NO AMBIGUITY CAN OCCUR, WE SHALL DENOTE                   *
C          *    2,1                                                         *
C          *G    ((Z/2)**2|1|A,B,C) BY (2,1 G 1,3)(Z).                      *
C          *    1,3                                                         *
C          *                                                                *
C          *                                                                *
C          *NOTE: SUBROUTINES CCOEF5 AND CHECK5 ARE THE SAME AS THE SUBROU- *
C          *TINES WITH THE SAME NAME FOR THE G  3,1 ((Z/2)**2|1|A,B,C) CASE,*
C          *                                      1,3                       *
C          *PROVIDED THAT THE PARAMETER VAL AS DEFINED IN THE COMMENTS OF   *
C          *THE SUBROUTINES IS CORRECTLY SPECIFIED.  THIS IS AUTOMATICALLY  *
C          *DONE IF ONE USES THE RESPECTIVE MAINLINE PROGRAMS.  IF THE      *
C          *SUBROUTINES ARE USED INDEPENDENTLY OF THE MAINLINE PROGRAM, VAL *
C          *MUST BE SPECIFIED BY THE USER. SEE CH. 8 FOR SUBROUTINE LINCOM. *
C          *                                                                *
C          *DESCRIPTION OF VARIABLES.                                       *
C          *                                                                *
C          *AP    -INPUT - PARAMETER A IN (2,1 G 1,3)(Z).                   *
C          *                                                                *
C          *BP    -INPUT - PARAMETER B IN (2,1 G 1,3)(Z).                   *
C          *                                                                *
C          *CP    -INPUT - PARAMETER C IN (2,1 G 1,3)(Z).                   *
C          *                                                                *
C          *W     -INPUT - THIS IS A PRESELECTED SCALE FACTOR SUCH THAT     *
C          *               0.LE.(W/Z).LE.1.  SEE TEXT.                      *
C          *                                                                *
C          *N     -INPUT - TWO LESS THAN THE NUMBER OF COEFFICIENTS TO BE   *
C          *               COMPUTED IN THE FIRST SET OF COEFFICIENTS.  MAX- *
C          *               IMUM VALUE IS 100.                               *
C          *                                                                *
C          *NN    -INPUT - TWO LESS THAN THE NUMBER OF COEFFICIENTS TO BE   *
C          *               COMPUTED IN THE SECOND SET OF COEFFICIENTS.  NN  *
C          *               MUST BE GREATER THAN N.  MAXIMUM VALUE IS 100.   *
C          *                                                                *
C          *T     -INPUT - THE VALUE OF (2,1 G 1,3)(Z) AT Z=W.             *
C          *                                                                *
C          *S1    -OUTPUT - THE SUM OF THE FIRST SET OF COEFFICIENTS.       *
C          *                                                                *
C          *S2    -OUTPUT - THE SUM OF THE SECOND SET OF COEFFICIENTS.      *
C          *                                                                *
C          *C1    -OUTPUT - THE FIRST VECTOR CONTAINING THE N+2 CHEBYSHEV   *
C          *               COEFFICIENTS FOR THE APPROXIMATION TO            *
C          *               (2,1 G 1,3)(Z).  DIMENSION OF THE VECTOR IS SET  *
C          *               AT 102.                                          *
C          *                                                                *
C          *C2    -OUTPUT - THE SECOND VECTOR CONTAINING THE NN+2 CHEBYSHEV *
C          *               COEFFICIENTS FOR THE APPROXIMATION TO            *
C          *               (2,1 G 1,3)(Z).  DIMENSION OF THE VECTOR IS SET  *
C          *               AT 102.                                          *
C          *                                                                *
C          *C     -OUTPUT - THE VECTOR CONTAINING THE FINAL NN+2 CHEBYSHEV  *
C          *               COEFFICIENTS FOR THE APPROXIMATION TO            *
C          *               (2,1 G 1,3)(Z).  DIMENSION OF THE VECTOR IS SET  *
C          *               AT 102.                                          *
C          *                                                                *
C          *TEST1 -OUTPUT - THE RESULT OF THE FIRST TEST PERFORMED ON THE   *
C          *               COEFFICIENTS.  SEE TEXT.                         *
C          *                                                                *
C          *TEST2 -OUTPUT - THE RESULT OF THE SECOND TEST PERFORMED ON THE  *
C          *               COEFFICIENTS.  SEE TEXT.                         *
C          *                                                                *
C          *SUMM  -OUTPUT - THE SUM OF THE COEFFICIENTS.                    *
C          *                                                                *
C          *ALL OTHER VARIABLES ARE FOR INTERNAL USE.                       *
C          *                                                                *
C          *SUBROUTINES CCOEF5, LINCOM AND CHECK5 ARE REQUIRED.            *
C          ******************************************************************
           IMPLICIT REAL*16 (A-H,O-Z)
           DIMENSION C1(102),C2(102),C(102)
           DATA ONE/1.Q0/
```

```
C
C
C          ---------- THE NEXT PROGRAM CARD IS USED TO INITIALIZE ------------
C          ---------- DEVICE NUMBERS FOR THE PARTICULAR INSTALL-
C          ---------- ATION OF THE OPERATING SYSTEM.  IF A SYSTEM
C          ---------- HAS DIFFERENT DEVICE NUMBERS THIS CARD CAN
C          ---------- BE CHANGED TO INITIALIZE THOSE VALUES AS
C          ---------- DEVICE NUMBERS.
           DATA NREAD,NWRITE,NPUNCH/5,6,7/
C
C          ---------- READ IN THE INPUT VARIABLES                ------------
           READ(NREAD,2) AP,BP,CP,W
           READ(NREAD,4) N
           N2=N+2
           READ(NREAD,4) NN
           NN2=NN+2
           READ(NREAD,2) T
C
C          ---------- CALL SUBROUTINE TO COMPUTE THE FIRST SET OF ------------
C          ---------- COEFFICIENTS
           CALL CCOEF5(AP,BP,CP,W,N,-ONE,C1,S1)
           WRITE(NWRITE,5) AP,BP,CP,W,N
           DO 14 K=1,N2
           K1=K-1
        14 WRITE(NWRITE,6) K1,C1(K)
           WRITE(NWRITE,18) S1
C
C          ---------- CALL SUBROUTINE TO COMPUTE THE SECOND SET   ------------
C          ---------- OF COEFFICIENTS
           CALL CCOEF5(AP,BP,CP,W,NN,-ONE,C2,S2)
           WRITE(NWRITE,5) AP,BP,CP,W,NN
           DO 15 K=1,NN2
           K1=K-1
        15 WRITE(NWRITE,6) K1,C2(K)
           WRITE(NWRITE,18) S2
C
C          ---------- CALL SUBROUTINE TO COMPUTE THE FINAL VALUES ------------
C          ---------- OF COEFFICIENTS
           CALL LINCOM(N,NN,C1,C2,S1,S2,T,C,SUMM)
C
C          ---------- WRITE OUT VALUES OF PARAMETERS AND COEFFI-  ------------
C          ---------- CIENTS
           WRITE(NWRITE,5) AP,BP,CP,W,NN
           DO 1 K=1,NN2
           K1=K-1
         1 WRITE(NWRITE,6) K1,C(K)
           WRITE(NWRITE,18) SUMM
C
C          ---------- CALL SUBROUTINE TO TEST THE COEFFICIENT     ------------
C          ---------- VALUES
           WRITE(NWRITE,19)
           CALL CHECK5(AP,BP,CP,W,NN,-ONE,C,TEST1,TEST2)
           WRITE(NWRITE,17) TEST1,TEST2
C
C          ---------- PUNCH THE VALUES OF COEFFICIENTS ON CARDS   ------------
           WRITE (NPUNCH,2) (C(K),K=1,NN2)
         2 FORMAT(Q39.32)
         4 FORMAT(I5)
         5 FORMAT('1',4X,'A=',Q39.32/5X,'B=',Q39.32/5X,'C=',Q39.32/5X,'W=',
          :Q39.32/5X,'N=',I5/////5X,'K',25X,'C(K)'/)
         6 FORMAT(1X,I5,8X,Q39.32)
        17 FORMAT(/1X,'TEST1=',Q12.5/1X,'TEST2=',Q12.5)
        18 FORMAT(/1X,'THE SUM OF THE COEFFICIENTS IS',Q39.32)
        19 FORMAT(/1X,'THE FINAL COEFFICIENTS ARE DIRECTLY ABOVE.')
           STOP
           END
```

```
A= 0.1000000000000000000000000000000000Q+01
B= 0.5000000000000000000000000000000000Q+00
C= 0.5000000000000000000000000000000000Q+00
W= 0.8000000000000000000000000000000000Q+01
N=    99
```

K	C(K)
0	0.1008885687017036760538337446126Q+01
1	0.9252867349956327363337589509858Q-02
2	0.3530050771986443872224736672735Q-03
3	-0.2994156368969133461452898749786Q-04
4	-0.1540931653189510966087947840686Q-04
5	0.1524623887183155987377318403079Q-05
6	0.9035895078867346677157754839799Q-06
7	-0.3216484524443618390467533070863Q-06
8	-0.2381604632353157234086249079261Q-08
9	0.4019805783515083230986177182568Q-07

```
 10        -0.16713319864278084785351863196416Q-07
 11         0.18713033997027045637415058077222Q-08
 12         0.18035898965937202853599923724126Q-08
 13        -0.13788655542846150045167702376966200-08
 14         0.50336686997054390195063747337401400-09
 15        -0.47292786216778402181598590598960Q-10
 16        -0.76784715201339966904367288619803Q-10
 17         0.66471264227700660476180280521299Q-10
 18        -0.31861850389083024393502978433918Q-10
 19         0.84163370798022836206535897501111Q-11
 20         0.14223331309510314601695203081597Q-11
 21        -0.34413526874056949429527622564704100-11
 22         0.25593044320897147723448713577340Q-11
 23        -0.12708254093231694520526436791835Q-11
 24         0.39395711075246725539919569545275Q-12
 25         0.20165312783495793306254459781804Q-13
 26        -0.14056031028574772046697259329809Q-12
 27         0.12808485667699386889394868381542100-12
 28        -0.79267386839712473150791076431507Q-13
 29         0.36178294353723729712158766476068Q-13
 30        -0.96803154888541369831182635961745Q-14
 31        -0.25345979651570391155543478385144Q-14
 32         0.59974403599477282971779385501532Q-14
 33        -0.53785212523143234977599753484828290-14
 34         0.35360425895945361198025483184268Q-14
 35        -0.18203382861857276420903400280108220-14
 36         0.65846176126649696836456411083160Q-15
 37        -0.34784441944575081078530145706346Q-16
 38        -0.21322919829773153772191704151416Q-15
 39         0.25167201030998018665428034486830Q-15
 40        -0.20010733255775364694721846613056Q-15
 41         0.12768769648597000442101222309206Q-15
 42        -0.66090285679272366487579174184907Q-16
 43         0.24561985273375346813856810881274Q-16
 44        -0.14904607258520769091642531351801Q-17
 45        -0.84380058522203028362549265904397Q-17
 46         0.10600912147330226891785752005346Q-16
 47        -0.90507590874847648297510652982748Q-17
 48         0.63197906794662211049501492049511Q-17
 49        -0.37222472947297314957420676610184Q-17
 50         0.17698673769059318035872365202894Q-17
 51        -0.52867263971672741188653191748419Q-18
 52        -0.13335288448376415851187081308911Q-18
 53         0.40034965979974232454262527789740Q-18
 54        -0.43720648898030621481238729091069Q-18
 55         0.36451070111151140615326131192836Q-18
 56        -0.25794063499311842232131809776091Q-18
 57         0.15779179225598028017696339835002Q-18
 58        -0.80595527225042759628152296052987Q-19
 59         0.29097037016415455490344855984637Q-19
 60         0.58204042569401978447225001244710-21
 61        -0.14478271277755865433016907438094Q-19
 62         0.18398060704800094368085227355961Q-19
 63        -0.16911586079344536710851051743471Q-19
 64         0.13149525572041971387554717076933Q-19
 65        -0.89915145634623994290495888145650Q-20
 66         0.53904844591023699535167609009420-20
 67        -0.26947513941245943291079343317846Q-20
 68         0.90773955151631905515837529719290Q-21
 69         0.12922688189030838524520785726890100-21
 70        -0.62295336399026599439843910201458Q-21
 71         0.76671275127744203696077789773266Q-21
 72        -0.71377852933490438090350946032566Q-21
 73         0.57196356889418045785008240543960Q-21
 74        -0.40889622870056003803230052184793Q-21
 75         0.26140185601402872742424230270613Q-21
 76        -0.14520008467465513340228143608112Q-21
 77         0.63099721116269994387127158463009Q-22
 78        -0.11098330130619687567204824198421Q-22
 79        -0.17536763236794011119402251702170Q-22
 80         0.29830859678340411850566269464753Q-22
 81        -0.31856283738332991381544743241795Q-22
 82         0.28267114833610536924979808793183Q-22
 83        -0.22289509579925124634288600987729Q-22
 84         0.15937408651351883293593565421590Q-22
 85        -0.10305925625953446479134934395078Q-22
 86         0.58580523434218405918132984815240-23
 87        -0.26643010313436674805223734111369Q-23
 88         0.58260385581976903725107781901243Q-24
 89         0.61871842869573312141073110157284Q-24
 90        -0.11849391082774175935492561141338Q-23
 91         0.13338629794111391292516671801689Q-23
 92        -0.12385275529909057780073869533427Q-23
 93         0.10245432668328861328892435398453Q-23
 94        -0.77532489286863554075631910227221Q-24
 95         0.54064338555912561946804555378897Q-24
 96        -0.34568482181835430776611773960815Q-24
 97         0.19909534132218937763210897382195Q-24
 98        -0.99364494557056313088295277280943Q-25
 99         0.39438711379763119293629265866477Q-25
100        -0.97526696619771061395448249293848Q-26
```

THE SUM OF THE COEFFICIENTS IS 0.1018448338950509494147099935937Q+01

```
A=  0.10000000000000000000000000000000Q+01
B=  0.50000000000000000000000000000000Q+00
C=  0.50000000000000000000000000000000Q+00
W=  0.80000000000000000000000000000000Q+01
N=  100
```

K	C(K)
0	0.10087681900286848255876179345904Q+01
1	0.90617401578503410828622950147586Q-02
2	0.25402647689159622262288017741818Q-03
3	-0.57170447076764270984967583320012Q-04
4	-0.15916567168766867341341095759618Q-04
5	0.29111181419086611394928627655916Q-05
6	0.78326783270324905394440350417427Q-06
7	-0.41679216641163428385416481766132Q-06
8	0.34426331024783295579479850020219Q-07
9	0.37356140548113795316889713234375Q-07
10	-0.21145703221548065920927646811061Q-07
11	0.44354255449988647663180137530720Q-08
12	0.11839194726194274548262838006049Q-08
13	-0.14851015703900686420136924398336Q-08
14	0.68459172681862938714934487003735Q-09
15	-0.14745811228209260676466495639136Q-09
16	-0.47806325628929510327569660571384Q-10
17	0.68339659470607280300089081313521Q-10
18	-0.40196767684833973955962534065597Q-10
19	0.14482237562341516589001970136156Q-10
20	-0.13355914917159839248825660263730Q-11
21	-0.27637142281321953162453182709114Q-11
22	0.27408706439776064739450238189920Q-11
23	-0.16238266442797970402801278049441Q-11
24	0.65864858847976924741593642335742Q-12
25	-0.11712118707860489393726712408524Q-12
26	-0.93379158558941632209597110547530Q-13
27	0.12633850156166044031423651545150Q-12
28	-0.92810304283933527523697460009480Q-13
29	0.50154234098449521579197965515018Q-13
30	-0.19166023245278583190270513424110Q-13
31	0.23437408081846523379176001949346Q-14
32	0.42665532543437504941075810165924Q-14
33	-0.52981853207667821276098292649090Q-14
34	0.40753175637539614054315974467608Q-14
35	-0.24279548149263882626714773564030Q-14
36	0.11165370672318485802944008946276Q-14
37	-0.30744143781095181967806778723452Q-15
38	-0.86657285201194006032749240531880Q-16
39	0.21587621215843283400995677672916Q-15
40	-0.20939026621919644709061400438827Q-15
41	0.15245366669350972839207594215611Q-15
42	-0.90766811200037063081364871744499Q-16
43	0.43035830935702300325354240801282Q-16
44	-0.12886574449471995117548980046316Q-16
45	-0.27227451570163432179847677388328Q-17
46	0.86181902335866443842324101531130Q-17
47	-0.91081929739068219901448939839520Q-17
48	0.72353609076103922442568452149305Q-17
49	-0.48018405309675556894170380956665Q-17
50	0.26820749005493544633961027158731Q-17
51	-0.11690306663537008595141187702050Q-17
52	0.24969386660620825001012867438103Q-18
53	0.21273984264450227023048157306298Q-18
54	-0.37633897691758230193102470511719Q-18
55	0.37346225195041082693042744959390Q-18
56	-0.29693063013912212236980810735101Q-18
57	0.20284122235332321011867976458428Q-18
58	-0.11963468036724628254976593867926Q-18
59	0.57807300548421944679227349321267Q-19
60	-0.17825037150671997170430097729399Q-19
61	-0.43861792548307092441480671697096Q-20
62	0.14122244510335124080763561173492Q-19
63	-0.16195643684871385946267151410173Q-19
64	0.14273826633942949955769185760288Q-19
65	-0.10805642311541792203739596404076Q-19
66	0.72215089159071579591226505303466Q-20
67	-0.42165444843174058555946903931001Q-20
68	0.20168756687725467041729613711110Q-20
69	-0.58830078987444498085711284609120Q-21
70	-0.21993879210374934986097129611860Q-21
71	0.58734004280006639909349655282277Q-21
72	-0.67594357299148938339014192275024Q-21
73	0.61163247324894467215353909670919Q-21
74	-0.48152807495122536657524143276503Q-21
75	0.33937399490116599576085400020317Q-21
76	-0.21375317893322747330622193692032Q-21
77	0.11624667936231894278434810988152Q-21
78	-0.48190918791641770201071203487539Q-22
79	0.56391043228002415954798044693430Q-23
80	0.17358409686578109147359492511808Q-22

```
 81        -0.26821139622010710050150318428829Q-22
 82         0.27872029005549128714367334481888Q-22
 83        -0.24397414188512057318733205926095Q-22
 84         0.19068326652056094011842963059190Q-22
 85        -0.13536889489787449967669635786652Q-22
 86         0.86893088592754905697055821343851Q-23
 87        -0.48884857813777417040120163250757Q-23
 88         0.21748063814752200499467762095848Q-23
 89        -0.41589445896137356339901935817490Q-24
 90        -0.59157126129055019707195012800793Q-24
 91         0.10592724625819092451646495913757Q-23
 92        -0.11737320699358592705590297018755Q-23
 93         0.10824786895728336079204847541364Q-23
 94        -0.89228772558861343295544277516040Q-24
 95         0.67389280825737413811890126643853Q-24
 96        -0.46937925435356173469830123391139Q-24
 97         0.29993337423321287966695115776691Q-24
 98        -0.17269442814454927005025772905825Q-24
 99         0.86180539675341283766107448585762Q-25
100        -0.34206886355271910557139385873540Q-25
101         0.84596912715628293851439769363551Q-26
```

THE SUM OF THE COEFFICIENTS IS 0.10180142045484878330474738795069Q+01

```
A=  0.10000000000000000000000000000000Q+01
B=  0.50000000000000000000000000000000Q+00
C=  0.50000000000000000000000000000000Q+00
W=  0.80000000000000000000000000000000Q+01
N=  100
```

K	C(K)
0	0.10090063050674650865582526462660Q+01
1	0.94490714367080076052863338208206Q-02
2	0.45461283728353557402861513642960Q-03
3	-0.19894018891553978556988954110987Q-05
4	-0.14888591841363153730402885958872Q-04
5	0.10130030858593557586059942683751Q-06
6	0.10271072770541932696815814603714Q-05
7	-0.22192431862228515438683237160834Q-06
8	-0.44273517863568409872667909597833Q-07
9	0.43115464718927524121606062561520Q-07
10	-0.12163199501162758080067129500055Q-07
11	-0.76092929463108606611585297827372Q-09
12	0.24397205710583014552123450708372Q-08
13	-0.12698075771286280303104244087010Q-08
14	0.31321232831760860883624104927870Q-09
15	0.55533222517400563534760703501913Q-10
16	-0.10653285516478542279106122173658Q-09
17	0.64553238967734549788058516280010Q-10
18	-0.23305533424740518998409854824865Q-10
19	0.21893086209568927393752838116679Q-11
20	0.42535162606664908614283610408181Q-11
21	-0.41369911969477831033764954040290Q-11
22	0.23729152930253415130410632061163Q-11
23	-0.90844743388326801956616382130694Q-12
24	0.12223465678767944077281435463776Q-12
25	0.16109854176036033112982193267370Q-12
26	-0.18899473075961240068577869717712Q-12
27	0.12987760006830636021004634774719Q-12
28	-0.65364730076470032622708926939528Q-13
29	0.21831112961607475047461682918257Q-13
30	0.57360281059301243574868646911325Q-16
31	-0.75425196184927278323496085764025Q-14
32	0.77743048661607653265713106502771Q-14
33	-0.54609911400835644955791340052538Q-14
34	0.29824429021449762070501455757973Q-14
35	-0.11966720289070647882561593140054Q-14
36	0.18821864159261509428369425345911Q-15
37	0.24511511751085598880860743456314Q-15
38	-0.34316322989255975918766818421460Q-15
39	0.28841864907175926654718924956664Q-15
40	-0.19057781718106777507600033002158Q-15
41	0.10226387000922735691529029513639Q-15
42	-0.40758279800583357599229762141667Q-16
43	0.55974204878561537017663117555018Q-17
44	0.10208366955460904200573034660202Q-16
45	-0.14305080510170787466341648219305Q-16
```
```

```
 46   0.12636300927346835130266215077214Q-16
 47  -0.89917995898336375365615104446218Q-17
 48   0.53799002405880151531945592374090Q-17
 49  -0.26139769191472145679508012636720Q-17
 50   0.83342896574976371178267743307496Q-18
 51   0.12869516036450612321141095767031Q-18
 52  -0.52657447243620015310651151147080Q-18
 53   0.59294293988194717543505617021342Q-18
 54  -0.49969081936533664222461545451534Q-18
 55   0.35532137102192606298742743747530Q-18
 56  -0.21791495206553202086054538519560Q-18
 57   0.11154571814408008686854519693556Q-18
 58  -0.40519380521765140212500285511973Q-19
 59  -0.37585469840490679394740166712190Q-21
 60   0.19478063570509054138009970443191Q-19
 61  -0.24838438645640889150235064985026Q-19
 62   0.22787455025456616796558544720561Q-19
 63  -0.17646545988118376084658864754132Q-19
 64   0.11995359800790018501812681281372Q-19
 65  -0.71291983415048563453428373317520Q-20
 66   0.35108226621135016472022971755309Q-20
 67  -0.11325350491017236543272615800494Q-20
 68  -0.23085937197281402279608342571600Q-21
 69   0.86581417753578968034740691929655Q-21
 70  -0.10366731761913721604366173125794Q-20
 71   0.95085012109482953193451018950023Q-21
 72  -0.75261849226667249707613913588311Q-21
 73   0.53124094295275908029209153264197Q-21
 74  -0.33433506916237260170893303535425Q-21
 75   0.18135855011989698689304234859107Q-21
 76  -0.74826020655829623701562692989606Q-22
 77   0.85410250280154749161291433490524Q-23
 78   0.26979545671876459310261823091774Q-22
 79  -0.41328249310039473108876144598290Q-22
 80   0.42634614252585933310415352228150Q-22
 81  -0.37025175941061934204709552350348Q-22
 82   0.28672695291467805065570063269326Q-22
 83  -0.20125612688249176381990989884570Q-22
 84   0.12723324353593156315785770684677Q-22
 85  -0.69891379556793723956921488767383Q-23
 86   0.29515894119475776774822591040732Q-23
 87  -0.38103545914055099469464179779945Q-24
 88  -0.10518922042526896148972867347199Q-23
 89   0.16808136466891719752042221049448Q-23
 90  -0.17940685327358154142545430744596Q-23
 91   0.16157474197752033397495557389300Q-23
 92  -0.13050441927132092415043645160326Q-23
 93   0.96506891059910614740186977699461Q-24
 94  -0.65525518680092401840412320326797Q-24
 95   0.40385447026861732615202208092943Q-24
 96  -0.21870470504679860612381847853560Q-24
 97   0.95578756690046062594068561134609Q-25
 98  -0.24086704450052565042596008261634Q-25
 99  -0.85447159535499522274166781684498Q-26
100   0.15351222195828103721430226707020Q-25
101  -0.86844053002030066815280476839439Q-26
```

THE SUM OF THE COEFFICIENTS IS 0.10188940052261324432338905125844Q+01

THE FINAL COEFFICIENTS ARE DIRECTLY ABOVE.

TEST1=-0.44266Q-20
TEST2=-0.45271Q-16

10 DIFFERENTIAL AND INTEGRAL PROPERTIES OF EXPANSIONS IN SERIES OF CHEBYSHEV POLYNOMIALS OF THE FIRST KIND

Given the coefficients in the expansion for $f(x)$ in series of the shifted polynomials $T_n^*(x)$, one can readily obtain the coefficients in the corresponding expansions for

$$f'(x) \ , \ [f(x) - f(0)]/x \ , \ \int_0^x f(t)dt \ , \ \text{etc.}$$ Such formulas

are presented herein. Similar results for expansions in series of the even and odd Chebyshev polynomials $T_{2n}(x)$ and $T_{2n+1}(x)$, respectively, are also given. Programs for these algorithms are omitted.

10.1. Series of Shifted Chebyshev Polynomials

We suppose that

$$f(x) = \sum_{n=0}^{\infty} b_n T_n^*(x) \ . \tag{1}$$

Throughout this chapter, we use the notation

$$\varepsilon_0 = 1 \quad \text{and} \quad \varepsilon_n = 2 \quad \text{if} \quad n > 0 \ . \tag{2}$$

If $f(0) = 0$, then

$$f(x) = x \sum_{n=0}^{\infty} c_n T_n^*(x) \ ,$$

$$c_n = 2\varepsilon_n \sum_{k=0}^{\infty} (-)^k (k+1) b_{k+n+1} \ ,$$

$$b_0 = \tfrac{1}{2} c_0 + \tfrac{1}{4} c_1 \ , \quad b_1 = \tfrac{1}{2} c_0 + \tfrac{1}{2} c_1 + \tfrac{1}{4} c_2 \ ,$$

$$b_n = \tfrac{1}{4}(c_{n-1} + 2c_n + c_{n+1}) \ , \quad n = 2,3,\ldots, \quad \sum_{k=0}^{\infty} (-)^k b_k = 0 \ . \tag{3}$$

116

$$xf(x) = \{(2b_0+b_1)T_0^*(x) + (2b_0+2b_1+b_2)T_1^*(x)$$

$$- \sum_{k=2}^{\infty} (b_{k-1}+2b_k+b_{k+1})T_k^*(x)\} . \tag{4}$$

$$x^2f(x) = \frac{1}{24} \{6b_0+4b_1+b_2)T_0^*(x) + (8b_0+7b_1+4b_2+b_3)T_1^*(x)$$

$$- (2b_0+4b_1+6b_2+4b_3+b_4)T_2^*(x)$$

$$- \sum_{k=3}^{\infty} (b_{k-2}+4b_{k-1}+6b_k+4b_{k+1}+b_{k+2})T_k^*(x)\} . \tag{5}$$

$$f'(x) = \sum_{n=0}^{\infty} d_n T_n^*(x) ,$$

$$d_n = 2\varepsilon_n \sum_{k=0}^{n} (n+2k+1)b_{n+2k+1} ,$$

$$2d_n/\varepsilon_n = d_{n+2} + 4(n+1)b_{n+1} . \tag{6}$$

$$\int_0^x f(t)dt = \left(\frac{b_0}{2} - \frac{b_1}{8} - \frac{1}{2}\sum_{n=2}^{\infty} \frac{(-)^n b_n}{n^2-1}\right) T_0^*(x) + \tfrac{1}{4}(2b_0-b_2)T_1^*(x)$$

$$+ \frac{1}{4}\sum_{n=2}^{\infty} \frac{(b_{n-1} - b_{n+1})}{n} T_n^*(x) ,$$

$$\int_0^1 f(t)dt = - \sum_{n=0}^{\infty} b_{2n}/(4n^2-1) . \tag{7}$$

If $f(0) = 1$,

$$\int_0^x t^{-1}[1-f(t)]dt = \sum_{n=0}^{\infty} d_n T_n^*(x) ,$$

$$d_0 = -b_1 + \sum_{n=2}^{\infty} (-)^n\{2\sum_{k=1}^{n-1} k^{-1} + n^{-1}\}b_n ,$$

$$d_n = - \frac{b_n}{n} + \frac{2(-)^{n-1}}{n} \sum_{k=n+1}^{\infty} (-)^k b_k , \quad n = 1,2,\ldots,$$

$$d_{n+1} = -\frac{nd_n}{n+1} + \frac{b_{n+1}-b_n}{n+1} \; , \; n = 0,1,\ldots . \tag{8}$$

$$\int_0^x tf(t)\,dt = \frac{1}{4}\left\{\frac{3b_0}{4} + \frac{b_1}{12} - \frac{19b_2}{48} + 3\sum_{k=3}^{\infty}\frac{(-)^k b_k}{(k^2-1)(k^2-4)}\right\}T_0^*(x)$$

$$+ \frac{1}{16}\{4b_0+b_1-2b_2-b_3\}T_1^*(x) + \frac{1}{32}\{2b_0+2b_1-2b_3-b_4\}T_2^*(x)$$

$$+ \frac{1}{16}\sum_{k=3}^{\infty}\left\{\frac{b_{k-2}+2b_{k-1}-2b_{k+1}-b_{k+2}}{k}\right\}T_k^*(x) \; . \tag{9}$$

If

$$\int_0^x f(t)(\ln t)\,dt = \sum_{n=0}^{\infty} a_n T_n^*(x) + (\ln x)\sum_{n=0}^{\infty} g_n T_n^*(x) \; ,$$

then

$$g_0 = \frac{b_0}{2} - \frac{b_1}{8} - \frac{1}{2}\sum_{n=2}^{\infty}\frac{(-)^n b_n}{n^2-1} \; , \; g_1 = \tfrac{1}{4}(2b_0-b_1) \; ,$$

$$g_n = (1/4n)(b_{n-1}-b_{n+1}) \; , \; n = 2,3,\ldots ,$$

$$a_0 = -\left\{\frac{h_0}{2} - \frac{h_1}{8} - \frac{1}{2}\sum_{n=2}^{\infty}\frac{(-)^n h_n}{n^2-1}\right\} \; ,$$

$$h_n = 2\varepsilon_n \sum_{k=0}^{\infty}(-)^k(k+1)g_{k+n+1} \; , \; a_1 = g_1 - 2g_0 \; ,$$

$$a_n = -(1/n)\{g_n - 2\sum_{k=0}^{\infty}(-)^k g_{k+n+1}\} \; , \; n = 2,3,\ldots . \tag{10}$$

For generalizations of (9) and (10), see Chapter 11.

$$\int_0^x \int_0^t f(u)\,du\,dt = \frac{1}{4}\left\{\frac{3b_0}{4} - \frac{b_1}{3} - \frac{3b_2}{16} - \sum_{k=3}^{\infty}\frac{(-)^k b_k}{k^2-4}\right\}T_0^*(x)$$

$$+ \frac{1}{4}\left\{b_0 - \frac{3b_1}{8} - \frac{b_2}{3} + \frac{b_3}{4} - \sum_{k=4}^{\infty}\frac{(-)^k b_k}{k^2-1}\right\}T_1^*(x)$$

(continued on next page)

$$+ \frac{1}{96} \{6b_0 - 4b_2 + b_4\} T_2^*(x)$$

$$+ \frac{1}{16} \sum_{k=3}^{\infty} \left\{ \frac{(k+1)b_{k-2} - 2kb_k + (k-1)b_{k+2}}{k(k^2-1)} \right\} T_k^*(x) \quad . \qquad (11)$$

Let

$$A(x) = \sum_{n=0}^{\infty} a_n T_n^*(x) \;, \; B(x) = \sum_{n=0}^{\infty} b_n T_n^*(x) \;, \; C(x) = \sum_{n=0}^{\infty} c_n T_n^*(x) \quad .$$

If

$$C(x) = A(x)B(x) \;,$$

then

$$c_0 = a_0 b_0 + \tfrac{1}{2} \sum_{m=1}^{\infty} a_m b_m$$

$$c_n = a_0 b_n + \sum_{m=1}^{\infty} \left(\frac{b_{n+m}}{2} + \frac{b_{|n-m|}}{\varepsilon_{|n-m|}} \right) a_m$$

$$= b_0 a_n + \sum_{m=1}^{\infty} \left(\frac{a_{n+m}}{2} + \frac{a_{|n-m|}}{\varepsilon_{|n-m|}} \right) b_m \;, \; n > 0 \;,$$

$$\varepsilon_0 = 1 \;, \; \varepsilon_n = 2 \quad \text{for} \quad n \geq 1 \quad . \qquad (12)$$

This result is also true if in $A(x)$, $B(x)$, and $C(x)$, $T_n^*(x)$ is replaced by $T_n(x)$ or $T_{2n}(x)$.

The above system of equations may be expressed in matrix form as follows. Let $A = (\alpha_{ij})$ and $B = (\beta_{ij})$ be infinite matrices where

$$\alpha_i = a_{i-1} \;, \qquad\qquad \alpha_{i1} = \alpha_i \quad \text{for} \quad i = 1,2,3,\ldots,$$

$$\alpha_{ij} = \tfrac{1}{2}\alpha_j \qquad\qquad \text{for} \quad j = 2,3,\ldots,$$

$$\alpha_{ii} = \alpha_1 + \tfrac{1}{2}\alpha_{2i-1} \qquad \text{for} \quad i = 2,3,\ldots,$$

$$\alpha_{ij} = \tfrac{1}{2}(\alpha_{i+j-1} + \alpha_{j-i+1}) \qquad \text{for} \quad j > i \geq 2 \;,$$

$$\alpha_{ij} = \alpha_{ji} , \qquad\qquad i,j, = 2,3,\ldots, \qquad\qquad (13)$$

and where (13) is also valid if a_i , α_i , and α_{ij} are replaced by b_i , β_i , and β_{ij} , respectively. Let α , β ,
and γ stand for the infinite vectors

$$\alpha = \begin{pmatrix} \alpha_1 \\ \alpha_2 \\ \alpha_3 \\ \vdots \end{pmatrix} , \quad \beta = \begin{pmatrix} \beta_1 \\ \beta_2 \\ \beta_3 \\ \vdots \end{pmatrix} , \quad \gamma = \begin{pmatrix} \gamma_1 \\ \gamma_2 \\ \gamma_3 \\ \vdots \end{pmatrix} . \qquad (14)$$

Then

$$A\beta = B\alpha = \gamma . \qquad\qquad (15)$$

10.2.   Series of Chebyshev Polynomials of Even Order

In this section we suppose that

$$f(x) = \sum_{n=0}^{\infty} b_n T_{2n}(x) . \qquad\qquad (1)$$

Some of the results of this section are essentially restate-
ments of results in 10.1 since $T_n^*(x^2) = T_{2n}(x)$.   For example, if $f(0) = 0$, then

$$f(x) = x^2 \sum_{n=0}^{\infty} c_n T_{2n}(x) , \qquad\qquad (2)$$

where c_n is given by 10.1(3).

$$xf(x) = \frac{1}{2} \sum_{n=0}^{\infty} (\frac{2b_n}{\varepsilon_n} + b_{n+1}) T_{2n+1}(x) . \qquad (3)$$

$$x^2 f(x) = \tfrac{1}{4}(2b_0 + b_1) T_0(x) + \tfrac{1}{4}(2b_0 + 2b_1 + b_2) T_2(x)$$

$$+ \frac{1}{4} \sum_{n=2}^{\infty} (b_{n-1} + 2b_n + b_{n+1}) T_{2n}(x) . \qquad (4)$$

$$f'(x) = \sum_{n=0}^{\infty} d_n T_{2n+1}(x) \ , \ d_n = 4 \sum_{k=0}^{\infty} (n+k+1) b_{n+k+1} \ ,$$

$$d_n = d_{n+1} + 4(n+1) b_{n+1} \ . \tag{5}$$

$$\int_{0}^{x} f(t) dt = (b_0 - \tfrac{1}{2} b_1) T_1(x) + \frac{1}{2} \sum_{n=1}^{\infty} \frac{(b_n - b_{n+1})}{2n+1} T_{2n+1}(x) \ ,$$

$$\int_{0}^{1} f(t) dt = - \sum_{n=0}^{\infty} \frac{b_n}{4n^2 - 1} \ . \tag{6}$$

$$x^{-1} \int_{0}^{x} f(t) dt = \sum_{n=0}^{\infty} e_n T_{2n}(x) \ ,$$

$$e_n = \frac{b_n}{2n+1} + 2(-)^n \varepsilon_n \sum_{k=n+1}^{\infty} \frac{(-)^k k b_k}{4k^2 - 1} \ ,$$

$$e_{n+1} = -e_n + \frac{b_n - b_{n+1}}{2n+1} \ , \ n = 1,2,3,\ldots ,$$

$$e_1 = -2e_0 + 2b_0 - b_1 \ . \tag{7}$$

If $f(0) = 1$, then

$$\int_{0}^{x} t^{-1} \{1 - f(t)\} dt = \sum_{n=0}^{\infty} q_n T_{2n}(x) \ ,$$

$$q_0 = \sum_{r=1}^{\infty} (-)^r b_r \left\{ \sum_{k=1}^{r-1} \frac{1}{k} + \frac{1}{2r} \right\} \ ,$$

$$q_n = - \frac{b_n}{2n} - \frac{(-)^n}{n} \sum_{k=n+1}^{\infty} (-)^k b_k \ , \ n = 1,2,3,\ldots ,$$

$$q_{n+1} = - \frac{n q_n}{n+1} + \frac{b_{n+1} - b_n}{2(n+1)} \ , \ n = 1,2,3,\ldots ,$$

$$\sum_{n=0}^{\infty} (-)^n b_n = 1 \ , \quad \sum_{n=0}^{\infty} (-)^n q_n = 0 \ . \tag{8}$$

$$\int_0^x tf(t)dt = \frac{1}{4} [b_0 - \frac{b_1}{4} - \sum_{n=2}^{\infty} \frac{(-)^n b_n}{n^2-1}] T_0(x)$$

$$+ \frac{1}{8}(2b_0 - b_2)T_2(x) + \frac{1}{8}\sum_{n=2}^{\infty} \frac{(b_{n-1}-b_{n+1})}{n} T_{2n}(x) \ . \tag{9}$$

$$\int_0^x f(t)\ln t \ dt = \sum_{n=0}^{\infty} a_n T_{2n+1}(x) + \ln x \sum_{n=0}^{\infty} g_n T_{2n+1}(x) \ ,$$

$$g_0 = b_0 - \tfrac{1}{2}b_1 \ , \quad g_n = \frac{1}{2}(\frac{b_n - b_{n+1}}{2n+1}) \ , \quad n = 1,2,\ldots,$$

$$a_n = \frac{g_n}{2n+1} + \frac{2}{2n+1} \sum_{k=1}^{\infty} (-)^k g_{n+k} \ . \tag{10}$$

$$\int_0^x \int_0^t f(u)du \ dt = \frac{1}{4} [b_0 - \frac{3b_1}{4} + \sum_{n=2}^{\infty} \frac{(-)^n b_n}{n^2-1}] T_0(x)$$

$$+ \frac{1}{24}[6b_0 - 4b_1 + b_2]T_2(x)$$

$$+ \frac{1}{8} \sum_{n=2}^{\infty} \frac{[(2n+1)b_{n-1} - 4nb_n + (2n-1)b_{n+1}]}{n(4n^2-1)} T_{2n}(x) \ . \tag{11}$$

$$\int_0^x \int_0^t uf(u)du \ dt = \frac{1}{4} [\frac{b_0}{2} - \frac{b_1}{4} - \frac{b_2}{12} - \sum_{n=3}^{\infty} \frac{(-)^n b_n}{n^2-1}] T_1(x)$$

$$+ \frac{1}{96}[4b_0 - b_1 - 2b_2 + b_3]T_3(x)$$

$$+ \frac{1}{16} \sum_{n=2}^{\infty} \frac{[(n+1)\{b_{n-1}-b_{n+1}\} - n\{b_n - b_{n+2}\}]}{n(n+1)(2n+1)} T_{2n+1}(x) \ . \tag{12}$$

10.3. Series of Chebyshev Polynomials of Odd Order

We suppose throughout that

$$f(x) = \sum_{n=0}^{\infty} b_n T_{2n+1}(x) \ . \tag{1}$$

If

$$f(x) = x \sum_{n=0}^{\infty} c_n T_{2n}(x) \ ,$$

then

$$c_n/\varepsilon_n + \tfrac{1}{2}c_{n+1} = b_n \ , \quad c_n = \varepsilon_n \sum_{k=0}^{\infty} (-)^k b_{n+k} \ ,$$

$$\sum_{n=0}^{\infty} (-)^n (2n+1) b_n = \sum_{n=0}^{\infty} (-)^n c_n \ . \tag{2}$$

$$xf(x) = \tfrac{1}{2}b_0 T_0(x) + \tfrac{1}{2} \sum_{n=1}^{\infty} (b_{n-1}+b_n) T_{2n}(x) \ . \tag{3}$$

$$x^2 f(x) = \tfrac{1}{4}(3b_0+b_1)T_1(x) + \tfrac{1}{4} \sum_{n=1}^{\infty} (b_{n-1}+2b_n+b_{n+1}) T_{2n+1}(x) \ . \tag{4}$$

$$f'(x) = \sum_{n=0}^{\infty} d_n T_{2n}(x) \ , \quad d_n = \varepsilon_n \sum_{k=0}^{\infty} (2n+2k+1) b_{n+k} \ ,$$

$$d_{n+1} = \frac{2d_n}{\varepsilon_n} + 2(2n+1)b_n \ . \tag{5}$$

$$\int_0^x f(t)dt = \frac{1}{4} \{ b_0 + \sum_{k=1}^{\infty} \frac{(-)^k(2k+1)b_k}{k(k+1)} \}$$

$$+ \frac{1}{4} \sum_{n=1}^{\infty} \frac{(b_{n-1}-b_n)}{n} T_{2n}(x) \ ,$$

$$\int_0^1 f(t)dt = \tfrac{1}{2} \sum_{k=0}^{\infty} \frac{(b_{2k}-b_{2k+1})}{2k+1} \ . \tag{6}$$

$$\int_0^x t^{-1}f(t)dt = \sum_{n=0}^{\infty} r_n T_{2n+1}(x) \; ,$$

$$r_n = \frac{b_n}{2n+1} + \frac{2}{2n+1} \sum_{k=1}^{\infty} (-)^k b_{n+k} \; ,$$

$$\sum_{n=0}^{\infty} (-)^n (2n+1) b_n = \sum_{n=0}^{\infty} (-)^n (2n+1) r_n \; . \tag{7}$$

$$\int_0^x t f(t)dt = \frac{(b_0 - b_1)}{4} T_1(x) + \frac{1}{4} \sum_{n=1}^{\infty} \frac{(b_{n-1} - b_{n+1})}{2n+1} T_{2n+1}(x). \tag{8}$$

Let

$$A(x^2) = \int_0^x f(t)(\ln t)dt = \int_0^x t(\ln t) \sum_{n=0}^{\infty} c_{2n} T_{2n}(t)dt$$

where c_n is defined in (2). Replace t by $\tau^{\frac{1}{2}}$ and x by $y^{\frac{1}{2}}$. So

$$A(y) = \frac{1}{4}\int_0^y (\ln \tau) \sum_{n=0}^{\infty} c_n T_n^*(\tau)d\tau$$

which can be evaluated by 10.1(10). Indeed, if we now write

$$f(x) = \sum_{n=0}^{\infty} B_n T_{2n+1}(x) \; ,$$

$$A(x) = \frac{1}{4} \sum_{n=0}^{\infty} a_n T_{2n}(x) + \frac{1}{2}(\ln x) \sum_{n=0}^{\infty} g_n T_{2n}(x) \; ,$$

then

$$g_0 = \frac{c_0}{2} - \frac{c_1}{8} - \frac{1}{2} \sum_{n=2}^{\infty} \frac{(-)^n c_n}{n^2 - 1} \; , \quad g_1 = B_0 + 2 \sum_{k=1}^{\infty} (-)^k B_k \; ,$$

$$g_n = \frac{1}{2n} (B_{n-1} - B_n) \; , \quad n = 2, 3, \ldots,$$

$$a_0 = - \left(\frac{h_0}{2} - \frac{h_1}{8} - \frac{1}{2} \sum_{n=2}^{\infty} \frac{(-)^n h_n}{n^2 - 1} \right) \; , \tag{9}$$

$$h_n = 2\varepsilon_n \sum_{k=0}^{\infty} (-)^k (k+1) g_{k+n+1} \;,$$

$$a_1 = g_1 - 2g_0 \;, \quad a_n = -(1/n)\{g_n - 2\sum_{k=0}^{\infty} (-)^k g_{k+n+1}\} \;,$$

$$n = 2,3,\ldots,$$

where c_n is defined in (2) with b_n replaced by B_n .

$$\int_0^x \int_0^t f(u)\,du\,dt = \frac{1}{4}\left(\frac{b_0}{2} - b_1 + \sum_{n=2}^{\infty} \frac{(-)^n (2n+1) b_n}{n(n+1)}\right) T_1(x)$$

$$+ \frac{1}{8} \sum_{n=1}^{\infty} \frac{[(n+1) b_{n-1} - (2n+1) b_n + n b_{n+1}]}{n(n+1)(2n+1)} T_{2n+1}(x) \;. \qquad (10)$$

XI EXPANSION OF EXPONENTIAL TYPE INTEGRALS IN SERIES OF CHEBYSHEV POLYNOMIALS OF THE FIRST KIND

11.1. Introduction

Suppose we are given the coefficients b_k in the expansion

$$f(x) = \sum_{k=0}^{\infty} b_k T_k^*(x/\lambda) , \quad 0 \leq x \leq \lambda . \tag{1}$$

We assume that (1) converges even though its power series counterpart might be divergent but asymptotic. In this chapter we show how to get the coefficients g_k for the expansion of

$$g(x) = e^{-ax} x^{-u-1} \int_0^x e^{at} t^u f(t) dt \tag{2}$$

in the form

$$g(x) = \sum_{k=0}^{\infty} g_k T_k^*(x/\lambda) , \quad 0 \leq x \leq \lambda , \tag{3}$$

which converges so long as $t^u f(t)$ is integrable. These developments are given in 11.2.

Similar results for

$$F(x) = \sum_{k=0}^{\infty} c_k T_k^*(\lambda/x) , \quad 0 < \lambda \leq x , \tag{4}$$

$$G(x) = e^{bx} x^{-u} \int_x^{\infty} e^{-bt} t^u F(t) dt , \quad R(b) > 0 , \tag{5}$$

$$G(x) = \sum_{k=0}^{\infty} h_k T_k^*(\lambda/x) , \quad 0 < \lambda \leq x , \tag{6}$$

are presented in 11.3. Here we assume that (4) is convergent even though its power series counterpart might be divergent but asymptotic. We further suppose that the parameters in (5) (and also in (2)) are such that the integrals have meaning.

In this connection, note that the restrictions on (6) might be less severe than those on (5) in view of analytic continuation. Also in 11.3, we consider (4-6) except that the limits of integration in (5) are 0 and x.   In 11.4, we take up the cases akin to those of 11.2 and 11.3 where the integrals also have (ln t) present in the integrand.   Some numerical examples are treated in 11.5.   The results given in this chapter are based on a paper by Luke (1976).   A number of errata have been found and for the convenience of the reader these are recorded in 11.6.   The chapter concludes with some programs.

11.2.   The Representation for g(x)

Let

$$f(x) = \sum_{k=0}^{\infty} b_k T_k^*(x/\lambda) , \quad 0 \le x \le \lambda , \tag{1}$$

$$g(x) = e^{-ax} x^{-u-1} \int_0^x e^{at} t^u f(t) dt , \tag{2}$$

$$g(x) = \sum_{k=0}^{\infty} g_k T_k^*(x/\lambda) , \quad 0 \le x \le \lambda . \tag{3}$$

We assume (formally at least) that

$$g(x) = \sum_{k=0}^{\infty} d_k x^k , \tag{4}$$

and that the Chebyshev series (1) is convergent. We also have (formally at least)

$$g(x) = \sum_{k=0}^{\infty} f_k x^k , \tag{5}$$

$$f_0 = d_0/(u+1) , \quad f_1 = (d_1 - af_0)/(u+2) , \quad f_2 = (d_2 - af_1)/(u+3) ,$$
$$(k+u+1) f_k = d_k - af_{k-1} , \quad k > 0 . \tag{6}$$

The g_k's are found by use of a recurrence formula used in the backward direction.   The details are as follows.   We first suppose that a $\neq$ 0. The recurrence formula for g_k is

$$L\{g_k\} = (4/a\lambda)(b_{k+1} - b_{k+2}) , \quad k \geq 0 , \tag{7}$$

$$L\{g_k\} \equiv (2g_k/\varepsilon_k) + \{(4/a\lambda)(k+u+2)+1\}g_{k+1}$$

$$+ \{(4/a\lambda)(k-u+1) -1\}g_{k+2} - g_{k+3} , \tag{8}$$

$$\varepsilon_0 = 1 , \quad \varepsilon_k = 2 \text{ for } k > 0 . \tag{9}$$

Let n be a large positive integer. Put $A_{k,n}$ = 0 for $k \geq n + 1$ and evaluate $A_{k,n}$ from

$$L\{A_{k,n}\} = (4/a\lambda)(b_{k+1}-b_{k+2}) , \quad 0 \leq k \leq n . \tag{10}$$

Put $B_{k,n}$ = 0 , $k \geq n + 2$, $B_{n+1,n}$ = 1 and evaluate $B_{k,n}$ from

$$L\{B_{k,n}\} = 0 , \quad 0 \leq k \leq n . \tag{11}$$

Now define

$$g_{k,n} = A_{k,n} + B_{k,n}\left(\{d_0/(u+1)\} - \sum_{m=0}^{n} (-)^m A_{m,n}\right)/ \sum_{m=0}^{n+1} (-)^m B_{m,n} .$$

Then
$$\tag{12}$$

$$g_k = \lim_{n\to\infty} g_{k,n} , \quad k = 0,1,\ldots . \tag{13}$$

However, we hasten to remark that though the algorithm posseses excellent theoretical convergence properties, its application can give rise to rather serious numerical instabilities since one may be in the position of determining a linear combination of two large quantities which is small. The details are quite lengthy and will not be taken up here. The reader will find the discussion in Wimp and Luke (1969) very illuminating. In this connection, the noted instabilities can be avoided if one determines the g_k's by solving systems of linear equations. Further details are given in the latter paper.

If f(x) = 1, the analysis simplifies. For then b_0 = 1 and b_k = 0 if k > 0, whence $A_{k,n}$ = 0 for all $k \geq 0$.

Next we consider the case a = 0. From (7) and (8),

$$(2g_k/\varepsilon_k) = \{(u-k)g_{k+1} + (2b_k/\varepsilon_k) - b_{k+1}\}/(k+u+1) . \qquad (14)$$

Let n be as before. Put $g_{n+1,n} = 0$ and evaluate $g_{k,n}$ from (14) with g_k replaced by $g_{k,n}$ for $0 \le k \le n$. Then the statement (13) is true also for $a = 0$.

The following equations can be used for checks on the b_k's and the g_k's.

$$\sum_{k=0}^{\infty} (-)^k b_k = d_0 , \qquad (15)$$

$$\sum_{k=1}^{\infty} (-)^k k^2 b_k = - \lambda d_1/2 , \qquad (16)$$

$$\sum_{k=2}^{\infty} (-)^k k^2 (k^2-1) b_k = 3\lambda^2 d_2/2 , \qquad (17)$$

$$2g_0 + g_1 + 2uG + a\lambda H = 2B , \quad G = g_0 - (g_1/4) - \sum_{n=2}^{\infty} (-)^n g_n/(n^2-1) ,$$

$$H = (3g_0/4) + (g_1/12) - (19g_2/48) + 3 \sum_{n=3}^{\infty} (-)^n g_n/(n^2-1)(n^2-4) ,$$

$$B = b_0 - (b_1/4) - \sum_{n=2}^{\infty} (-)^n b_n/(n^2-1) , \qquad (18)$$

$$2(1+u)g_0 + 2g_1 + (1-u)g_2 + (a\lambda/4)(4g_0+g_1-2g_2-g_3) = 2b_0 - b_2 , \quad (19)$$

$$(k+u)g_{k-1} + 2kg_k + (k-u)g_{k+1} + (a\lambda/4)\{(2g_{k-2}/\varepsilon_{k-2}) + 2g_{k-1}$$

$$- 2g_{k+1} - g_{k+2}\} = b_{k-1} - b_{k+1} , \quad k \ge 2 , \qquad (20)$$

$$\sum_{k=0}^{\infty} (-)^k g_k = f_0 , \qquad (21)$$

$$\sum_{k=1}^{\infty} (-)^k k^2 g_k = - \lambda f_1/2 , \qquad (22)$$

$$\sum_{k=2}^{\infty} (-)^k k^2 (k^2-1) g_k = 3\lambda^2 f_2/2 . \qquad (23)$$

11.3. The Representation for G(x).

Let

$$F(x) = \sum_{k=0}^{\infty} c_k T_k^*(\lambda/x) , \quad 0 < \lambda \le x , \qquad (1)$$

$$G(x) = e^{bx} x^{-u} \int_x^\infty e^{-bt} t^u F(t) \, dt \, , \tag{2}$$

$$G(x) = \sum_{k=0}^\infty h_k T_k^*(\lambda/x) \, , \quad 0 < \lambda \le x \, . \tag{3}$$

It is convenient to suppose that $b \ne 0$ and that u is not a positive integer or zero. If $b = 0$, then upon replacing x, λ, and t by their reciprocals, (2) with $b = 0$ becomes 11.1(2) if in the latter u and $f(t)$ are replaced by $-(u+2)$ and $F(1/t)$, respectively. The situation when u is a positive integer or zero is treated at the end of this section. We also assume that (formally at least)

$$F(x) = \sum_{k=0}^\infty t_k x^{-k} \, , \tag{4}$$

whence (formally at least),

$$G(x) = \sum_{k=0}^\infty q_k x^{-k} \, , \tag{5}$$

$$bq_0 = t_0 \, , \quad bq_1 = t_1 + uq_0 \, , \quad bq_2 = t_2 - (1-u)q_1 \, ,$$

$$bq_k = t_k - (k-1-u)q_{k-1} \, , \quad k > 0 \, . \tag{6}$$

The h_k's are found by use of a recurrence formula in the backward direction. The recurrence formula for h_k is

$$M\{h_k\} = 4\lambda(c_{k+1} - c_{k+2}) \, , \quad k \ge 0 \, , \tag{7}$$

$$M\{h_k\} \equiv (k-u)(2h_k/\varepsilon_k) + (3k+3+4b\lambda-u)h_{k+1}$$

$$+ (3k+6-4b\lambda+u)h_{k+2} + (k+3+u)h_{k+3} \, , \tag{8}$$

with ε_k as in 11.2(9). Let n be a large positive integer. Put $C_{k,n} = 0$ for $k \ge n+1$ and compute $C_{k,n}$ from

$$M\{C_{k,n}\} = 4\lambda(c_{k+1} - c_{k+2}) \, , \quad 0 \le k \le n \, . \tag{9}$$

Again, let

$$D_{k,n} = 0 \, , \; k \ge n + 2, \; D_{n+1,n} = 1 \, ,$$

$$M\{D_{k,n}\} = 0 \, , \quad 0 \le k \le n \, . \tag{10}$$

Define

$$h_{k,n} = C_{k,n} + D_{k,n}\{(t_0/b) - \sum_{k=0}^{n} (-)^k C_{k,n}\}/\sum_{k=0}^{n+1} (-)^k D_{k,n} . \qquad (11)$$

Then

$$h_k = \lim_{n \to \infty} h_{k,n} , \quad |arg\ b| < \pi . \qquad (12)$$

As a function of arg b, the convergence of the backward recurrence scheme weakens as $|arg\ b| \to \pi$ and fails when $|arg\ b| = \pi$. In these situations, the h_k's can be efficiently obtained provided the backward recurrence scheme is modified as outlined in the discussion surrounding 1.3(36-40). For the benefit of the reader, we briefly describe this procedure. Let n_1 , n_2 be two large distinct positive integers. Evaluate h_{k,n_i} , i = 1 , 2 from (11) with $n = n_i$. Let $N = N\{h_k\}$ be an operator acting on the h_k's which produces a known numerical value. Compute

$$h_{k,n_1,n_2} = h_{k,n_1} + (N - N_1)(h_{k,n_2} - h_{k,n_1})/(N_2 - N_1) \qquad (13)$$

or the right hand side of this formula with the subscripts 1 and 2 interchanged. Then under certain conditions noted in Wimp (1969), we have

$$\lim_{n_1,n_2 \to \infty} h_{k,n_1,n_2} = h_k , \quad k = 0,1,\ldots ., n_1 \neq n_2 . \qquad (14)$$

These conditions are very difficult to apply in terms of the general formulation given here as they depend on the growth properties of the c_k's and the nature of the operator N. A common choice for N is often based on knowledge of G(x) for a particular value of x, say $x = \lambda$. Then in this event,

$$N = G(\lambda) = \sum_{k=0}^{\infty} h_k . \qquad (15)$$

Other possible choices of N could be based on (19), (20), (22) or (23). As remarked, in practice, the hypotheses noted might be difficult to verify, and some experimentation might be necessary. Indeed, in some recent exploratory work, we found

that the scheme worked if N is given by (15), but failed when
N is based on any of the four possible choices named above.
In particular, when (22) was employed, we found that N_1 and N_2
were much nearer to each other than h_{k,n_1} and h_{k,n_2}. Notice
that if $N_1 = N_2$, (13) is not defined. The advantage of having several possible forms for N is that those not used in
(13) are available as checks. The relations analogous to
11.2(15-23) are

$$\sum_{k=0}^{\infty} (-)^k b_k = t_0 , \tag{16}$$

$$\sum_{k=1}^{\infty} (-)^k k^2 b_k = - t_1/2\lambda , \tag{17}$$

$$\sum_{k=2}^{\infty} (-)^k k^2 (k^2-1) b_k = 3t_2/2\lambda^2 , \tag{18}$$

$$b\lambda G - (u+2)H/2 + (6h_0+4h_1+h_2)/8 = \lambda B , \tag{19}$$

where G, H and B are defined in 11.2(18) with b_k and g_k replaced by c_k and h_k, respectively,

$$(2b\lambda-u)h_0 + (5-u)h_1/4 + (2+\tfrac{1}{2}u-b\lambda)h_2 + (u+3)h_3/4 = \lambda(2c_0-c_2) , \tag{20}$$

$$\sum_{k=0}^{\infty} (-)^k h_k = q_0 , \tag{21}$$

$$\sum_{k=1}^{\infty} (-)^k k^2 h_k = - q_1/2\lambda , \tag{22}$$

$$\sum_{k=2}^{\infty} (-)^k k^2 (k^2-1) h_k = 3q_2/2\lambda^2 . \tag{23}$$

We now consider the situation in (2) when u is a positive
integer or zero. First suppose u = 0. Then use of (7) to
compute the elements outlined in (10) and (11) will fail for
k = 0. In this case, designate the left hand side of (20) by
$L^*(h)$. Then in (10) and (11) replace the k = 0 computations
by $L^*(C) = \lambda(2c_0 - c_2)$ and $L^*(D) = 0$ and proceed in the usual
fashion. Next suppose that u is a positive integer. Then use
of (7) to compute the coefficients outlined in (10) and (11)

will fail for k = u. In this event, notice from (4) that

$$x^u F(x) = F_1(x) + F_2(x) ,$$

$$F_1(x) = \sum_{k=0}^{u-1} t_{u-1-k} x^{k+1} , \quad F_2(x) = \sum_{k=0}^{\infty} t_{k+u} x^{-k} . \qquad (24)$$

Now $e^{bx} \int_x^{\infty} e^{-bt} F_1(t)dt$ is a polynomial in x of degree u-1 and

no Chebyshev expansions are needed. We can easily convert $F_2(x)$ into a form like (1), see Chapter 12. So the problem is reduced to the form (2) with u = 0 and F(t) replaced by $F_2(t)$.

In some applications, we need

$$G(x) = e^{ax} x^{-u} \int_0^x e^{-at} t^u F(t)dt \qquad (25)$$

where F(x) and G(x) are represented by (1) and (3), respectively. Here (4) is an asymptotic expansion for F(x) in some domain. The corresponding asymptotic expansion for G(x) is given by (5) with t_k replaced by $- t_k$ for all k. For the present situation, (7)-(13), (19) and (20) are valid provided we replace c_k by $- c_k$ for all k.

11.4. Exponential Type Integrals Involving Logarithms

From 11.2(2)

$$\frac{\partial g(x)}{\partial u} = - (\ln x) g(x) + h(x) ,$$

$$h(x) = e^{-ax} x^{-u-1} \int_0^x e^{at} t^u (\ln t) f(t)dt , \qquad (1)$$

$$\frac{\partial g(x)}{\partial u} = \sum_{k=0}^{\infty} p_k T_k^*(x/\lambda) , \quad p_k = \frac{\partial g_k}{\partial u} . \qquad (2)$$

Partial integration shows that

$$\frac{\partial g(x)}{\partial u} = - e^{-ax} x^{-u-1} \int_0^x t^{-1} v(t)dt ,$$

$$v(x) = e^{ax} x^{u+1} g(x) . \qquad (3)$$

For the present situation, in place of 11.2(7-9), we have

$$L\{p_k\} = - (4/a\lambda)(g_{k+1} - g_{k+2}) .$$ (4)

Here $L\{p_k\}$ is the same as $L\{g_k\}$ with g replaced by p. Also 11.2(18,19) remain valid provided that the right hand sides of these equations are replaced by

$$- 2G \text{ and } - 2g_0 + g_2 ,$$ (5)

respectively, G as in 11.2(18). The analogs of equations 11.2(5,21-23) are

$$\frac{\partial g(x)}{\partial u} = \sum_{k=0}^{\infty} w_k x^k , \quad w_k = \frac{\partial f_k}{\partial u} ,$$ (6)

$$\sum_{k=0}^{\infty} (-)^k p_k = w_0 , \quad \sum_{k=1}^{\infty} (-)^k k^2 p_k = - \lambda w_1/2 ,$$

$$\sum_{k=2}^{\infty} (-)^k k^2 (k^2-1) p_k = 3\lambda^2 w_2/2 .$$ (7)

The discussion surrounding equations 11.2(10-12) can stand except that for 11.2(10), put

$$L\{A_{k,n}\} = - (4/a\lambda)(g_{k+1} - g_{k+2}) ,$$ (8)

and in 11.2(12), replace $g_{k,n}$ by $p_{k,n}$.   Then

$$p_k = \lim_{n \to \infty} p_{k,n} .$$ (9)

Analysis of $\partial G(x)/\partial u$ where $G(x)$ is given by 11.3(2) or 11.3(25) is much akin to the above developments and we dispense with specific results, save that the backward recursion process converges so long as $|\arg b| < \pi$. It will converge more slowly as $|\arg b| \to \pi$ and will fail when $|\arg b| = \pi$. Here, the pertinent coefficients can be easily determined if the backward recurrence scheme is modified after the manner of our previous discussion.

11.5. Numerical Examples

In this section, we describe developments to get Chevy-shev coefficients for the two integrals

$$x^{-1}F(x) = x^{-1}\int_0^x t^{-1}K(t)dt , \tag{1}$$

$$x^{-1}e^x P(x) = x^{-1}e^x \int_0^x t^{-1}e^{-t}K(t)dt , \tag{2}$$

where

$$K(t) = Ei(t) - (\gamma + \ln t) , \tag{3}$$

see 2.2(45). In all of the numerics, we take $\lambda = 8$.

To get the coefficients for $x^{-1}F(x)$, we obtain in all three sets of coefficients based on the theory in 11.2. Pass 1 is the nomenclature employed to describe the input and output data for the computer to achieve the first set of coefficients by use of the backward recursion process described by 11.2(10-14). Similarly, we speak of Pass 2 and Pass 3. We also need three passes for (2), but the first two passes to get this function and (1) are identical.

The data for the passes are based on 11.2(2) and are given in the following table.

Pass No.	a	u	f(x)	g(x)
1	-1	0	1	$U(x) = x^{-1}(e^x-1)$
2	0	0	$U(x)$	$V(x) = x^{-1}\int_0^x U(t)dt$
3A	0	0	$V(x)$	$x^{-1}F(x) = x^{-1}\int_0^x V(t)dt$
3B	-1	0	$V(x)$	$x^{-1}e^x P(x) = x^{-1}e^x\int_0^x e^{-t}V(t)dt$

$$(4)$$

To describe the Chebyshev expansions, some notation changes are necessary to avoid confusion with the notation of

11.2 and to correspond to the notation of the computer.   We use the nomenclature of (4) and write

$$U(x) = \sum_{K=0}^{\infty} B(K)T_K^*(x/8) , \qquad (5)$$

$$V(x) = \sum_{K=0}^{\infty} G(K) \; T_K^*(x/8) , \qquad (6)$$

$$x^{-1}F(x) = \sum_{K=0}^{\infty} H(K)T_K^*(x/8) , \qquad (7)$$

$$x^{-1}e^x P(x) = \sum_{K=0}^{\infty} E(K)T_K^*(x/8) , \qquad (8)$$

where the range of validity is $0 \leq x \leq 8$.   The coefficients $B(K)$ and $G(K)$ are presented to 20D in Table 1, and the coefficients $H(K)$ and $E(K)$ are presented to 20D in Table 2.

In the numerics quadruple precision was employed.   Several values of n (see 11.2(10)) were used to insure that the final coefficients would be accurate to about 28D.   For each n, the coefficients were checked using 11.2(18,19,22).   These are called Test 1, Test 2 and Test 3, respectively.   Test 1, for example, is the difference between the left and right hand sides of 11.2(18) with g_k replaced by the computed g_k. Numerous other checks were made.   The data in Tables 1 and 2 are rounded 20D values from the n = 40 computations.   The test values for the n = 40 unrounded data are presented below each corresponding set of rounded coefficients.

Using the data in Table 2, we calculated $F(x)$ and $P(x)$ for x = 0(0.1)2(0.2)5(0.5)8.   Chipman (1972) tabulated these functions to 12 significant figures for the same values of x as well as for many values of x > 8.   For the x values stated, the data given by Chipman are in perfect agreement with our findings.

Table I

Chebyshev Coefficients For U(x) and V(x)

K	B(K)	G(K)
0	84.21528770042713795950	14.41456779944782689066
1	139.60143980195862213768	21.39840922840560632498
2	85.21528770042713795950	11.58933364472027152821
3	40.60764385021356897975	4.91982145802637719756
4	15.77850825131791519843	1.71661658893004833037
5	5.16137322191878098368	0.50851302118722314073
6	1.45576180748035323106	0.13090665746301778165
7	0.36059548129911692234	0.02980328732335197286
8	0.07958677802143021666	0.00608320067012441754
9	0.01583486248656648906	0.00112538868796799622
10	0.00286744579127871872	0.00019039220173420091
11	0.00047637622313130745	0.00002967553490712013
12	0.00007309908702441011	0.00000428824702013234
13	0.00001042060996351576	0.00000057760548326449
14	0.00000138697475244150	0.00000007285834195164
15	0.00000017311660000426	0.00000000864164451161
16	0.00000002034127261190	0.00000000096726768044
17	0.00000000225773521050	0.00000000010249917713
18	0.00000000023743852051	0.00000000001031244128
19	0.00000000002372498310	0.00000000000098761962
20	0.00000000000225795513	0.00000000000009024398
21	0.00000000000020514367	0.00000000000000788439
22	0.00000000000001782899	0.00000000000000065990
23	0.00000000000000148504	0.00000000000000005301
24	0.00000000000000011875	0.00000000000000000409
25	0.00000000000000000913	0.00000000000000000030
26	0.00000000000000000068	0.00000000000000000002
27	0.00000000000000000005	

```
        TEST1 = 0.31554Q-29     TEST1 = 0.13805Q-29
        TEST2 = 0.12572Q-29     TEST2 = 0.17218Q-30
        TEST3 = 0.29582Q-30     TEST3 = 0.49304Q-31
```

Table II

Chebyshev Coefficients For $x^{-1}F(x)$ And $x^{-1}e^{x}P(x)$

K	H(K)	E(K)
0	3.71536318524502372817	125.71108767544436730357
1	4.01002445709397614210	210.08614599238009058502
2	1.78902666949738251257	129.42645086068939103174
3	0.65121608910087339647	62.18703685169731545791
4	0.19944683756427842710	24.33360295198771131866
5	0.05271734498035826353	8.00700906357096171078
6	0.01226045876841115558	2.26965701687427594241
7	0.00254669312679795329	0.56458844764792477821
8	0.00047807737697770414	0.12506441586340564631
9	0.00008188944866963551	0.02496184587178997702
10	0.00001290022217082669	0.00453266875813334169
11	0.00000188142229479871	0.00075485244962925663
12	0.00000025547457721847	0.00011608009582794006
13	0.00000003245600275231	0.00001657953104894205
14	0.00000000387408482927	0.00000221052600142585
15	0.00000000043610178578	0.00000027633809425030
16	0.00000000004644988391	0.00000003251563439163
17	0.00000000000469502980	0.00000000361363776047
18	0.00000000000045154114	0.00000000038048072047
19	0.00000000000004141888	0.00000000003805883127
20	0.00000000000000363147	0.00000000000362573457
21	0.00000000000000030494	0.00000000000032971390
22	0.00000000000000002457	0.00000000000002867974
23	0.00000000000000000190	0.00000000000000239073
24	0.00000000000000000014	0.00000000000000019132
25	0.00000000000000000001	0.00000000000000001472
26		0.00000000000000000109
27		0.00000000000000000008
28		0.00000000000000000001

TEST1 = 0.89363Q-31   TEST1 = 0.17392Q-28
TEST2 =-0.26193Q-31   TEST2 = 0.17873Q-30
TEST3 = 0.0   TEST3 = 0.24652Q-30

11.6. Errata

In this section, we record errata found in Luke (1976).

Page

189 The right hand side of (3.23) should read
$$3\{t_2 + (u-1)(t_1/b) + u(u-1)(t_0/b^2)\}/2b\lambda^2 .$$

191 Line 16. Read-(3.9), (3.10), ...

192 Line 5. Read-and (3.3), respectively.

193 Second line after (4.3). Read-equations (2.19)-
(2.22) are ... .

193 In (4.4), w_1 and w_2 should read as follows.

$$w_1 = - \{d_1 - ad_0(2u+3)/(u+1)^2\}/(u+2)^2 .$$

$$w_2 = - \{d_2 - [ad_1(2u+5)/(u+2)^2]$$
$$+ a^2 d_0(3u^2+12u+11)/(u+1)^2(u+2)^2\}/(u+3)^2 .$$

194 Eq. (4.6) should read $L\{p_k\} = - (4/a\lambda)(g_{k+1} - g_{k+2})$.

194 For line 6, read
and in (2.26) replace $g_{k,n}$ by $p_{k,n}$ and $d_0/(u+1)$
by $- d_0/(u+1)^2$. Then

196 For lines 11 and 12, read
coefficients were checked using (2.7), (2.8) and
(2.21). These are caleed Test 1, Test 2 and Test

```
C         ----------- IBM S/370 ----------- MULTIPLE PRECISION ---------------
C
C
C         ***********************************************************************
C         *THIS MAINLINE PROGRAM UTILIZES THE SUBROUTINES CCOEF7 AND CHECK7******
C         *TO COMPUTE AND TEST THE COEFFICIENTS IN THE CHEBYSHEV EXPANSION    *
C         *OF EXP(-A*X)*(X**-(U+1))*(INTEGRAL FROM 0 TO X OF EXP(A*T)*        *
C         *(T**U)*F(T)DT) .                                                   *
C         *                                                                   *
C         *WHERE NO AMBIGUITY CAN OCCUR, WE SHALL DENOTE THIS FUNCTION BY     *
C         *INT1(X).                                                           *
C         *                                                                   *
C         *DESCRIPTION OF VARIABLES.                                          *
C         *                                                                   *
C         *FA     -INPUT  - PARAMETER A IN INT1(X).                           *
C         *                                                                   *
C         *FU     -INPUT  - PARAMETER U IN INT1(X).                           *
C         *                                                                   *
C         *FLAM   -INPUT  - THIS IS A PRESELECTED SCALE FACTOR SUCH THAT      *
C         *                 0.LE.(Z/W).LE.1.  SEE TEXT.                       *
C         *                                                                   *
C         *N      -INPUT  - TWO LESS THAN THE NUMBER OF COEFFICIENTS TO BE    *
C         *                 GENERATED.  MAXIMUM VALUE IS 100.                 *
C         *                                                                   *
C         *B      -INPUT  - A VECTOR CONTAINING THE N+3 COEFFICIENTS IN THE   *
C         *                 EXPANSION OF F(X) INTO SHIFTED CHEBYSHEV POLY-     *
C         *                 NOMIALS.  THESE ARE TO BE READ SEQUENTIALLY.      *
C         *                 DIMENSION OF THE VECTOR IS SET AT 103.            *
C         *                                                                   *
C         *G      -OUTPUT - A VECTOR CONTAINING THE N+2 CHEBYSHEV COEFFI-     *
C         *                 CIENTS FOR THE APPROXIMATION TO INT1(X).          *
C         *                 DIMENSION OF THE VECTOR IS SET AT 103.            *
C         *                                                                   *
C         *TEST1  -OUTPUT - THE RESULT OF THE FIRST TEST PERFORMED ON THE     *
C         *                 COEFFICIENTS.  SEE TEXT.                          *
C         *                                                                   *
C         *TEST2  -OUTPUT - THE RESULT OF THE SECOND TEST PERFORMED ON THE    *
C         *                 COEFFICIENTS.  SEE TEXT.                          *
C         *                                                                   *
C         *SUM    -OUTPUT - THE SUM OF THE COEFFICIENTS.                      *
C         *                                                                   *
C         *ALL OTHER VARIABLES ARE FOR INTERNAL USE.                         *
C         *                                                                   *
C         *SUBROUTINES CCOEF7 AND CHECK7 ARE REQUIRED.                       *
C         ***********************************************************************
          IMPLICIT REAL*16 (A-H,O-Z)
          DIMENSION A(103),B(103),G(103)
          REAL*16 C(103)/103*0.Q0/,D(103)/103*0.Q0/
C         ---------- THE NEXT PROGRAM CARD IS USED TO INITIALIZE -------------
C         ---------- DEVICE NUMBERS FOR THE PARTICULAR INSTALL-  -------------
C         ---------- ATION OF THE OPERATING SYSTEM.  IF A SYSTEM -------------
C         ---------- HAS DIFFERENT DEVICE NUMBERS THIS CARD CAN  -------------
C         ---------- BE CHANGED TO INITIALIZE THOSE VALUES AS    -------------
C         ---------- DEVICE NUMBERS.                             -------------
          DATA NREAD,NWRITE,NPUNCH,ZERO,ONE/5,6,7,0.Q0,1.Q0/
C         ---------- READ IN THE INPUT VARIABLES                 -------------
          READ(NREAD,2) FA,FU,FLAM
          READ(NREAD,1) N
          N1=N+1
          N2=N+2
          N3=N+3
          READ(NREAD,2) (B(I),I=1,N3)
C         ---------- CALL SUBROUTINE TO COMPUTE THE COEFFICIENTS -------------
          CALL CCOEF7(FA,FU,FLAM,N,B,ZERO,A)
          IF(FA.EQ.ZERO) GO TO 9
          CALL CCOEF7(FA,FU,FLAM,N,D,ONE,C)
          P=ONE
          SA=A(1)
          SB=B(1)
          SC=C(1)
          DO 3 I=2,N1
          P=-P
          SA=SA+P*A(I)
          SB=SB+P*B(I)
        3 SC=SC+P*C(I)
          SC=SC-P*C(N2)
          SB=SB-P*(B(N2)-B(N3))
          FAC=(SB/(ONE+FU)-SA)/SC
          GO TO 10
        9 FAC=ZERO
       10 SUM=ZERO
C         ---------- WRITE OUT VALUES OF PARAMETERS AND COEFFI- -------------
C         ---------- CIENTS                                      -------------
          WRITE(NWRITE,6) FA,FU,FLAM,N
          DO 4 I=1,N2
          I1=I-1
          G(I)=A(I)+FAC*C(I)
          SUM=SUM+G(I)
        4 WRITE(NWRITE,5) I1,G(I)
```

```
C
C          ---------- CALL SUBROUTINE TO TEST THE COEFFICIENT      ----------
C          ---------- VALUES                                       ----------
      CALL CHECK7(FA,FU,FLAM,N,B,G,TEST1,TEST2)
      WRITE(NWRITE,7) TEST1,TEST2
      WRITE(NWRITE,8) SUM
C
C          ---------- PUNCH THE VALUES OF COEFFICIENTS ON CARDS    ----------
      WRITE(NPUNCH,2) (G(K),K=1,N2)
    1 FORMAT(I5)
    2 FORMAT(Q39,32)
    5 FORMAT(1X,I5,8X,Q39,32)
    6 FORMAT('1',4X,'A=',Q39,32/5X,'U=',Q39,32/3X,'LAM=',Q39,32/5X,'N=',
     1I5////5X,'K',25X,'G(K)'/)
    7 FORMAT(/1X,'TEST1=',Q39,32/1X,'TEST2=',Q39,32)
    8 FORMAT(/1X,'THE SUM OF THE COEFFICIENTS IS',Q39,32)
      STOP
      END
C
C          ----------- IBM S/370 ----------- MULTIPLE PRECISION ------------
C
C     ****************************************************************
C     *THIS SUBROUTINE, CCOEF7, COMPUTES MULTIPLE PRECISION VALUES OF*
C     *THE COEFFICIENTS IN THE CHEBYSHEV EXPANSION OF EXP(-A*X)*     *
C     *(X**-(U+1))*(INTEGRAL FROM 0 TO X OF EXP(A*T)*(T**U)*F(T)DT). *
C     *                                                              *
C     *WHERE NO AMBIGUITY CAN OCCUR, WE SHALL DENOTE THIS FUNCTION BY*
C     *INT1(X).                                                      *
C     *                                                              *
C     *DESCRIPTION OF VARIABLES.                                     *
C     *                                                              *
C     *FA      -INPUT  - PARAMETER A IN INT1(X).                     *
C     *                                                              *
C     *FU      -INPUT  - PARAMETER U IN INT1(X).                     *
C     *                                                              *
C     *FLAM    -INPUT  - THIS IS A PRESELECTED SCALE FACTOR SUCH THAT*
C     *                  0.LE.(Z/W).LE.1.  SEE TEXT.                 *
C     *                                                              *
C     *N       -INPUT  - TWO LESS THAN THE NUMBER OF COEFFICIENTS TO BE*
C     *                  GENERATED.  MAXIMUM VALUE IS 100.           *
C     *                                                              *
C     *VIN     -INPUT  - A VECTOR CONTAINING THE N+3 COEFFICIENTS IN THE*
C     *                  EXPANSION OF F(X) INTO SHIFTED CHEBYSHEV POLY-*
C     *                  NOMIALS.                                    *
C     *                                                              *
C     *START   -INPUT  - THE DESIRED VALUE OF VOUT(N+2).            *
C     *                                                              *
C     *VOUT    -OUTPUT - A VECTOR CONTAINING THE N+2 CHEBYSHEV COEFFI-*
C     *                  CIENTS FOR THE APPROXIMATION TO INT1(X).    *
C     *                                                              *
C     *ALL OTHER VARIABLES ARE FOR INTERNAL USE.                    *
C     ****************************************************************
      SUBROUTINE CCOEF7(FA,FU,FLAM,N,VIN,START,VOUT)
      IMPLICIT REAL*16 (A-H,O-Z)
      DIMENSION VIN(1),VOUT(1)
      DATA ZERO,ONE,TWO,FOUR/0.Q0,1.Q0,2.Q0,4.Q0/
      N2=N+2
      N1=N+1
      NCOUNT=N2
      VOUT(NCOUNT)=START
      Z=N2
      IF(FA.EQ.ZERO) GO TO 2
C
C          ---------- START COMPUTING COEFFICIENTS BY MEANS OF      ----------
C          ---------- BACKWARD RECURRENCE SCHEME                    ----------
      A3=ZERO
      A2=ZERO
      A1=START
      B2=VIN(N2+1)
      B1=VIN(N2)
      FMUL=FOUR/(FA*FLAM)
      F1=FU+ONE+Z
      F2=Z-FU
      DO 1 I=1,N1
      NCOUNT=NCOUNT-1
      F1=F1-ONE
      F2=F2-ONE
      VOUT(NCOUNT)=-(FMUL*F1+ONE)*A1-(FMUL*F2-ONE)*A2+A3+FMUL*(B1-B2)
      A3=A2
      A2=A1
      A1=VOUT(NCOUNT)
      B2=B1
    1 B1=VIN(NCOUNT)
      VOUT(1)=VOUT(1)/TWO
      RETURN
```

```
    2 B2=VIN(N2)
      B1=VIN(N1)
      F1=FU-Z
      F2=FU+Z+ONE
      DO 3 I=1,N
      NCOUNT=NCOUNT-1
      F1=F1+ONE
      F2=F2-ONE
      VOUT(NCOUNT)=(F1*VOUT(NCOUNT+1)+B1-B2)/F2
      B2=B1
    3 B1=VIN(NCOUNT-1)
      VOUT(1)=(FU*VOUT(2)+TWO*VIN(1)-VIN(2))/(TWO*(FU+ONE))
      RETURN
      END
```

```
C
C     ------------ IBM S/370 ------------ MULTIPLE PRECISION ------------
C
C     ***********************************************************************
C     *THIS SUBROUTINE, CHECK7, RUNS CHECKS ON THE COEFFICIENTS           *
C     *GENERATED BY SUBROUTINE CCOEF7.  EXCEPT FOR ROUND-OFF ERROR AND    *
C     *TRUNCATION ERROR, BOTH TEST RESULTS SHOULD BE ZERO.  SEE TEXT.     *
C     *                                                                   *
C     *WHERE NO AMBIGUITY CAN OCCUR, WE SHALL DENOTE EXP(-A*X)*           *
C     *(X**-(U+1))*(INTEGRAL FROM 0 TO X OF EXP(A*T)*(T**U)*F(T)DT) BY    *
C     *INT1(X).                                                           *
C     *                                                                   *
C     *DESCRIPTION OF VARIABLES.                                          *
C     *                                                                   *
C     *FA      -INPUT  - PARAMETER A IN INT1(X).                          *
C     *                                                                   *
C     *FU      -INPUT  - PARAMETER U IN INT1(X).                          *
C     *                                                                   *
C     *FLAM    -INPUT  - THIS IS A PRESELECTED SCALE FACTOR SUCH THAT     *
C     *                  0.LE.(Z/W).LE.1.  SEE TEXT.                      *
C     *                                                                   *
C     *N       -INPUT  - TWO LESS THAN THE NUMBER OF COEFFICIENTS         *
C     *                  GENERATED.  MAXIMUM VALUE IS 100.                *
C     *                                                                   *
C     *B       -INPUT  - A VECTOR CONTAINING THE N+3 COEFFICIENTS IN THE  *
C     *                  EXPANSION OF F(X) INTO SHIFTED CHEBYSHEV POLY-    *
C     *                  NOMIALS.                                         *
C     *                                                                   *
C     *G       -INPUT  - A VECTOR CONTAINING THE N+2 CHEBYSHEV COEFFI-    *
C     *                  CIENTS FOR THE APPROXIMATION TO INT1(X).         *
C     *                                                                   *
C     *TEST1   -OUTPUT - THE RESULT OF THE FIRST TEST PERFORMED ON THE    *
C     *                  COEFFICIENTS.  SEE TEXT.                         *
C     *                                                                   *
C     *TEST2   -OUTPUT - THE RESULT OF THE SECOND TEST PERFORMED ON THE   *
C     *                  COEFFICIENTS.  SEE TEXT.                         *
C     *                                                                   *
C     *ALL OTHER VARIABLES ARE FOR INTERNAL USE.                         *
C     ***********************************************************************
      SUBROUTINE CHECK7(FA,FU,FLAM,N,B,G,TEST1,TEST2)
      IMPLICIT REAL*16 (A-H,O-Z)
      DIMENSION B(1),G(1)
      DATA ONE,TWO,THREE,FOUR,TWELVE/1.Q0,2.Q0,3.Q0,4.Q0,12.Q0/
C
C     ---------- COMPUTATION OF LEFT SIDES OF TEST EQUATIONS -----------
      SB=B(1)-B(2)+B(3)
      SB1=-B(2)+FOUR*B(3)
      SB2=TWELVE*B(3)
      SG1=-G(2)+FOUR*G(3)
      SG2=TWELVE*G(3)
      P=-ONE
      COUNT=TWO
      DO 1 I=2,N
      COUNT=COUNT+ONE
      CSQ=COUNT*COUNT
      SB=SB+P*B(I+2)
      SB1=SB1+P*CSQ*B(I+2)
      SB2=SB2+P*CSQ*(CSQ-ONE)*B(I+2)
      SG1=SG1+P*CSQ*G(I+2)
      SG2=SG2+P*CSQ*(CSQ-ONE)*G(I+2)
    1 P=-P
C
C     --------- COMPUTATION OF TEST VALUES ---------        ----------
      TEST1=SG1+FLAM/(TWO*(FU+TWO))*(SB1-FA*SB/(FU+ONE))
      TEST2=SG2-THREE*FLAM*FLAM/(TWO*(FU+THREE))*(SB2-FA*SB1/(FU+TWO)+FA
     :*FA*SB/((FU+TWO)*(FU+ONE)))
      RETURN
      END
```

```
A=-0.1000000000000000000000000000000000000Q+01
U= 0.0
LAM= 0.8000000000000000000000000000000000000Q+01
N=    40
```

K	G(K)
0	0.8421528770042713795950388664549000Q+02
1	0.1396014398019586221376838603259300Q+03
2	0.8521528770042713795950388664549100Q+02
3	0.4060764385021356897975194332274500Q+02
4	0.1577850825131791519842803035769300Q+02
5	0.5161373221918780983681867428514000Q+01
6	0.1455761807480353231064295500664300Q+01
7	0.3605954812991169223392612483398000Q+00
8	0.7958677780214302166640836834941540Q-01
9	0.1583486248656664890581357992118100Q-01
10	0.2867445791278718724292456504350600Q-02
11	0.4763762231313074479344530355002600Q-03
12	0.7309908702441011121240743860215000Q-04
13	0.1042060996351576173954297593534000Q-04
14	0.1386974752441504089411291641058000Q-05
15	0.1731166000426010456489568217690000Q-06
16	0.2034127261190432185041481047526000Q-07
17	0.2257735210496621728495971191571400Q-08
18	0.2374385205080951303208520790322000Q-09
19	0.2372498310444428439894097100809700Q-10
20	0.2257955133370276779319533859310400Q-11
21	0.2051436677086645475282344365657300Q-12
22	0.1782899263134497259482054224895000Q-13
23	0.1485041023164119196753255766810400Q-14
24	0.1187528015729049215796719026158100Q-15
25	0.9131356878349910788030207282848300Q-17
26	0.6761656571637583322361920515593900Q-18
27	0.4828266099525462852082252557625000Q-19
28	0.3328909869366836586291280288077200Q-20
29	0.2218711196290309642756548907770800Q-21
30	0.1431096436687014175541832564317200Q-22
31	0.8942468739158892231277697239849000Q-24
32	0.5418632675968125504485336752501000Q-25
33	0.3186866597062749761699223671435800Q-26
34	0.1820769907630704836557529168918490Q-27
35	0.1011386832352239369052805296331200Q-28
36	0.5466196486683684950187430940934100Q-30
37	0.2876576254143433828479613474303400Q-31
38	0.1474983270479634836884936857263240Q-32
39	0.7374916352398174184424684286311220Q-34
40	0.3588767081458965539866026416969690Q-35
41	0.1794383540729482769933011320834850Q-36

```
TEST1=-0.674845852513287442910814671081760Q-30
TEST2=-0.616297582203915472977912941627180Q-31
```

THE SUM OF THE COEFFICIENTS IS 0.37249474838021603434294901243164Q+03

```
C
C          ---------- IBM S/370 ---------- MULTIPLE PRECISION ------------
C
C      **********************************************************************
C      *THIS MAINLINE PROGRAM UTILIZES THE SUBROUTINES CCOEF8 AND CHECK8 *
C      *TO COMPUTE AND TEST THE COEFFICIENTS IN THE CHEBYSHEV EXPANSION  *
C      *OF EXP(B*X)*(X**-U)*(INTEGRAL FROM X TO INFINITY OF EXP(-B*T)*   *
C      *(T**U)*F(T)DT).                                                  *
C      *                                                                 *
C      *WHERE NO AMBIGUITY CAN OCCUR, WE SHALL DENOTE THIS FUNCTION BY   *
C      *INT2(X).                                                         *
C      *                                                                 *
C      *DESCRIPTION OF VARIABLES.                                        *
C      *                                                                 *
C      *FB     -INPUT  - PARAMETER B IN INT2(X).                         *
C      *                                                                 *
C      *FU     -INPUT  - PARAMETER U IN INT2(X).                         *
C      *                                                                 *
C      *FLAM   -INPUT  - THIS IS A PRESELECTED SCALE FACTOR SUCH THAT    *
C      *                 0.LE.(Z/W).LE.1.  SEE TEXT.                     *
C      *                                                                 *
C      *N      -INPUT  - TWO LESS THAN THE NUMBER OF COEFFICIENTS TO BE  *
C      *                 GENERATED.  MAXIMUM VALUE IS 100.               *
C      *                                                                 *
C      *B      -INPUT  - A VECTOR CONTAINING THE N+3 COEFFICIENTS IN THE *
C      *                 EXPANSION OF F(X) INTO SHIFTED CHEBYSHEV POLY-  *
C      *                 NOMIALS.  THESE ARE TO BE READ SEQUENTIALLY.    *
C      *                                                                 *
C      *NN     -INPUT  - TWO LESS THAN THE NUMBER OF COEFFICIENTS TO BE  *
C      *                 COMPUTED IN THE SECOND SET OF COEFFICIENTS.     *
C      *                 NN IS TO BE ENTERED ONLY IF FB IS NEGATIVE.  NN *
C      *                 MUST BE GREATER THAN N.  MAXIMUM VALUE IS 100.  *
C      *                                                                 *
C      *T      -INPUT  - THE VALUE OF INT2(FLAM).  THIS IS TO BE ENTERED *
C      *                 ONLY IF FB IS NEGATIVE.                         *
C      *                                                                 *
C      *S1     -OUTPUT - THE SUM OF THE FIRST SET OF COEFFICIENTS.       *
C      *                                                                 *
C      *S2     -OUTPUT - THE SUM OF THE SECOND SET OF COEFFICIENTS.      *
C      *                                                                 *
C      *H1     -OUTPUT - THE FIRST VECTOR CONTAINING THE N+2 CHEBYSHEV   *
C      *                 COEFFICIENTS FOR THE APPROXIMATION TO INT2(X).  *
C      *                 DIMENSION OF THE VECTOR IS SET AT 103.          *
C      *                                                                 *
C      *H2     -OUTPUT - THE SECOND VECTOR CONTAINING THE NN+2 CHEBYSHEV *
C      *                 COEFFICIENTS FOR THE APPROXIMATION TO INT2(X) IF*
C      *                 FB IS NEGATIVE.  DIMENSION OF THE VECTOR IS SET *
C      *                 103.                                            *
C      *                                                                 *
C      *H      -OUTPUT - THE VECTOR CONTAINING THE FINAL NN+2 CHEBYSHEV  *
C      *                 COEFFICIENTS FOR THE APPROXIMATION TO INT2(X).  *
C      *                 DIMENSION OF THE VECTOR IS SET AT 103.          *
C      *                                                                 *
C      *TEST1  -OUTPUT - THE RESULT OF THE FIRST TEST PERFORMED ON THE   *
C      *                 COEFFICIENTS.  SEE TEXT.                        *
C      *                                                                 *
C      *TEST2  -OUTPUT - THE RESULT OF THE SECOND TEST PERFORMED ON THE  *
C      *                 COEFFICIENTS.  SEE TEXT.                        *
C      *                                                                 *
C      *SUM    -OUTPUT - THE SUM OF THE COEFFICIENTS.                    *
C      *                                                                 *
C      *ALL OTHER VARIABLES ARE FOR INTERNAL USE.                       *
C      *                                                                 *
C      *SUBROUTINES CCOEF8 AND CHECK8 ARE REQUIRED.                     *
C      **********************************************************************
       IMPLICIT REAL*16 (A-H,O-Z)
       DIMENSION A(103),B(103),H1(103),H2(103),H(103)
       REAL*16 C(103)/103*0.Q0/,D(103)/103*0.Q0/
C
C          ---------- THE NEXT PROGRAM CARD IS USED TO INITIALIZE ----------
C          ---------- DEVICE NUMBERS FOR THE PARTICULAR INSTALL-  ----------
C          ---------- ATION OF THE OPERATING SYSTEM.  IF A SYSTEM ----------
C          ---------- HAS DIFFERENT DEVICE NUMBERS THIS CARD CAN  ----------
C          ---------- BE CHANGED TO INITIALIZE THOSE VALUES AS    ----------
C          ---------- DEVICE NUMBERS.                             ----------
       DATA NREAD,NWRITE,NPUNCH,ZERO,ONE/5,6,7,0.Q0,1.Q0/
C
C          ---------- READ IN THE INPUT VARIABLES                 ----------
       READ(NREAD,2) FB,FU,FLAM
       READ(NREAD,1) N
       N1=N+1
       N2=N+2
       N3=N+3
       READ(NREAD,2) (B(I),I=1,N3)
       NUM=1
       IF(FB) 9,11,11
```

```
    9 READ(NREAD,1) NN
      NN1=NN+1
      NN2=NN+2
      NN3=NN+3
      READ(NREAD,2) T
      NUM=2
      NDIF=NN-N
      DO 25 KK=1,NDIF
      KKN=KK+N+3
   25 READ(NREAD,2) B(KKN)
C         ----------- CALL SUBROUTINE TO COMPUTE THE FIRST OR      -----------
C         ----------- ONLY SET OF COEFFICIENTS                     -----------
   11 CALL CCOEF8(FB,FU,FLAM,N,B,ZERO,A)
      CALL CCOEF8(FB,FU,FLAM,N,D,ONE,C)
      P=ONE
      SA=A(1)
      SB=B(1)
      SC=C(1)
      DO 3 I=2,N1
      P=-P
      SA=SA+P*A(I)
      SB=SB+P*B(I)
    3 SC=SC+P*C(I)
      SC=SC-P*C(N2)
      SB=SB-P*(B(N2)-B(N3))
      FAC=(SB/FB-SA)/SC
      S1=ZERO
      WRITE(NWRITE,6) FB,FU,FLAM,N
      DO 4 I=1,N2
      I1=I-1
      H1(I)=A(I)+FAC*C(I)
      S1=S1+H1(I)
    4 WRITE(NWRITE,5) I1,H1(I)
      WRITE(NWRITE,8) S1
      IF(NUM.EQ.2) GO TO 12
      DO 13 K=1,N2
   13 H(K)=H1(K)
      NN=N
      NN2=N2
      GO TO 14
C         ----------- CALL SUBROUTINE TO COMPUTE THE SECOND SET    -----------
C         ----------- OF COEFFICIENTS                              -----------
   12 CALL CCOEF8(FB,FU,FLAM,NN,B,ZERO,A)
      CALL CCOEF8(FB,FU,FLAM,NN,D,ONE,C)
      P=ONE
      SA=A(1)
      SB=B(1)
      SC=C(1)
      DO 23 I=2,NN1
      P=-P
      SA=SA+P*A(I)
      SB=SB+P*B(I)
   23 SC=SC+P*C(I)
      SC=SC-P*C(NN2)
      SB=SB-P*(B(NN2)-B(NN3))
      FAC=(SB/FB-SA)/SC
      S2=ZERO
      WRITE(NWRITE,6) FB,FU,FLAM,NN
      DO 24 I=1,NN2
      I1=I-1
      H2(I)=A(I)+FAC*C(I)
      S2=S2+H2(I)
   24 WRITE(NWRITE,5) I1,H2(I)
      WRITE(NWRITE,8) S2
C         ----------- CALL SUBROUTINE TO COMPUTE THE FINAL VALUES  -----------
C         ----------- OF COEFFICIENTS                              -----------
      CALL LINCOM(N,NN,H1,H2,S1,S2,T,H,SUM)
C         ----------- WRITE OUT VALUES OF PARAMETERS AND COEFFI-   -----------
C         ----------- CIENTS                                       -----------
      WRITE(NWRITE,6) FB,FU,FLAM,NN
      DO 15 K=1,NN2
      K1=K-1
   15 WRITE(NWRITE,5) K1,H(K)
      WRITE(NWRITE,8) SUM
C         ----------- CALL SUBROUTINE TO TEST THE COEFFICIENT      -----------
C         ----------- VALUES                                       -----------
   14 WRITE(NWRITE,16)
      CALL CHECK8(FB,FU,FLAM,NN,B,H,TEST1,TEST2)
      WRITE(NWRITE,7) TEST1,TEST2
C         ----------- PUNCH THE VALUES OF COEFFICIENTS ON CARDS    -----------
      WRITE(NPUNCH,2) (H(K),K=1,NN2)
```

```
  1 FORMAT(I5)
  2 FORMAT(Q39,32)
  5 FORMAT(1X,I5,8X,Q39,32)
  6 FORMAT('1','4X,'B=',Q39,32/5X,'U=',Q39,32/3X,'LAM=',Q39,32/5X,'N=',
   :I5/////5X,'K',25X,'H(K)'/)
  7 FORMAT(/1X,'TEST1=',Q39,32/1X,'TEST2=',Q39,32)
  8 FORMAT(/1X,'THE SUM OF THE COEFFICIENTS IS',Q39,32)
 16 FORMAT(/1X,'THE FINAL COEFFICIENTS ARE DIRECTLY ABOVE.')
    STOP
    END
C
C
C      ----------- IBM S/370 ----------- MULTIPLE PRECISION -----------
C
C      ************************************************************************
C      *THIS SUBROUTINE, CCOEF8, COMPUTES MULTIPLE PRECISION VALUES OF   *
C      *THE COEFFICIENTS IN THE CHEBYSHEV EXPANSION OF EXP(B*X)*(X**-U)* *
C      *(INTEGRAL FROM X TO INFINITY OF EXP(-B*T)*(T**U)*F(T)DT).        *
C      *                                                                *
C      *WHERE NO AMBIGUITY CAN OCCUR, WE SHALL DENOTE THIS FUNCTION BY   *
C      *INT2(X).                                                         *
C      *                                                                *
C      *                                                                *
C      *DESCRIPTION OF VARIABLES.                                       *
C      *                                                                *
C      *FB     -INPUT  - PARAMETER B IN INT2(X).                        *
C      *                                                                *
C      *FU     -INPUT  - PARAMETER U IN INT2(X).                        *
C      *                                                                *
C      *FLAM   -INPUT  - THIS IS A PRESELECTED SCALE FACTOR SUCH THAT   *
C      *                 0.LE.(Z/W).LE.1.  SEE TEXT.                    *
C      *                                                                *
C      *N      -INPUT  - TWO LESS THAN THE NUMBER OF COEFFICIENTS TO BE *
C      *                 GENERATED.  MAXIMUM VALUE IS 100.              *
C      *                                                                *
C      *VIN    -INPUT  - A VECTOR CONTAINING THE N+3 COEFFICIENTS IN THE*
C      *                 EXPANSION OF F(X) INTO SHIFTED CHEBYSHEV POLY- *
C      *                 NOMIALS.                                       *
C      *                                                                *
C      *START  -INPUT  - THE DESIRED VALUE OF VOUT(N+2).                *
C      *                                                                *
C      *VOUT   -OUTPUT - A VECTOR CONTAINING THE N+2 CHEBYSHEV COEFFI-  *
C      *                 CIENTS FOR THE APPROXIMATION TO INT2(X).       *
C      *                                                                *
C      *ALL OTHER VARIABLES ARE FOR INTERNAL USE.                       *
C      ************************************************************************
       SUBROUTINE CCOEF8(FB,FU,FLAM,N,VIN,START,VOUT)
       IMPLICIT REAL*16 (A-H,O-Z)
       DIMENSION VIN(1),VOUT(1)
       DATA ZERO,ONE,TWO,THREE,FOUR/0.Q0,1.Q0,2.Q0,3.Q0,4.Q0/
       N2=N+2
       N1=N+1
       NCOUNT=N2
       VOUT(NCOUNT)=START
       Z=N2
C
C      ---------- START COMPUTING COEFFICIENTS BY MEANS OF    -----------
C      ---------- BACKWARD RECURRENCE SCHEME                  -----------
       A3=ZERO
       A2=ZERO
       A1=START
       B2=VIN(N2+1)
       B1=VIN(N2)
       FMUL=FOUR*FLAM
       C=FMUL*FB-FU
       F0=Z-ONE-FU
       F1=THREE*Z+C
       F2=THREE*Z+THREE-C
       F3=Z+TWO+FU
       DO 1 I=1,N1
       NCOUNT=NCOUNT-1
       F0=F0-ONE
       F1=F1-THREE
       F2=F2-THREE
       F3=F3-ONE
       VOUT(NCOUNT)=-(F1*A1+F2*A2+F3*A3-FMUL*(B1-B2))/F0
       A3=A2
       A2=A1
       A1=VOUT(NCOUNT)
       B2=B1
  1    B1=VIN(NCOUNT)
       VOUT(1)=VOUT(1)/TWO
       RETURN
       END
```

```
C
C
C        ------------ IBM S/370 ----------- MULTIPLE PRECISION ------------
C
C        ********************************************************************
C        *THIS SUBROUTINE, LINCOM, FORMS A LINEAR COMBINATION OF TWO SETS  *
C        *OF CHEBYSHEV COEFFICIENTS IN ORDER TO SATISFY A GIVEN VALUE OF   *
C        *THE FUNCTION THAT IS TO BE APPROXIMATED.                         *
C        *                                                                 *
C        *DESCRIPTION OF VARIABLES.                                        *
C        *                                                                 *
C        *N      -INPUT  - TWO LESS THAN THE NUMBER OF COEFFICIENTS        *
C        *                 COMPUTED IN THE FIRST SET OF COEFFICIENTS.      *
C        *                                                                 *
C        *NN     -INPUT  - TWO LESS THAN THE NUMBER OF COEFFICIENTS        *
C        *                 COMPUTED IN THE SECOND SET OF COEFFICIENTS.  NN *
C        *                 MUST BE GREATER THAN N.                         *
C        *                                                                 *
C        *C1     -INPUT  - THE FIRST VECTOR CONTAINING THE N+2 CHEBYSHEV   *
C        *                 COEFFICIENTS FOR THE APPROXIMATION TO THE       *
C        *                 FUNCTION.                                       *
C        *                                                                 *
C        *C2     -INPUT  - THE SECOND VECTOR CONTAINING THE NN+2 CHEBYSHEV *
C        *                 COEFFICIENTS FOR THE APPROXIMATION TO THE       *
C        *                 FUNCTION.                                       *
C        *                                                                 *
C        *S1     -INPUT  - THE SUM OF THE FIRST SET OF COEFFICIENTS.       *
C        *                                                                 *
C        *S2     -INPUT  - THE SUM OF THE SECOND SET OF COEFFICIENTS.      *
C        *                                                                 *
C        *T      -INPUT  - THE VALUE OF THE FUNCTION AT Z=W.               *
C        *                                                                 *
C        *C      -OUTPUT - THE FINAL VECTOR CONTAINING THE NN+2 CHEBYSHEV  *
C        *                 COEFFICIENTS FOR THE APPROXIMATION TO THE       *
C        *                 FUNCTION.                                       *
C        *                                                                 *
C        *SUM    -OUTPUT - THE SUM OF THE COEFFICIENTS.                    *
C        *                                                                 *
C        *ALL OTHER VARIABLES ARE FOR INTERNAL USE.                       *
C        ********************************************************************
         SUBROUTINE LINCOM(N,NN,C1,C2,S1,S2,T,C,SUM)
         IMPLICIT REAL*16 (A-H,O-Z)
         DIMENSION C(1),C1(1),C2(1)
         N2=N+2
         N3=N2+1
         NN2=NN+2
         FAC1=(T-S2)/(S1-S2)
         FAC2=(S1-T)/(S1-S2)
         C(1)=C2(1)+FAC1*(C1(1)-C2(1))
         SUM=C(1)
         DO 1 K=2,N2
         C(K)=C2(K)+FAC1*(C1(K)-C2(K))
       1 SUM=SUM+C(K)
         DO 2 K=N3,NN2
         C(K)=FAC2*C2(K)
       2 SUM=SUM+C(K)
         RETURN
         END
C        ------------ IBM S/370 ----------- MULTIPLE PRECISION ------------
C
C        ********************************************************************
C        *THIS SUBROUTINE, CHECK8, RUNS CHECKS ON THE COEFFICIENTS         *
C        *GENERATED BY SUBROUTINE CCOEF8.  EXCEPT FOR ROUND-OFF ERROR AND  *
C        *TRUNCATION ERROR, BOTH TEST RESULTS SHOULD BE ZERO.  SEE TEXT.   *
C        *                                                                 *
C        *WHERE NO AMBIGUITY CAN OCCUR, WE SHALL DENOTE EXP(B*X)*(X**-U)*  *
C        *(INTEGRAL FROM X TO INFINITY OF EXP(-B*T)*(T**U)*F(T)DT) BY      *
C        *INT2(X).                                                         *
C        *                                                                 *
C        *DESCRIPTION OF VARIABLES.                                        *
C        *                                                                 *
C        *FB     -INPUT  - PARAMETER B IN INT2(X).                         *
C        *                                                                 *
C        *FU     -INPUT  - PARAMETER U IN INT2(X).                         *
C        *                                                                 *
C        *FLAM   -INPUT  - THIS IS A PRESELECTED SCALE FACTOR SUCH THAT    *
C        *                 0.LE.(Z/W).LE.1.  SEE TEXT.                     *
C        *                                                                 *
C        *N      -INPUT  - TWO LESS THAN THE NUMBER OF COEFFICIENTS        *
C        *                 GENERATED.  MAXIMUM VALUE IS 100.               *
C        *                                                                 *
C        *B      -INPUT  - A VECTOR CONTAINING THE N+3 COEFFICIENTS IN THE *
C        *                 EXPANSION OF F(X) INTO SHIFTED CHEBYSHEV POLY-  *
C        *                 NOMIALS.                                        *
```

```
C
C     *
C     *H      -INPUT  - A VECTOR CONTAINING THE N+2 CHEBYSHEV COEFFI-      *
C     *                 CIENTS FOR THE APPROXIMATION TO INT2(X).           *
C     *                                                                    *
C     *TEST1 -OUTPUT - THE RESULT OF THE FIRST TEST PERFORMED ON THE       *
C     *                 COEFFICIENTS.  SEE TEXT.                           *
C     *                                                                    *
C     *TEST2 -OUTPUT - THE RESULT OF THE SECOND TEST PERFORMED ON THE      *
C     *                 COEFFICIENTS.  SEE TEXT.                           *
C     *                                                                    *
C     *                                                                    *
C     *ALL OTHER VARIABLES ARE FOR INTERNAL USE.                          *
C     **********************************************************************
      SUBROUTINE CHECK8(FB,FU,FLAM,N,B,H,TEST1,TEST2)
      IMPLICIT REAL*16 (A-H,O-Z)
      DIMENSION B(1),H(1)
      DATA ONE,TWO,THREE,FOUR,TWELVE/1.Q0,2.Q0,3.Q0,4.Q0,12.Q0/
C
C     ----------- COMPUTATION OF LEFT SIDES OF TEST EQUATIONS -----------
      SB=B(1)-B(2)+B(3)
      SB1=-B(2)+FOUR*B(3)
      SB2=TWELVE*B(3)
      SH1=-H(2)+FOUR*H(3)
      SH2=TWELVE*H(3)
      P=ONE
      COUNT=TWO
      DO 1 I=2,N
      COUNT=COUNT+ONE
      CSQ=COUNT*COUNT
      SB=SB+P*B(I+2)
      SB1=SB1+P*CSQ*B(I+2)
      SB2=SB2+P*CSQ*(CSQ-ONE)*B(I+2)
      SH1=SH1+P*CSQ*H(I+2)
      SH2=SH2+P*CSQ*(CSQ-ONE)*H(I+2)
    1 P=-P
C
C     ----------- COMPUTATION OF TEST VALUES            -----------
      TEST1=SH1+(-TWO*FLAM*SB1+FU*SB/FB)/(TWO*FLAM*FB)
      TEST2=SH2-(SB2-THREE/(TWO*FLAM*FLAM)*(FU-ONE)/FB*(TWO*FLAM*SB1-FU*
     1SB/FB))/FB
      RETURN
      END
```

```
B= 0.100000000000000000000000000000000000Q+01
U=-0.500000000000000000000000000000000000Q+00
LAM= 0.900000000000000000000000000000000000Q+01
N=  45
```

K	H(K)
0	0.9750834237085559285373955580671130+00
1	-0.2404939385041460496480699369385Q-01
2	0.8204522408804319851858882992549980-03
3	-0.4342930813034275689534498126458880-04
4	0.3018447034034931913045528846864Q-05
5	-0.2544733192508205188574764112291515Q-06
6	0.2485835302051093671155318394888870-07
7	-0.2731720132382240374017035312794220-08
8	0.330847222796561028522769189302011Q-09
9	-0.4350549079673218279647253168821440-10
10	0.6141214566159980240622764968614Q-11
11	-0.9223692795989471709835577307964Q-12
12	0.146356647010933518837628571075190-12
13	-0.2433277512379979070103644755410Q-13
14	0.4249758881830505728550638529502Q-14
15	-0.77084411309328062407819718294069Q-15
16	-0.14506883283159871467300605026364Q-15
17	-0.28242433151649893456671259497888810-16
18	0.5673294811993644768916444088555340-17
19	-0.11732702853472893014932694159535Q-17
20	0.249304605375757421984642821275430-18
21	-0.5433369342356253544197913614986909-19
22	0.121264661039980161308450449273090-19
23	-0.27676739833998377808081162044528Q-20
24	0.6451505279698972964765148433087809-21
25	-0.15341738098791147321116441763820Q-21
26	0.3717943110576168234975295626312Q-22
27	-0.9173447018847710118661463370774609-23
28	0.230242712439101956794233736894940Q-23
29	-0.5873750174263885194032428046246809-24
30	0.152195352855145442576350893947254Q-24
31	-0.4002654986017563009478565475699409-25
32	0.1067780777785669640770003637686109-25
33	-0.288768055297431807884218848650309-26
34	0.79124849523626728039006299296557Q-27
35	-0.2195598345013924763096728772863Q-27
36	0.6166846754259216716625902163007Q-28
37	-0.17524710493137681178994002887830Q-28
38	0.503657324356327323437631329826330-29
39	-0.14633426998170994379351466034240Q-29
40	0.4296571562447418274663103411157909-30
41	-0.12744044448635057762666597168528Q-30
42	0.38175665988979681076321180679013Q-31
43	-0.1155496560830137264148451769785809-31
44	0.35417710745339716308451460149745Q-32
45	-0.10866996399904574479183684141406309-32
46	0.28335148203815552322554514861883Q-33

THE SUM OF THE COEFFICIENTS IS 0.95181383918392523181467

THE FINAL COEFFICIENTS ARE DIRECTLY ABOVE.

```
TEST1=-0.41288326889016610622295687130378Q-30
TEST2=-0.93075490128345942954624324739005Q-27
```

```
C
C           ----------- IBM S/370 ----------- MULTIPLE PRECISION --------------
C
CCCCC
CCCCC     *******************************************************************
CCCCC     *THIS MAINLINE PROGRAM UTILIZES THE SUBROUTINES CCOEF9 AND CHECK9 *
CCCCC     *TO COMPUTE AND TEST THE COEFFICIENTS IN THE CHEBYSHEV EXPANSION  *
CCCCC     *OF THE PARTIAL DERIVATIVE WITH RESPECT TO U OF EXP(-A*X)*        *
CCCCC     *(X**-(U+1))*(INTEGRAL FROM 0 TO X OF EXP(A*T)*(T**U)*F(T)DT).    *
CCCCC     *                                                                *
CCCCC     *WHERE NO AMBIGUITY CAN OCCUR, WE SHALL DENOTE THIS FUNCTION BY   *
CCCCC     *INT3(X).                                                        *
CCCCC     *                                                                *
CCCCC     *DESCRIPTION OF VARIABLES.                                       *
CCCCC     *                                                                *
CCCCC     *FA     -INPUT  - PARAMETER A IN INT3(X).                        *
CCCCC     *                                                                *
CCCCC     *FU     -INPUT  - PARAMETER U IN INT3(X).                        *
CCCCC     *                                                                *
CCCCC     *FLAM   -INPUT  - THIS IS A PRESELECTED SCALE FACTOR SUCH THAT    *
CCCCC     *                 0.LE.(Z/W).LE.1.  SEE TEXT.                     *
CCCCC     *                                                                *
CCCCC     *N      -INPUT  - TWO LESS THAN THE NUMBER OF COEFFICIENTS TO BE  *
CCCCC     *                 GENERATED.  MAXIMUM VALUE IS 100.               *
CCCCC     *                                                                *
CCCCC     *G      -INPUT  - A VECTOR CONTAINING THE N+3 COEFFICIENTS IN THE *
CCCCC     *                 EXPANSION OF INT1 INTO SHIFTED CHEBYSHEV POLY-  *
CCCCC     *                 NOMIALS.  THESE ARE TO BE READ SEQUENTIALLY.    *
CCCCC     *                                                                *
CCCCC     *P      -OUTPUT - A VECTOR CONTAINING THE N+2 CHEBYSHEV COEFFI-   *
CCCCC     *                 CIENTS FOR THE APPROXIMATION TO INT3(X).        *
CCCCC     *                 DIMENSION OF THE VECTOR IS SET AT 103.          *
CCCCC     *                                                                *
CCCCC     *TEST1  -OUTPUT - THE RESULT OF THE FIRST TEST PERFORMED ON THE   *
CCCCC     *                 COEFFICIENTS.  SEE TEXT.                        *
CCCCC     *                                                                *
CCCCC     *TEST2  -OUTPUT - THE RESULT OF THE SECOND TEST PERFORMED ON THE  *
CCCCC     *                 COEFFICIENTS.  SEE TEXT.                        *
CCCCC     *                                                                *
CCCCC     *SUM    -OUTPUT - THE SUM OF THE COEFFICIENTS.                    *
CCCCC     *                                                                *
CCCCC     *ALL OTHER VARIABLES ARE FOR INTERNAL USE.                       *
CCCCC     *SUBROUTINES CCOEF9 AND CHECK9 ARE REQUIRED.                     *
CCCCC     *******************************************************************
          IMPLICIT REAL*16 (A-H,O-Z)
          DIMENSION A(103),G(103),P(103)
          REAL*16 C(103)/103*0.Q0/,D(103)/103*0.Q0/
C
C           ----------- THE NEXT PROGRAM CARD IS USED TO INITIALIZE ----------
C           ----------- DEVICE NUMBERS FOR THE PARTICULAR INSTALL- ----------
C           ----------- ATION OF THE OPERATING SYSTEM.  IF A SYSTEM ----------
C           ----------- HAS DIFFERENT DEVICE NUMBERS THIS CARD CAN ----------
C           ----------- BE CHANGED TO INITIALIZE THOSE VALUES AS   ----------
C           ----------- DEVICE NUMBERS.                            ----------
          DATA NREAD,NWRITE,NPUNCH,ZERO,ONE/5,6,7,0.Q0,1.Q0/
C
C           ----------- READ IN THE INPUT VARIABLES              ----------
          READ(NREAD,2) FA,FU,FLAM
          READ(NREAD,1) N
          N1=N+1
          N2=N+2
          N3=N+3
          READ(NREAD,2) (G(I),I=1,N3)
C
C           ----------- CALL SUBROUTINE TO COMPUTE THE COEFFICIENTS ----------
          CALL CCOEF9(FA,FU,FLAM,N,G,ZERO,A)
          IF(FA.EQ.ZERO) GO TO 9
          CALL CCOEF9(FA,FU,FLAM,N,D,ONE,C)
          Q=ONE
          SA=A(1)
          SB=G(1)
          SC=C(1)
          DO 3 I=2,N1
          Q=-Q
          SA=SA+Q*A(I)
          SB=SB+Q*G(I)
        3 SC=SC+Q*C(I)
          SC=SC-Q*C(N2)
          SB=SB-Q*(G(N2)-G(N3))
          FAC=-(SB/(FU+ONE)+SA)/SC
          GO TO 10
        9 FAC=ZERO
       10 SUM=ZERO
C
C           ----------- WRITE OUT VALUES OF PARAMETERS AND COEFFI- ----------
C           ----------- CIENTS                                    ----------
          WRITE(NWRITE,6) FA,FU,FLAM,N
          DO 4 I=1,N2
          I1=I-1
          P(I)=A(I)+FAC*C(I)
          SUM=SUM+P(I)
        4 WRITE(NWRITE,5) I1,P(I)
```

```
C
C             ---------- CALL SUBROUTINE TO TEST THE COEFFICIENT     ----------
C             ---------- VALUES                                      ----------
        CALL CHECK9(FA,FU,FLAM,N,G,P,TEST1,TEST2)
        WRITE(NWRITE,7) TEST1,TEST2
        WRITE(NWRITE,8) SUM
C
C             ---------- PUNCH THE VALUES OF COEFFICIENTS ON CARDS   ----------
        WRITE(NPUNCH,2) (P(K),K=1,N2)
      1 FORMAT(I5)
      2 FORMAT(Q39.32)
      5 FORMAT(1X,I5,8X,Q39.32)
      6 FORMAT('1',4X,'A=',Q39.32/5X,'U=',Q39.32/3X,'LAM=',Q39.32/5X,'N=',
       1I5/////5X,'K',25X,'P(K)'/)
      7 FORMAT(/1X,'TEST1=',Q39.32/1X,'TEST2=',Q39.32)
      8 FORMAT(/1X,'THE SUM OF THE COEFFICIENTS IS',Q39.32)
        STOP
        END
C
C             ----------- IBM S/370 ----------- MULTIPLE PRECISION -----------
C
C     ****************************************************************
C     *THIS SUBROUTINE, CCOEF9, COMPUTES MULTIPLE PRECISION VALUES OF *
C     *THE COEFFICIENTS IN THE CHEBYSHEV EXPANSION OF THE PARTIAL DERI*
C     *VATIVE WITH RESPECT TO U OF EXP(-A*X)*(X**-(U+1))*(INTEGRAL FROM*
C     *0 TO X OF EXP(A*T)*(T*U)*F(T)DT).                             *
C     *                                                              *
C     *WHERE NO AMBIGUITY CAN OCCUR, WE SHALL DENOTE THIS FUNCTION BY *
C     *INT3(X).                                                      *
C     *                                                              *
C     *                                                              *
C     *DESCRIPTION OF VARIABLES.                                     *
C     *                                                              *
C     *FA     -INPUT  - PARAMETER A IN INT3(X).                      *
C     *                                                              *
C     *FU     -INPUT  - PARAMETER U IN INT3(X).                      *
C     *                                                              *
C     *FLAM   -INPUT  - THIS IS A PRESELECTED SCALE FACTOR SUCH THAT  *
C     *                 0.LE.(Z/W).LE.1.  SEE TEXT.                  *
C     *                                                              *
C     *N      -INPUT  - TWO LESS THAN THE NUMBER OF COEFFICIENTS TO BE*
C     *                 GENERATED.  MAXIMUM VALUE IS 100.            *
C     *                                                              *
C     *VIN    -INPUT  - A VECTOR CONTAINING THE N+3 COEFFICIENTS IN THE*
C     *                 EXPANSION OF INT1 INTO SHIFTED CHEBYSHEV POLY-*
C     *                 NOMIALS.                                     *
C     *                                                              *
C     *START  -INPUT  - THE DESIRED VALUE OF VOUT(N+2).              *
C     *                                                              *
C     *VOUT   -OUTPUT - A VECTOR CONTAINING THE N+2 CHEBYSHEV COEFFI- *
C     *                 CIENTS FOR THE APPROXIMATION TO INT3(X).     *
C     *                                                              *
C     *ALL OTHER VARIABLES ARE FOR INTERNAL USE.                     *
C     ****************************************************************
        SUBROUTINE CCOEF9(FA,FU,FLAM,N,VIN,START,VOUT)
        IMPLICIT REAL*16 (A-H,O-Z)
        DIMENSION VIN(1),VOUT(1)
        DATA ZERO,ONE,TWO,FOUR/0.Q0,1.Q0,2.Q0,4.Q0/
        N2=N+2
        N1=N+1
        NCOUNT=N2
        VOUT(NCOUNT)=START
        Z=N2
        IF(FA.EQ.ZERO) GO TO 2
C
C             ---------- START COMPUTING COEFFICIENTS BY MEANS OF    ----------
C             ---------- BACKWARD RECURRENCE SCHEME                  ----------
        A3=ZERO
        A2=ZERO
        A1=START
        B2=VIN(N2+1)
        B1=VIN(N2)
        FMUL=FOUR/(FA*FLAM)
        F1=FU+ONE+Z
        F2=Z-FU
        DO 1 I=1,N1
        NCOUNT=NCOUNT-1
        F1=F1-ONE
        F2=F2-ONE
        VOUT(NCOUNT)=-(FMUL*F1+ONE)*A1-(FMUL*F2-ONE)*A2+A3+FMUL*(B2-B1)
        A3=A2
        A2=A1
        A1=VOUT(NCOUNT)
        B2=B1
      1 B1=VIN(NCOUNT)
        VOUT(1)=VOUT(1)/TWO
        RETURN
```

```
    2 B2=VIN(N2)
      B1=VIN(N1)
      F1=FU-Z
      F2=FU+Z+ONE
      DO 3 I=1,N
      NCOUNT=NCOUNT-1
      F1=F1+ONE
      F2=F2-ONE
      VOUT(NCOUNT)=(F1*VOUT(NCOUNT+1)+B2-B1)/F2
      B2=B1
    3 B1=VIN(NCOUNT-1)
      VOUT(1)=(FU*VOUT(2)-TWO*VIN(1)+VIN(2))/(TWO*(FU+ONE))
      RETURN
      END
C
C
C     ----------- IBM S/370 ----------- MULTIPLE PRECISION ------------
C
C     ***********************************************************************
C     *THIS SUBROUTINE, CHECK9, RUNS CHECKS ON THE COEFFICIENTS           *
C     *GENERATED BY SUBROUTINE CCOEF9.  EXCEPT FOR ROUND-OFF ERROR AND    *
C     *TRUNCATION ERROR, BOTH TEST RESULTS SHOULD BE ZERO.  SEE TEXT.     *
C     *                                                                   *
C     *WHERE NO AMBIGUITY CAN OCCUR, WE SHALL DENOTE THE PARTIAL DERI-    *
C     *VATIVE WITH RESPECT TO U OF EXP(-A*X)*(X**-(U+1))*(INTEGRAL FROM   *
C     *0 TO X OF EXP(A*T)*(T**U)*F(T)DT) BY INT3(X).                      *
C     *                                                                   *
C     *                                                                   *
C     *DESCRIPTION OF VARIABLES.                                          *
C     *                                                                   *
C     *FA    -INPUT  - PARAMETER A IN INT3(X).                            *
C     *                                                                   *
C     *FU    -INPUT  - PARAMETER U IN INT3(X).                            *
C     *                                                                   *
C     *FLAM  -INPUT  - THIS IS A PRESELECTED SCALE FACTOR SUCH THAT       *
C     *                0.LE.(Z/W).LE.1.  SEE TEXT.                        *
C     *                                                                   *
C     *N     -INPUT  - TWO LESS THAN THE NUMBER OF COEFFICIENTS           *
C     *                GENERATED.  MAXIMUM VALUE IS 100.                  *
C     *                                                                   *
C     *G     -INPUT  - A VECTOR CONTAINING THE N+3 COEFFICIENTS IN THE    *
C     *                EXPANSION OF INT1 INTO SHIFTED CHEBYSHEV POLY-     *
C     *                NOMIALS.                                           *
C     *                                                                   *
C     *P     -INPUT  - A VECTOR CONTAINING THE N+2 CHEBYSHEV COEFFI-      *
C     *                CIENTS FOR THE APPROXIMATION TO INT3(X).           *
C     *                                                                   *
C     *TEST1 -OUTPUT - THE RESULT OF THE FIRST TEST PERFORMED ON THE      *
C     *                COEFFICIENTS.  SEE TEXT.                           *
C     *                                                                   *
C     *TEST2 -OUTPUT - THE RESULT OF THE SECOND TEST PERFORMED ON THE     *
C     *                COEFFICIENTS.  SEE TEXT.                           *
C     *                                                                   *
C     *ALL OTHER VARIABLES ARE FOR INTERNAL USE.                         *
C     ***********************************************************************
      SUBROUTINE CHECK9(FA,FU,FLAM,N,G,P,TEST1,TEST2)
      IMPLICIT REAL*16 (A-H,O-Z)
      DIMENSION G(1),P(1)
      DATA ONE,TWO,THREE,FOUR,FIVE,ELEVEN,TWELVE/1.Q0,2.Q0,3.Q0,4.Q0,
     :5.Q0,11.Q0,12.Q0/
C
C     ----------- COMPUTATION OF LEFT SIDES OF TEST EQUATIONS ------------
      SB=G(1)-G(2)+G(3)
      SB1=-G(2)+FOUR*G(3)
      SB2=TWELVE*G(3)
      SG1=-P(2)+FOUR*P(3)
      SG2=TWELVE*P(3)
      Q=-ONE
      COUNT=TWO
      DO 1 I=2,N
      COUNT=COUNT+ONE
      CSQ=COUNT*COUNT
      SB=SB+Q*G(I+2)
      SB1=SB1+Q*CSQ*G(I+2)
      SB2=SB2+Q*CSQ*(CSQ-ONE)*G(I+2)
      SG1=SG1+Q*CSQ*P(I+2)
      SG2=SG2+Q*CSQ*(CSQ-ONE)*P(I+2)
    1 Q=-Q
C
C     ----------- COMPUTATION OF TEST VALUES                 ------------
      TEST1=SG1+(SB1+FA*FLAM*SB/(TWO*(FU+ONE)))/(FU+TWO)
      D0=SB/(FU+ONE)
      D1=-TWO*(FU+TWO)/FLAM*SB1+FA*SB
      D2=TWO*(FU+THREE)/(THREE*FLAM*FLAM)*SB2+FA/(FU+TWO)*(D1-FA*D0/
     :(FU+ONE))
      TEST2=SG2+THREE*FLAM*FLAM/(TWO*(FU+THREE)*(FU+THREE))*(D2-FA/
     :((FU+TWO)*(FU+TWO))*((TWO*FU+FIVE)*D1-FA*(ELEVEN+FU*(TWELVE+THREE*
     :FU))/((FU+ONE)*(FU+ONE))))
      RETURN
      END
```

```
   A=-0.10000000000000000000000000000000000000Q+01
   U= 0.0
 LAM= 0.80000000000000000000000000000000000000Q+01
   N=    39
```

K	P(K)
0	-0.2154426198834685123799501171866620Q+03
1	-0.3646728008667210105622981197722210Q+03
2	-0.2298571876829163392706120736691510Q+03
3	-0.1132925342933282579430721107097040Q+03
4	-0.4546296036371310953072383317952300Q+02
5	-0.1532182027975313439139233649312000Q+02
6	-0.4441200423198757580182432925689800Q+01
7	-0.1127924030230403093609430917294100Q+01
8	-0.2547062639252462208610476358812200Q+00
9	-0.5175449648278365831701459476470400Q-01
10	-0.9555581845554892813665469672811400Q-02
11	-0.1616286191203562284750535330360500Q-02
12	-0.2521958146389725851957709893143400Q-03
13	-0.3651666243720825450130438891471000Q-04
14	-0.4931812159202459401096398062079600Q-05
15	-0.6240664272417415047875125076208200Q-06
16	-0.7428106067695834830089942966678900Q-07
17	-0.8345816452300798383405792130856600Q-08
18	-0.8878927945505571410740775508096990Q-09
19	-0.8969585531593169253693521135610900Q-10
20	-0.8625914116357118186775375248316100Q-11
21	-0.7915119752115024400549233932090000Q-12
22	-0.6944476750974406856600551521527960Q-13
23	-0.5836922632760466161591248384314600Q-14
24	-0.4708205701545528712829583219021800Q-15
25	-0.3650553324263703510123283365523800Q-16
26	-0.2724866176104190916621692587954300Q-17
27	-0.1960734760949489220684757175829700Q-18
28	-0.1361887408961558636070102263972600Q-19
29	-0.9141881333728234686090537437837800Q-21
30	-0.5937325941505698088391848855742600Q-22
31	-0.3734779093755007733651936679857200Q-23
32	-0.2277651008942483194499507830092000Q-24
33	-0.1347911755487158550670861835986500Q-25
34	-0.7747615426510719552301036311034000Q-27
35	-0.4328782305903226510782366037095550Q-28
36	-0.2352847286091518453683074080349000Q-29
37	-0.1245001622497063741348453218948400Q-30
38	-0.6418842757401365759120251941352200Q-32
39	-0.3217814629544386769689723834907000Q-33
40	-0.1658903217016059926155942004600000Q-34

```
TEST1=-0.485642494776685392706595398002220Q-29
TEST2=-0.631088724176809444329382852226230Q-29
```

THE SUM OF THE COEFFICIENTS IS-0.98993697480374393470837373621675Q+03

XII CONVERSION OF A POWER SERIES INTO A SERIES OF CHEBYSHEV POLYNOMIALS OF THE FIRST KIND

Let

$$f(z) = \sum_{k=0}^{\infty} a_k z^k , \quad |z| < r ,$$ (1)

$$f(z) = \sum_{n=0}^{\infty} S_n(w) T_n^*(z/w) , \quad 0 \leq z/w \leq 1 .$$ (2)

Then

$$S_n(w) = \varepsilon_n (w/4)^n \sum_{k=0}^{\infty} \frac{(n+1)_k (n+\frac{1}{2})_k a_{k+n} w^k}{(2n+1)_k k!} ,$$ (3)

$$\varepsilon_0 = 1 , \quad \varepsilon_n = 2 \text{ if } n > 0 .$$ (4)

Now consider

$$z^\delta f(z^2) = \sum_{k=0}^{\infty} a_k z^{2k+\delta} , \quad |z| < r , \quad \delta = 0 \text{ or } \delta = 1 ,$$ (5)

$$z^\delta f(z^2) = \sum_{n=0}^{\infty} V_n(w) T_{2n+\delta}(z/w) , \quad -1 \leq z/w \leq 1 .$$ (6)

Then

$$V_n(w) = \{\varepsilon_n(1-\delta)+\delta\}(w^2/4)^n w^\delta \sum_{k=0}^{\infty} \frac{(n+\frac{1}{2}+\delta)_k (n+1)_k a_{k+n} w^{2k}}{(2n+1+\delta)_k k!} .$$ (7)

Notice that (1-3) and (5-7) are the same if in the former we replace z and w by z^2 and w^2, respectively, and if in the latter, we put $\delta = 0$. This follows since $T_n^*(x^2) = T_{2n}(x)$. Thus it is sufficient to have a machine program which pertains only to equations (5-7).

Next we consider the convergence of (3) and the error in $S_n(w)$ when only a finite number of the Taylor series coefficients are used in its determination. Once this information is at hand, the corresponding results for $V_n(w)$ are direct and we omit the details.

If the series for $f(w)$ converges, then likewise for the

series for $S_n(w)$. Indeed the series for $S_n(w)$ converges for $|w| < r$.

Suppose we approximate $S_n(w)$ by

$$S_{n,N}(w) = \varepsilon_n(w/4)^n \sum_{k=0}^{N-1} \frac{(n+1)_k (n+\frac{1}{2})_k a_{k+n} w^k}{(2n+1)_k k!} , \tag{8}$$

and let the error be given by

$$\delta_{n,N}(w) = S_n(w) - S_{n,N}(w) , \tag{9}$$

$$\delta_{n,N}(w) = q_{n,N} \sum_{k=0}^{\infty} \frac{u_k a_{k+n+N} w^k}{a_{n+N}} , \tag{10}$$

$$q_{n,N} = \frac{\varepsilon_n w^{n+N} (N+n)! \, \Gamma(N+\frac{1}{2}+n) a_{n+N}}{\Gamma(\frac{1}{2}) N! (2n+N)!} , \tag{11}$$

$$u_k = \frac{(n+N+1)_k (n+N+\frac{1}{2})_k}{(2n+N+1)_k (N+1)_k} . \tag{12}$$

Let n and N be fixed with N large and $N \gg n$. Then

$$\{u_{k+1}/u_k\} \le 1 \text{ if } n^2 \le \frac{1}{2}(n+N+k+1) , \tag{13}$$

$$q_{n,N} = \frac{\varepsilon_n w^{n+N} a_{n+N}}{(\pi N)^{\frac{1}{2}}} \{1 + O(N^{-1})\} . \tag{14}$$

Further, let

$$|a_{k+n+N+1} w / a_{k+n+N}| \le \rho < 1 \tag{15}$$

uniformly in n+N for all k. Then

$$|\delta_{n,N}(w)| \le \frac{\varepsilon_n |a_{n+N} w^{n+N}|}{(1-\rho)(\pi N)^{\frac{1}{2}}} = t_n . \tag{16}$$

For a numerical example let

$$f(z) = \{2/(2+z)\} = \sum_{n=0}^{\infty} b_n T_n^*(z) . \tag{17}$$

Then from 4(11),

$$b_n = 2\varepsilon_n (-)^n u^n / 6^{\frac{1}{2}} , \quad u = 5 - 2(6)^{\frac{1}{2}} . \tag{18}$$

For application of the theory of this chapter, we have

$$a_k = (-)^k/2^k , \quad w = 1, \quad \rho = \tfrac{1}{2}, \quad N = 36 . \tag{19}$$

The values of $S_{n,N}$ (w) are given at the end of the chapter. Using (16), we find

$$t_0 = 2.737(-12) , \quad t_4 = 0.342(-12) . \tag{20}$$

From (18) and the noted values of $S_{n,N}(w)$, we have

$$\delta_{0,36}(1) = 0.914(-12) , \quad \delta_{4,36}(1) = 1.124(-13) . \tag{21}$$

Thus the tabulated values are correct to about 12 decimals.

```
C        ----------- IBM S/370 ----------- MULTIPLE PRECISION -------------
C
C        *****************************************************************
C        *THIS SUBROUTINE CONVERTS A TAYLOR'S SERIES TO A SERIES OF CHEB- *
C        *YSHEV POLYNOMIALS.  MORE SPECIFICALLY, IT FINDS THE COEFFICIENTS *
C        *V(I) SUCH THAT (Z**DELTA)*F(Z**2)=SUM FROM I=0 TO N OF V (W)*   *
C        *                                                          I     *
C        *T                                                               *
C        * 2*I+DELTA    (Z/W), WHERE DELTA=0 OR 1.                        *
C        *                                                               *
C        *                                                               *
C        *DESCRIPTION OF VARIABLES.                                       *
C        *                                                               *
C        *N       -INPUT  - THE ORDER OF THE TAYLOR'S SERIES.             *
C        *                                                               *
C        *DELTA  -INPUT  - THE PARAMETER DELTA IN THE ABOVE EQUATION.     *
C        *                                                               *
C        *W       -INPUT  - THE PARAMETER W IN THE ABOVE EQUATION.        *
C        *                                                               *
C        *A       -INPUT  - A VECTOR CONTAINING THE N+1 COEFFICIENTS OF THE*
C        *                  TAYLOR'S SERIES.                              *
C        *                                                               *
C        *B       -INPUT  - THIS IS A VECTOR USED FOR COMPUTATION AND MUST *
C        *                  BE DIMENSIONED LIKE A.                        *
C        *                                                               *
C        *S       -OUTPUT - A VECTOR CONTAINING THE VALUES OF V(I).  V(I) IS*
C        *                  PLACED IN S(I+1).                             *
C        *                                                               *
C        *                                                               *
C        *ALL OTHER VARIABLES ARE FOR INTERNAL USE.                       *
C        *****************************************************************
         SUBROUTINE CONV(N,DELTA,W,A,B,S)
         IMPLICIT REAL*16 (A-H,O-Z)
         DIMENSION A(1),B(1),S(1)
         DATA ZERO,ONE,TWO,THREE/0.Q0,1.Q0,2.Q0,3.Q0/
         N1=N+1
         B(1)=(TWO-DELTA)*(ONE+(W-ONE)*DELTA)
         F1=DELTA-ONE/TWO
         F2=DELTA
         DO 1 I=2,N1
         F1=F1+ONE
         F2=F2+ONE
       1 B(I)=F1/F2*W*W*B(I-1)
         SUM=A(1)*B(1)
         DO 2 I=2,N1
       2 SUM=SUM+A(I)*B(I)
         S(1)=SUM/(TWO-DELTA)
         F1=ZERO
         F2=DELTA-ONE
         DO 4 I=2,N
         F1=F1+ONE
         F2=F2+ONE
         DO 5 J=I,N1
         COUNT=J
       5 B(J)=(COUNT-F1)/(COUNT+F2)*B(J)
         IUP=I+1
         SUM=A(I)*B(I)
         DO 6 K=IUP,N1
       6 SUM=SUM+A(K)*B(K)
       4 S(I)=SUM
         S(N1)=A(N1)*B(N1)/(F1+F2+THREE)
         RETURN
         END
```

THIS IS THE REPRESENTATION OF 2/(2+X) IN TERMS OF
SHIFTED CHEBYSHEV POLYNOMIALS OF THE FIRST KIND
BASED ON THE FIRST 35 TERMS OF THE TAYLOR'S SERIES.
THIS IS NOT THE TRUE EXPANSION.

```
 0      0.8164965809268128180846265646693Q+00
 1     -0.1649658092790369571929667555328Q+00
 2      0.1666493091551604665332040774174Q-01
 3     -0.1683499895675512465192335380619Q-02
 4      0.1700680241937419409498573756202Q-03
 5     -0.1718036032430919187003654386950Q-04
 6      0.1735568073816516237299736607107Q-05
 7     -0.1753285148449540008050631658332Q-06
 8      0.1771142278024458620840135522315Q-07
 9     -0.1789438042278142000139413308757Q-08
10      0.1806398120031480109196055645877Q-09
11     -0.1832054350137106809959129571471Q-10
12      0.1813024937182735286590947379564Q-11
13     -0.2016840336750008026288266121138Q-12
14      0.1182609513361590558598068548549Q-13
15     -0.4891997700640396922195748540993Q-14
16     -0.1002819320714037564155379048003Q-14
17     -0.4650153434412298439961486409306Q-15
18     -0.1527654528819667912588657358832Q-15
19     -0.5008986759387544919554386450868Q-16
20     -0.1487542240447722049562830188264Q-16
21     -0.4105414983287605902974109673787Q-17
22     -0.1038710323189951367206318715313Q-17
23     -0.2406583090901302207525388609693Q-18
24     -0.5071457470885545509704178601773Q-19
25     -0.9664341396039502641597054575219Q-20
26     -0.1652187903056979694345754601590Q-20
27     -0.2510689652517015968960276360841Q-21
28     -0.3351758599124261954327510537813Q-22
29     -0.3873241618483841196904975920300Q-23
30     -0.3799137356250156908362019771169Q-24
31     -0.3079392499240084052281439804134Q-25
32     -0.1983393530951928931756438545262Q-26
33     -0.9540286572516611521698092336388Q-28
34     -0.3056683600773142074597044819047Q-29
35     -0.4930380065763132378382330353301720Q-31
```

XIII   RATIONAL APPROXIMATIONS FOR $_2F_1(A,B;C;-Z)$

See Luke (1977b) for a short version of this chapter.   Let

$$E(z) = {}_2F_1(a,b;c;-z) \ . \tag{1}$$

The rational approximations given below follow from 1.5(1-6)
with a = 0, f = g = 0, $\alpha = \beta = 0$, p = 2, q = 1, α_1 = a, α_2 =
b and ρ_1 = c.   The a in 1.5(3-6) has nothing to do with the a
in (1).   The rational approximations are not of the Padé class.
We suppose that c $\neq$ a, c $\neq$ b.   If c = a, E(z) is the binomial
function $_1F_0(b;-z)$.   In this event, use the developments in
Chapter 14 with c = 1.   We have

$$E(z) = \{A_n(z)/B_n(z)\} + R_n(z) \ , \tag{2}$$

$$B_n(z) = L_n z^n {}_3F_2(-n,n+1,c; \ a+1,b+1;-1/z) \ , \tag{3}$$

$$A_n(z) = L_n z^n \sum_{k=0}^{n} \frac{(-n)_k(n+1)_k(a)_k(b)_k}{(a+1)_k(b+1)_k(k!)^2}$$

$$\times \ {}_4F_3 \left(\begin{array}{c} -n+k,n+1+k,c+k,1 \\ 1+k,a+1+k,b+1+k \end{array} \ \bigg| \ -\frac{1}{z} \right) \ , \tag{4}$$

$$L_n = \frac{(a+1)_n(b+1)_n}{(n+1)_n(c)_n} \ . \tag{5}$$

Here $R_n(z)$ is the remainder which we discuss later.

For the polynomials $B_n(z)$ and $A_n(z)$, we have

$$B_0(z) = 1, \ B_1(z) = 1 + \frac{(a+1)(b+1)z}{2c} \ ,$$

$$B_2(z) = 1 + \frac{(a+2)(b+2)z}{2(c+1)} + \frac{(a+1)_2(b+1)_2 z^2}{12(c)_2} \ ,$$

$$B_3(z) = 1 + \frac{(a+3)(b+3)z}{2(c+2)} + \frac{(a+2)_2(b+2)_2 z^2}{10(c+1)_2}$$

$$+ \frac{(a+1)_3(b+1)_3 z^3}{120(c)_3} \ ,$$

$$A_0(z) = 1, \ A_1(z) = B_1(z) - abz/c \ , \tag{6}$$

$$A_2(z) = B_2(z) - \frac{abz}{c}\left[1 + \frac{(a+2)(b+2)z}{2(c+1)}\right] + \frac{(a)_2(b)_2 z^2}{2(c)_2} ,$$

$$A_3(z) = B_3(z) - \frac{abz}{c}\left[1 + \frac{(a+3)(b+3)z}{2(c+2)} + \frac{(a+2)_2(b+2)_2 z^2}{10(c+1)_2}\right]$$

$$+ \frac{(a)_2(b)_2 z^2}{2(c)_2}\left[1 + \frac{(a+3)(b+3)z}{2(c+2)}\right] - \frac{(a)_3(b)_3 z^3}{6(c)_3} .$$

Both $A_n(z)$ and $B_n(z)$ satisfy the same recurrence formula

$$B_n(z) = (1 + F_1 z)B_{n-1}(z) + (E + F_2 z)zB_{n-2}(z)$$

$$+ F_3 z^3 B_{n-3}(z), \quad n \geq 3,$$

$$F_1 = \frac{3n^2 + (a+b-6)n + 2 - ab - 2(a+b)}{2(2n-3)(n+c-1)} ,$$

$$F_2 = -\frac{\{3n^2 - (a+b+6)n + 2 - ab\}(n+a-1)(n+b-1)}{4(2n-1)(2n-3)(n+c-2)_2} , \tag{7}$$

$$F_3 = \frac{(n+a-2)_2(n+b-2)_2(n-a-2)(n-b-2)}{8(2n-3)^2(2n-5)(n+c-3)_3} ,$$

$$E = -\frac{(n+a-1)(n+b-1)(n-c-1)}{2(2n-3)(n+c-2)_2} .$$

The recurrence formula is stable in the forward direction.

We now turn to the remainder. Several forms are presented according to the nature of the parameters and variable.
Although the representations are general, it is convenient to
make certain assumptions to simplify the discussion. If a
numerator parameter takes on a specialized form, then the numerator parameter is a. We assume that a is not a negative
integer for otherwise the $_2F_1$ is a polynomial. Similarly,
neither c-a nor c-b is allowed to be a negative integer, for
otherwise the $_2F_1$ is a polynomial in the variable z or z/(z+1)
except for a binomial multiplier in view of the Kummer relations 2.2(13-15). We forbid c=a for the reason given in the
discussion following 1.5(7). If c=a and we let a→∞, then (1)
becomes the binomial function $(1+z)^{-b}$ and the rational approx-

imations defined by (2)-(4) are the same as those of 14(1-5)
with c=1. Finally, if c and a are positive integers, then
c>a; and if b also is a positive integer, then c>b≥a.

The remainder is written in the form

$$R_n(z) = S_n(z)/B_n^*(z) \; , \; B_n^*(z) = z^{-n}L_n^{-1}B_n(z) \; . \tag{8}$$

We first consider $B_n^*(z)$. From the work of Luke (1969,1975a),
we have

$$B_n^*(z) = \frac{\Gamma(a+1)\Gamma(b+1)z^{\frac{1}{2}}N^{2\sigma}e^{N\zeta}}{2\Gamma(c)\Gamma(\frac{1}{2})(1+z)^{\sigma+\frac{1}{2}}}\left[1+O(N^{-1})\right] \; , \tag{9}$$

$$N^2 = n(n+1), \; 2\sigma = c - a - b - 3/2 \; ,$$
$$e^{-\zeta} = \{2 + z \mp 2(1 + z)^{\frac{1}{2}}\}/z \; , \tag{10}$$

and the sign is chosen so that $|e^{-\zeta}| < 1$. This is possible
for all z, z ≠ 1, $|\arg(1 + z)| < \pi$. (9) is used only for
$|\arg z| < \pi$. We also have

$$B_n^*(z) = \frac{(-)^n\Gamma(a+1)\Gamma(b+1)x^{\frac{1}{2}}N^{2\sigma}(-e^{-\zeta})^N}{2\Gamma(c)\Gamma(\frac{1}{2})(1-x)^{\sigma+\frac{1}{2}}}\left[1+O(N^{-1})\right] \; ,$$

$$z = -x, \; 0 < x < 1. \tag{11}$$

The expression

$$B_n^*(z) = \frac{\Gamma(a+1)\Gamma(b+1)}{2\Gamma(c)\Gamma(\frac{1}{2})(1+z)^{\sigma+\frac{1}{2}}} \frac{z^{n+1}(n+\frac{1}{2})^{2\sigma}}{(ze^{-\zeta})^{n+\frac{1}{2}}}\left[1+O(n^{-1})\right] \; ,$$

$$|\arg(1+z)| < \pi \tag{12}$$

is often most convenient. Frequently, it is desirable to have
the form for $e^{-\zeta}$ with z replaced by 1/z. In this case, we put

$$e^{-\xi} = 2z + 1 \mp 2(z^2 + z)^{\frac{1}{2}} \tag{13}$$

where the sign is chosen so that $|e^{-\xi}| < 1$ which is possible
for all z, z ≠ 1, $|\arg(1 + 1/z)| < \pi$. Tables of $|e^{-\zeta}|$ and
$|e^{-\xi}|$ for complex z are given in 1.9.

We now take up the structure of $S_n(z)$ which follows from the
works of Fields (1972,1976) and Luke (1969,1975a). We write

$$S_n(z) = F(z)M_n(z) + H(z)G_n(z) \; , \tag{14}$$

$$F(z) = - \frac{abz\Gamma(c)}{\Gamma(2-c)} \; _2F_1 \left(\begin{array}{c} a+1-c,b+1-c \\ \\ 2-c \end{array} \right| \left. -z \right) , \qquad (15)$$

or

$$F(z) = - \frac{abz\Gamma(c)(1+z)^{c-a-b}}{\Gamma(2-c)} \; _2F_1 \left(\begin{array}{c} 1-a,1-b \\ \\ 2-c \end{array} \right| \left. -z \right) , \qquad (16)$$

$$M_n(z) = \frac{\Gamma(n+1-c)}{\Gamma(n+1+c)} \; _3F_2 \left(\begin{array}{c} c,c-a,c-b \\ \\ c-n,n+1+c \end{array} \right| \left. -z \right) , \qquad (17)$$

$$H(z) = \frac{\Gamma(1-c)}{\Gamma(-a)\Gamma(-b)} \; _2F_1 \left(\begin{array}{c} a,b \\ \\ c \end{array} \right| \left. -z \right) , \qquad (18)$$

$$G_n(z) = \frac{n!\,\Gamma(n+1-a)\Gamma(n+1-b)z^{n+1}}{\Gamma(n+2-c)(2n+1)!} \; _3F_2 \left(\begin{array}{c} n+1,n+1-a,n+1-b \\ \\ 2n+2,n+2-c \end{array} \right| \left. -z \right). \quad (19)$$

We have need for asymptotic expressions for $M_n(z)$ and $G_n(z)$. For the latter, with all parameters and z fixed and $|\arg(1+z)| < \pi$,

$$G_n(z) = \frac{(\pi z)^{\frac{1}{2}} N^{2\sigma} e^{-N\zeta}}{(1+z)^{\sigma+\frac{1}{2}}} \; [1 + O(N^{-1})] \qquad (20)$$

where N, σ and $e^{-\zeta}$ are as in (10). Another form for (20) is

$$G_n(z) = \frac{(\pi z)^{\frac{1}{2}} (n+\frac{1}{2})^{2\sigma} e^{-(n+\frac{1}{2})\zeta}}{(1+z)^{\sigma+\frac{1}{2}}} \; [1 + O(n^{-1})] . \qquad (21)$$

Finally, if a, b, c and z are fixed and $|\arg(1+z)| < \pi$, then

$$M_n(z) = \frac{\Gamma(n+1-c)}{\Gamma(n+1+c)} \left[\; _3F_2^r \left(\begin{array}{c} c,c-a,c-b \\ c-n,n+c+1 \end{array} \right| \left. -z \right) + O(n^{-2r-2}) \right] + O(e^{-n\zeta}). \qquad (22)$$

If none of the numbers a, b or c is an integer, then from (14), (21) and (22), since $G_n(z)$ is subdominant to $M_n(z)$, we have with $|\arg(1+z)| < \pi$,

$$S_n(z) = \frac{F(z)\,(n+1-c)}{(n+1+c)} \left[\; _3F_2^r \left(\begin{array}{c} c,c-a,c-b \\ c-n,n+c+1 \end{array} \right| \left. -z \right) + O(n^{-2r-2}) \right]. \qquad (23)$$

If either a or b is a positive integer, and c is not a positive integer, then $H(z) = 0$, because of the factor $[\Gamma(-a)\Gamma(-b)]^{-1}$. In this event (23) is valid.

If c is a positive integer, but neither a nor b is a positive integer, it can be shown that

$$S_n(z) = \frac{(-z)^c \Gamma(a+1)\Gamma(b+1)\Gamma(n+1-c)E(z)}{\Gamma(a+1-c)\Gamma(b+1-c)\Gamma(n+1-c)}$$

$$\times \left[{_3F_2^r}\left(\begin{array}{c} c,c-a,c-b \\ c-n,n+c+1 \end{array} \middle| -z \right) + O(n^{-2r-2}) \right] , \qquad (24)$$

where $E(z)$ is defined by (1) and again $|\arg(1+z)| < \pi$.

Now suppose first that either a or b is a positive integer, but c is not a positive integer. If both a and b are positive integers, let $a \le b$. Then $H(z) = 0$ and it is convenient to put (14) in the form

$$S_n(z) = F^*(z)M_n^*(z), \qquad (25)$$

$$F^*(z) = -abz\Gamma(c)(1+z)^{c-a-b} \; {_2F_1^{a-1}}\left(\begin{array}{c} 1-a,1-b \\ 2-c \end{array} \middle| -z \right) , \qquad (26)$$

$$M_n^*(z) = \frac{\Gamma(n+1-c)\Gamma(c-n)}{\Gamma(n+1+c)\Gamma(2-c)}\left[\frac{1}{\Gamma(c-n)} \; {_3F_2}\left(\begin{array}{c} c,c-a,c-b \\ c-n,n+1+c \end{array} \middle| -z \right) \right]. \qquad (27)$$

Now let c approach a positive integer such that $c>b\ge a$. Then

$$\frac{\Gamma(c-n)}{\Gamma(2-c)} \rightarrow \frac{(-)^{n+1}\Gamma(c-1)}{\Gamma(n+1-c)} , \qquad (28)$$

$$\frac{1}{\Gamma(c-n)} \; {_3F_2}\left(\begin{array}{c} c,c-a,c-b \\ c-n,n+1+c \end{array} \middle| -z \right) \rightarrow \frac{(-)^{n+1-c}z^{-c}\Gamma(n+1+c)G_n(z)}{\Gamma(c)\Gamma(c-a)\Gamma(c-b)} . \qquad (29)$$

It follows that

$$S_n(z) = \frac{(-z)^{1-c}ab\Gamma(c-1)(1+z)^{c-a-b}}{\Gamma(c-a)\Gamma(c-b)} \; {_2F_1^{a-1}}\left(\begin{array}{c} 1-a,1-b \\ 2-c \end{array} \middle| -z \right) G_n(z). \qquad (30)$$

Under these same conditions, an alternative but less attractive form for $S_n(z)$ is

$$S_n(z) = \frac{(-z)^{1-c}ab\Gamma(c-1)}{\Gamma(c-a)\Gamma(c-b)}\left[{_2F_1^{c-1-a}}\left(\begin{array}{c} a+1-c,b+1-c \\ 2-c \end{array} \middle| -z \right) \right.$$

$$\left. + \frac{\Gamma(a)\Gamma(b)\Gamma(c-a)(-)^{c-a}z^{c-1}E(z)}{\Gamma(c-1)\Gamma(c)\Gamma(b+1-c)} \right] G_n(z). \qquad (31)$$

Asymptotic forms for $S_n(z)$ readily follow from the forms for $G_n(z)$. Asymptotic forms for $R_n(z)$ follow from the appropriate forms for $S_n(z)$ and use of (8) and (12). Thus if neither c nor a or if neither c nor both a and b are simultaneously positive integers, then

$$R_n(z) = W(n^{2\sigma+2c}e^{n\zeta})^{-1}\{1 + O(n^{-1})\} ,$$

$$|\arg(1 + z)| < \pi , \tag{32}$$

where W is free of n. Clearly

$$\lim_{n\to\infty} R_n(z) = 0, \quad |\arg(1 + z)| < \pi . \tag{33}$$

For a measure of the rate of convergence, we have

$$R_{n+1}(z)/R_n(z) = e^{-\zeta}[1 + O(n^{-1})] . \tag{34}$$

If c and a are both positive integers, c > a, and in the event that b is also a positive integer, $c > b \geq a$, then from (8), (12) and (21), we find

$$R_n(z) = \frac{2\pi\Gamma(c)\Gamma(c-1)(-z)^{1-c}(1+z)^{c-a-b}}{\Gamma(c-a)\Gamma(c-b)\Gamma(a)\Gamma(b)} \, _2F_1^{a-1}\left(\begin{array}{c}1-a,1-b\\2-c\end{array}\Bigg| -z\right)$$

$$\times e^{-(2n+1)\zeta}[1+O(n^{-1})] , \quad |\arg(1+z)| < \pi , \tag{35}$$

$$\lim_{n\to\infty} R_n(z) = 0, \quad |\arg(1+z)| < \pi , \tag{36}$$

$$R_{n+1}(z)/R_n(z) = e^{-2\zeta}[1+O(n^{-1})] . \tag{37}$$

It is of interest to compare the error $R_n(z)$ when a = 1 with the corresponding error, call it $R_{n,p}(z)$, for the Páde approximation in the next chapter. If c is not a positive integer, or if c is a positive integer but b is not a positive integer, then

$$R_{n,p}(z)/R_n(z) = O(n^{2\sigma+2c}e^{-n\zeta}) \tag{38}$$

whence the Padé approximation is superior.

Let us now return to (35). If z is small, $e^{-\zeta} \sim z/4$. So

$$R_n(z) = O(z^{2n+2-c}) . \tag{39}$$

Our rational approximation is a ratio of polynomials each of degree n. If we have a rational approximation of the Padé

class with numerator and denominator polynomials of degree n
+ 1 - c and n respectively, then the error would be as in
(39). Thus under the assumptions given for (35), our rational
approximations though not of the Padé class, are very much
akin to this class. Indeed, we find

$$R_{n,p}(z)/R_n(z) = (-)^c e^{-\zeta(c-1)}[1 + O(n^{-1})] . \qquad (40)$$

If a = 1, we forbid c = 1 for the reasons given in the discussion following 1.5(7). So c $\geq$ 2 and from (40), the Padé
approximation is superior.

Numerical Examples

1. Let

$$a = 0.8, \ b = 0.6, \ c = 1.5, \ z = 0.8 .$$

We compare values of $B_n^*(z)$, $B_n(z)$, $S_n(z)$ and $R_n(z)$ based on
(8) and (14) and the asymptotic representations (9) and (23)
with r = 1 and both without order terms with values determined
by the machine using the FORTRAN program delineated herein.
The latter values are called true. To simplify the discussion,
unless a quantity is labelled true, then it is derived from
the asymptotic representations noted.

n	$B_n^*(z)$	$B_n(z)$	True $B_n(z)$	$-S_n(z)$	$-R_n(z)$	True $-R_n(z)$
6	5.093(3)	8.776	8.733	7.046(-2)	1.383(-5)	1.393(-5)
7	2.868(4)	12.02	11.97	4.535(-2)	1.581(-6)	1.588(-5)
8	1.655(5)	16.46	16.41	3.092(-2)	1.868(-7)	1.874(-7)
9	9.733(5)	22.56	22.50	2.204(-2)	2.264(-8)	2.270(-8)
10	5.811(6)	30.92	30.85	1.626(-2)	2.798(-9)	2.804(-9)

2. Let

$$a = 0.8, \ b = 0.6, \ c = 2.0, \ z = 0.8 .$$

Again we compare asymptotic values with true values as in the
above example, where in the present instance we use (24) instead of (23) with r = 1 and without the order term. If n =
10, the asymptotic values for $B_n^*(z)$, $B_n(z)$, $S_n(z)$ and $R_n(z)$

are $1.440(7)$, 25.77, $1.822(-4)$ and $1.265(-13)$, respectively, while the true values of $B_n(z)$ and $R_n(z)$ are 26.71 and $1.226(-13)$, respectively.

3. Let

$a = 2.0$, $b = 3.0$, $c = 5.0$, $z = 2.0$.

Again we compare asymptotic values with true values as in the previous examples, but in the present case we estimate $R_n(z)$ using (35) without the order term. Thus for $n = 10$ and 20, we get the respective values $3.22(-11)$ and $1.17(-22)$. The respective values found from the machine program are $1.656(-11)$ and $0.862(-22)$. If we neglect the order term in (37), then for any n, $R_{n+1}(z)/R_n(z) = 0.718(-1)$. For $n = 10$ and 20, the values determined from the machine run are $0.772(-1)$ and $0.729(-1)$, respectively.

4. Let

$a = 1.5$, $b = 2.0$, $c = 4.0$, $z = 0.5$.

We estimate the error using (35) without the order term. When $n = 8$ and 10, we get $0.838(-14)$ and $0.873(-18)$, respectively. The corresponding values determined from the machine run are $0.530(-14)$ and $0.607(-18)$.

```
C       ...............................................................
C       .                                                             .
C       .        THIS MAINLINE PROGRAM ILLUSTRATES THE SYNTAX OF THE SUB- .
C       .   ROUTINE 'R2F1' FOR GENERATING VALUES OF THE NUMERATOR AND  .
C       .   DENOMINATOR POLYNOMIALS IN THE RATIONAL APPROXIMATION OF   .
C       .   2F1( AP , BP ; CP ; -Z ) .                                 .
C       ...............................................................
        IMPLICIT REAL*16(A-H,O-Z)
        DIMENSION A(26),B(26),R(26),D(26),E(26)
        DATA ZERO/0.Q0/
   10   READ(5,1,END=999) N,M
        READ(5,2) AP,BP,CP
        N1=N+1
        D(N1)=ZERO
        E(N1)=ZERO
        DO 100 I=1,M
          READ(5,2) Z
C------------------------------------------------------------------------
          CALL R2F1(AP,BP,CP,Z,A,B,N)
C------------------------------------------------------------------------
C       IN THE ABOVE :
C
C  AP AND BP   ARE THE NUMERATOR PARAMETERS OF THE 2F1
C  CP          IS THE DENOMINATOR PARAMETER OF THE 2F1
C  Z           IS THE VALUE OF THE ARGUMENT
C  A AND B     WILL CONTAIN THE VALUES OF THE NUMERATOR AND DENOMINATOR
C              POLYNOMIALS, RESPECTIVELY, FOR ALL DEGREES FROM 0 TO
C              N INCLUSIVE
C  N           IS THE MAXIMUM DEGREE FOR WHICH VALUES OF THE POLYNOMIALS
C              ARE TO BE CALCULATED
C
C       NOTE : VALUES OF THE K-TH DEGREE POLYNOMIALS WILL BE PLACED IN
C  A(K+1) AND B(K+1) RESPECTIVELY.
C------------------------------------------------------------------------
          R(N1)=A(N1)/B(N1)
          DO 50 J=1,N
            J1=N1-J
            R(J1)=A(J1)/B(J1)
            D(J1)=R(J1+1)-R(J1)
   50       E(J1)=R(N1)-R(J1)
          WRITE(6,3) N,AP,BP,CP,Z
          DO 60 J=1,N1
            J1=J-1
   60       WRITE(6,4) J1,A(J),B(J)
          WRITE(6,5)
          DO 70 J=1,N1
            J1=J-1
   70       WRITE(6,6) J1,R(J),D(J),E(J)
  100   CONTINUE
        GOTO 10
  999   STOP
    1   FORMAT(2I2)
    2   FORMAT(Q39.32)
    3   FORMAT('1','VALUES OF THE POLYNOMIALS IN THE RATIONAL APPROXIMATIO
       :N OF 2F1(AP,BP;CP;-Z)'//' ','N = ',I2,T20,'AP = ',Q39.32/' ',T20,
       :'BP = ',Q39.32//' ',T20,'CP = ',Q39.32//' ','      Z = ',Q39.32//
       :' ',' I',T24,'A(I)',T65,'B(I)'/)
    4   FORMAT(' ',I2,2X,Q39.32,2X,Q39.32)
    5   FORMAT('0','VALUES OF THE APPROXIMATION, 1ST DIFFERENCES AND APPRO
       :XIMATE ERRORS'//' ',' I',T12,'I-TH APPROXIMATION -- F(I)',T47,
       :'1ST DIFF''S.',T60,'F(N)-F(I)'/)
    6   FORMAT(' ',I2,2X,Q39.32,2X,Q10.3,2X,Q10.3)
        END
        SUBROUTINE R2F1(AP,BP,CP,Z,A,B,N)
C       ****************************************************************
C       *        THIS SUBROUTINE RETURNS VALUES A(I) AND B(I), I=1,...,N+1 *
C       *  OF THE NUMERATOR AND DENOMINATOR POLYNOMIALS IN THE RATIONAL  *
C       *  APPROXIMATION OF 2F1( AP , BP ; CP ; -Z )                   *
C       *                                                              *
C       *        NO OTHER SUBROUTINES ARE CALLED BY THIS ONE.          *
C       ****************************************************************
        IMPLICIT REAL*16(A-H,O-Z)
        DIMENSION A(1),B(1)
        DATA ONE/1.Q0/,TWO/2.Q0/,THREE/3.Q0/,FOUR/4.Q0/,SIX/6.Q0/,
       :HALF7/3.5Q0/,HALF3/1.5Q0/.FRTH3/0.75Q0/
```

```
C
C   INITIALIZATION :
C
      SABZ=(AP+BP)*Z
      AB=AP*BP
      ABZ=AB*Z
      ABZ1=Z+ABZ+SABZ
      ABZ2=ABZ1+SABZ+Z+Z+Z
      XI=THREE
      B(1)=ONE
      A(1)=ONE
      CP1=CP+ONE
      B(2)=ONE+ABZ1/(CP+CP)
      A(2)=B(2)-ABZ/CP
      CT1=CP1+CP1
      B(3)=ONE+ABZ2/CT1*(ONE+ABZ1/((-SIX)+CT1+CT1+CT1))
      A(3)=B(3)-ABZ/CP*(ONE+(ABZ2-ABZ1)/CT1)
      SABZ=SABZ/FOUR
      Z2=Z/TWO
      D1=(HALF7-AB)*Z2-SABZ
      D2=ABZ1/FOUR
      D3=D2-SABZ-SABZ
      D4=CP1+ONE
      D5=CP1*D4
      D6=CP*D5
      D7=HALF3
      D8=FRTH3
      D9=FRTH3*Z
C
C   FOR I=3,...,N , THE VALUES A(I+1) AND B(I+1) ARE CALCULATED
C   USING THE RECURRENCE RELATIONS BELOW.
C
      DO 100 I=3,N
C
C   CALCULATION OF THE MULTIPLIERS FOR THE RECURSION
C
      G3=D3/D8*D2/D6
      D2=D2+D9+SABZ
      D3=D3+D9-SABZ
      G3=G3*D2/D7
      G1=ONE+(D2+D1)/D7/D4
      G2=D2/D5/D7
      D8=D8+D7+D7
      D7=D7+ONE
      G2=G2*(CP1-XI-(D3+D1)/D7)
C-----------------------------------------------------------------
C   THE RECURRENCE RELATIONS FOR A(I+1) AND B(I+1) ARE AS FOLLOWS
C-----------------------------------------------------------------
C
      B(I+1)=G1*B(I)+G2*B(I-1)+G3*B(I-2)
      A(I+1)=G1*A(I)+G2*A(I-1)+G3*A(I-2)
C
      D9=D9+Z2
      D1=D1+D9+D9
      D6=D6+D5+D5+D5
      D5=D5+D4+D4
      D4=D4+ONE
  100 XI=XI+ONE
      RETURN
      END
```

VALUES OF THE POLYNOMIALS IN THE RATIONAL APPROXIMATION OF 2F1(AP,BP;CP;-Z)

N = 15   AP = 0.80000000000000000000000000000000000Q+00
        BP = 0.60000000000000000000000000000000000Q+00

        CP = 0.15000000000000000000000000000000000Q+01

  Z = 0.80000000000000000000000000000000000Q+00

I	A(I)	B(I)
0	0.10000000000000000000000000000000000Q+01	0.10000000000000000000000000000000000Q+01
1	0.15120000000000000000000000000000000Q+01	0.17680000000000000000000000000000000Q+01
2	0.20267648000000000000000000000000000Q+01	0.24629888000000000000000000000000000Q+01
3	0.27793473155657142857142857142857Q+01	0.33851005893485714285714285714286Q+01
4	0.38125855768609436734693877551020Q+01	0.46442737166717736054421768707483Q+01
5	0.52283361840981937389898989898989Q+01	0.63689492357696628536026936026935Q+01
6	0.71686640698506085451904895104894Q+01	0.87325909380120047006851282051281Q+01
7	0.98282944393655113505484656783215Q+01	0.11972452192313215030148012456876Q+02
8	0.13474072650678424083396888965586Q+02	0.16413603092244326991295912965045Q+02
9	0.18471762989352346740972298577828Q+02	0.22501595254349517344811320098459Q+02
10	0.25322724305588574795529125147622Q+02	0.30847174334598266844571842786682Q+02
11	0.34714234353232356944559856979948Q+02	0.42287552717300641093594613009363Q+02
12	0.47588422215064136268004683053801Q+02	0.57970396024961914852223494790017Q+02
13	0.65236780067557815161371080887525Q+02	0.79468950637206342496807997996671Q+02
14	0.89429717456501201871547030985462Q+02	0.10893986175728935488377113775259Q+03
15	0.12259415677642049949403598580130Q+03	0.14933951343378561165478260026070Q+03

VALUES OF THE APPROXIMATION, 1ST DIFFERENCES AND APPROXIMATE ERRORS

I	I-TH APPROXIMATION -- F(I)	1ST DIFF'S.	F(N)-F(I)
0	0.10000000000000000000000000000000000Q+01	-0.145Q+00	-0.179Q+00
1	0.85520361990950226244343891402715Q+00	-0.323Q-01	-0.343Q-01
2	0.82288835418171613285452215569051Q+00	-0.184Q-02	-0.198Q-02
3	0.82105309493936537065140617495423Q+00	-0.131Q-03	-0.144Q-03
4	0.82092180811279945470107636148100Q+00	-0.115Q-04	-0.128Q-04
5	0.82091032453744617671685600414576Q+00	-0.114Q-05	-0.128Q-05
6	0.82090918041817405190700980285576Q+00	-0.123Q-06	-0.139Q-06
7	0.82090905701637821127716258413638Q+00	-0.140Q-07	-0.159Q-07
8	0.82090904300902359137921864566204Q+00	-0.165Q-08	-0.187Q-08
9	0.82090904136149140923944307083607Q+00	-0.199Q-09	-0.227Q-09
10	0.82090904116253347504397182332365Q+00	-0.245Q-10	-0.280Q-10
11	0.82090904113801090819200409254136Q+00	-0.307Q-11	-0.352Q-11
12	0.82090904113493850691953033593299Q+00	-0.390Q-12	-0.447Q-12
13	0.82090904113454836810432009183813Q+00	-0.501Q-13	-0.566Q-13
14	0.82090904113449826865987733106322Q+00	-0.650Q-14	-0.650Q-14
15	0.82090904113449177360427967970090Q+00	0.0	0.0

VALUES OF THE POLYNOMIALS IN THE RATIONAL APPROXIMATION OF 2F1(AP,BP;CP;-Z)

N = 25   AP = 0.20000000000000000000000000000000000000Q+01
                      BP = 0.30000000000000000000000000000000000000Q+01

                      CP = 0.50000000000000000000000000000000000000Q+01

     Z = 0.20000000000000000000000000000000000000Q+01

I	A(I)	B(I)
0	0.10000000000000000000000000000000000000Q+01	0.10000000000000000000000000000000000000Q+0
1	0.10000000000000000000000000000000000000Q+01	0.34000000000000000000000000000000000000Q+0
2	0.14000000000000000000000000000000000000Q+01	0.70000000000000000000000000000000000000Q+0
3	0.31142857142857142857142857142857142857Q+01	0.13285714285714285714285714285714285714Q+0
4	0.75000000000000000000000000000000000000Q+01	0.24750000000000000000000000000000000000Q+0
5	0.10659259259259259259259259259259259259Q+02	0.45888888888888888888888888888888888888Q+0
6	0.19744242424242424242424242424242424259Q+02	0.84999999999999999999999999999999999999Q+0
7	0.36578978597160415342233524051705Q+02	0.15747425301970756516211061665607Q+0
8	0.67780380730380730380730380730380606Q+02	0.29189860139860139860139860139860Q+0
9	0.12575699109789155038023816304535930Q+03	0.54140236686390532544378698224851Q+0
10	0.23333290451555051347001811399809620Q+03	0.10047674956622325043377674956622Q+0
11	0.43337898374071748684751780727012Q+03	0.18657170278637770897832817337461Q+0
12	0.80511335685044528765083348472750Q+03	0.34660509994615695248351056669807Q+0
13	0.14963399446965373394391629764928Q+04	0.64418144679435914896312543054641Q+0
14	0.27820513337137396005794623623070Q+04	0.11976862608496233479407527058624Q+0
15	0.51741834775844802182456233058435Q+04	0.22275104657920347305571590935833Q+0
16	0.96259414646374371954506085982239Q+04	0.41440133401669284203399149321772Q+0
17	0.17912432163168711382078217390128Q+05	0.77113867886793930223828178076856Q+0
18	0.33339812856784071688878967487450Q+05	0.14352947163115323170193095934037Q+0
19	0.62066661848681052766996440189289Q+05	0.26719972891304278079333749254748Q+0
20	0.11556595469731796138248513624945Q+06	0.49751690231460455134893847791414Q+0
21	0.21521398047800833421061659148537Q+06	0.92550636757729661367850141284403Q+0
22	0.40084152906935623574400220954603Q+06	0.17256417461659136865072461347790Q+0
23	0.74667297770819225571986086819428Q+06	0.32144642493598237031782550205890Q+0
24	0.13910365654853060890190955028980Q+07	0.59884743030108454782694133291043Q+0
25	0.25917401933553475541729619744415Q+07	0.11157564145769409535710605908141Q+0

VALUES OF THE APPROXIMATION, 1ST DIFFERENCES AND APPROXIMATE ERRORS

I	I-TH APPROXIMATION -- F(I)	1ST DIFF'S.	F(N)-F(I)
0	0.10000000000000000000000000000000000Q+01	-0.706Q+00	-0.768Q+00
1	0.29411764705882352941176470588235Q+00	-0.941Q-01	-0.618Q-01
2	0.20000000000000000000000000000000Q+00	0.344Q-01	0.323Q-01
3	0.23440860215053763440860215053763Q+00	-0.209Q-02	-0.212Q-02
4	0.23232323232323232323232323232323Q+00	-0.391Q-04	-0.377Q-04
5	0.23228410008071025020177562550044Q+00	0.110Q-05	-0.138Q-05
6	0.23228520499108734402852049910873Q+00	0.251Q-06	0.280Q-06
7	0.23228545553146795743896350661601Q+00	0.264Q-07	0.299Q-07
8	0.23228548194110045025800329275288Q+00	0.234Q-08	0.255Q-08
9	0.23228548428294610908119933488171Q+00	0.193Q-09	0.209Q-09
10	0.23228548447586866370105574794708Q+00	0.153Q-10	0.166Q-10
11	0.23228548449114550909592814480518Q+00	0.118Q-11	0.128Q-11
12	0.23228548449232689090673928636202Q+00	0.899Q-13	0.972Q-13
13	0.23228548449242168060655891878337Q+00	0.677Q-14	0.731Q-14
14	0.23228548449242354052220761211Q+00	0.505Q-15	0.546Q-15
15	0.23228548449242407967868562520903Q+00	0.375Q-16	0.299Q-17
16	0.23228548449242411718463452032803Q+00	0.277Q-17	0.299Q-17
17	0.23228548449242411995496603163258Q+00	0.204Q-18	0.220Q-18
18	0.23228548449242412015882589058149Q+00	0.150Q-19	0.161Q-19
19	0.23228548449242412017378180283168Q+00	0.109Q-20	0.118Q-20
20	0.23228548449242412014876291827790Q+00	0.999Q-22	0.862Q-22
21	0.23228548449242412014956221686120Q+00	0.583Q-23	0.628Q-23
22	0.23228548449242412014960204873135Q+00	0.424Q-24	0.457Q-24
23	0.23228548449242412014962479042420Q+00	0.308Q-25	0.331Q-25
24	0.23228548449242412014962503742090Q+00	0.224Q-26	0.224Q-26
25	0.23228548449242412014962505981570Q+00	0.0	0.0

VALUES OF THE POLYNOMIALS IN THE RATIONAL APPROXIMATION OF 2F1(AP,BP;CP;-Z)

N = 12

AP = 0.15000000000000000000000000000Q+01
BP = 0.20000000000000000000000000000Q+01

CP = 0.40000000000000000000000000000Q+00

Z = 0.50000000000000000000000000000Q+00

I	A(I)	B(I)
0	0.10000000000000000000000000000Q+01	0.10000000000000000000000000000Q+01
1	0.10937500000000000000000000000Q+01	0.14687500000000000000000000000Q+01
2	0.13125000000000000000000000000Q+01	0.18093750000000000000000000000Q+01
3	0.16132812500000000000000000000Q+01	0.22205078125000000000000000000Q+01
4	0.19828404017857142857142857Q+01	0.27292181521045918367346938775Q+01
5	0.24405568440755208333333333Q+01	0.33552287699381510416666666666Q+01
6	0.30071286288174715909090909Q+01	0.41390689503062855113636363636Q+01
7	0.37081487068763146035384615846Q+01	0.51039662984701033353536346363Q+01
8	0.45752990072423761541193181818Q+01	0.62975284392184772283806818181Q+01
9	0.56477963543115540200626148897058Q+01	0.77737341546179617152494542738970Q+01
10	0.69741725018361078100165857477225Q+01	0.95993834721633575576905779510361Q+01
11	0.86144761557985401275206585319673Q+01	0.11857128571102401570162481912121Q+02
12	0.10642984999096790409606436024540Q+02	0.14649225247496329045490078304125Q+02

VALUES OF THE APPROXIMATION, 1ST DIFFERENCES AND APPROXIMATE ERRORS

I	I-TH APPROXIMATION -- F(I)	1ST DIFF'S.	F(N)-F(I)
0	0.10000000000000000000000000000Q+01	-0.255Q+00	-0.273Q+00
1	0.74446808510632978728340425531490Q+00	-0.193Q-01	-0.182Q-01
2	0.72538860103626943005181347150259Q+00	-0.115Q-02	-0.113Q-02
3	0.72653707450083560559415955668924Q+00	-0.138Q-04	-0.141Q-04
4	0.72652323532901883106806751070258Q+00	-0.271Q-06	-0.275Q-06
5	0.72652296441258911993882701662310Q+00	-0.359Q-08	-0.363Q-08
6	0.72652296080206812148748344451910Q+00	-0.426Q-10	-0.430Q-10
7	0.72652296077813053801925876158867Q+00	-0.479Q-12	-0.484Q-12
8	0.72652296077761443337458637301660Q+00	-0.525Q-14	-0.530Q-14
9	0.72652296077764619746353264644Q+00	-0.565Q-16	-0.571Q-16
10	0.72652296077764614101608580699994Q+00	-0.601Q-18	-0.607Q-18
11	0.72652296077764614040897400747330Q+00	-0.634Q-20	-0.634Q-20
12	0.72652296077764614040897400747330Q+00	0.0	0.0

171

```
C     ................................................................
C     .         THIS MAINLINE PROGRAM ILLUSTRATES THE SYNTAX OF THE SUB- .
C     .  ROUTINE 'C2F1' WHEN USED TO GENERATE COEFFICIENTS IN THE POLY- .
C     .  NOMIALS FOR THE RATIONAL APPROXIMATION OF 2F1(AP,BP;CP;-Z).     .
C     ................................................................
      IMPLICIT REAL*16(A-H,O-Z)
      DIMENSION CA(26),CB(26),NO(25)
   10 READ(5,2,END=999) AP,BP,CP
      WRITE(6,3) AP,BP,CP
      READ(5,1) M,(NO(J),J=1,M)
      DO 100 I=1,M
      N=NO(I)
C-----------------------------------------------------------------------
      CALL C2F1(AP,BP,CP,CA,CB,N)
C-----------------------------------------------------------------------
C     IN THE ABOVE:
C
C  AP AND BP   ARE THE NUMERATOR PARAMETERS OF THE 2F1
C  CP          IS THE DENOMINATOR PARAMETER OF THE 2F1
C  N           IS THE DEGREE OF THE POLYNOMIALS IN THE RATIONAL
C                 APPROXIMATION
C  CA AND CB   WILL CONTAIN THE COEFFICIENTS IN THE NUMERATOR AND
C                 DENOMINATOR POLYNOMIALS, RESPECTIVELY
C
C     NOTE : THE COEFFICIENTS OF THE K-TH POWER OF Z WILL BE PLACED
C IN CA(K+1) OR CB(K+1) AS APPROPRIATE
C-----------------------------------------------------------------------
      N1=N+1
  100 WRITE(6,4) N,(CA(J),CB(J),J=1,N1)
      GOTO 10
  999 STOP
    1 FORMAT(26I2)
    2 FORMAT(Q39.32)
    3 FORMAT('1','COEFFICIENTS FOR THE RATIONAL APPROXIMATION OF 2F1( AP
     : , BP ; CP ; -Z )'//' ',T20,'AP = ',Q39.32/' ',T20,'BP = ',Q39.32/
     :' ',T20,'CP = ',Q39.32/)
    4 FORMAT(' ','N = ',I2,T18,'CA(I)',T58,'CB(I)'//
     :26(1X,Q39.32,2X,Q39.32/)/)
      END
      SUBROUTINE C2F1(AP,BP,CP,A,B,N)
C     ****************************************************************
C     *        THIS SUBROUTINE RETURNS COEFFICIENTS A(I) AND B(I) ,   *
C     *  I = 1,2,...,N+1 , OF THE NUMERATOR AND DENOMINATOR POLYNOMIALS *
C     *  RESPECTIVELY, IN THE RATIONAL APPROXIMATION OF ORDER  N  FOR  *
C     *  2F1( AP , BP ; CP ; -Z ).                                    *
C     *                                                               *
C     *        NO OTHER SUBROUTINES ARE CALLED BY THIS ONE.           *
C     ****************************************************************
      IMPLICIT REAL*16(A-H,O-Z)
      DIMENSION A(1),B(1)
      DATA ONE/1.Q0/,ZERO/0.Q0/
C
C  INITIALIZATION :
C
      B(1)=ONE
      A(1)=ONE
      XI=ONE
      CP1=CP-ONE
      XIJ=ZERO
      XN=N
      XN1I=XN
      DO 100 I=1,N
      I1=I+1
C-----------------------------------------------------------------------
C  FOR I = 1,2,...,N , B(I+1) IS CALCULATED AS FOLLOWS
C-----------------------------------------------------------------------
C
      B(I1)=(AP+XN1I)/(CP1+XN1I)*(BP+XN1I)/(XN+XN1I)*XN1I/XI*B(I)
C
      AI1=ONE
      DO 50 J=1,I
C-----------------------------------------------------------------------
C  TO CALCULATE A(I+1), WE EMPLOY B(J), J = 1,2,...,I+1 AS FOLLOWS
C-----------------------------------------------------------------------
C
      AI1=B(J+1)-(AP+XIJ)/(CP+XIJ)*(BP+XIJ)/(ONE+XIJ)*AI1
C
   50 XIJ=XIJ-ONE
C
      A(I1)=AI1
      XIJ=XI
      XN1I=XN-XI
  100 XI=XI+ONE
      RETURN
      END
```

COEFFICIENTS FOR THE RATIONAL APPROXIMATION OF 2F1(AP   BP ; CP ; -Z)

AP = 0.15000000000000000000000000000000Q+01
BP = 0.20000000000000000000000000000000Q+01

CP = 0.40000000000000000000000000000000Q+01

N = 3   CA(I)   CB(I)

0.10000000000000000000000000000Q+01   0.10000000000000000000000000000Q+01
0.11250000000000000000000000000Q+01   0.18750000000000000000000000000Q+01
0.20625000000000000000000000000Q+00   0.10500000000000000000000000000Q+01
-0.62499999999999999999999999997Q-02   0.16406250000000000000000000000Q+00

N = 4   CA(I)   CB(I)

0.10000000000000000000000000000Q+01   0.10000000000000000000000000000000Q+01
0.16071428571428571428571429Q+01   0.23571428571428571428571429Q+01
0.68877551020408163265306122448979Q+00   0.18941326530612244897959183673469Q+01
0.57079081632653061224489795918369Q-01   0.58928571428571428571428571Q+00
-0.95663265306122448979591836734702Q-03   0.55245535714285714285714285Q-01

N = 5   CA(I)   CB(I)

0.10000000000000000000000000000Q+01   0.10000000000000000000000000000000Q+01
0.20937500000000000000000000000Q+01   0.28437500000000000000000000000Q+01
0.14088541666666666666666667Q+01   0.29791666666666666666666667Q+01
0.32421875000000000000000000000Q+00   0.13964843750000000000000000Q+01
0.15136718749999999999999999Q-01   0.27929687500000000000000000Q+00
-0.16276041666666666666666399Q-03   0.17456054687500000000000000Q-01

N = 6   CA(I)   CB(I)

0.10000000000000000000000000000Q+01   0.10000000000000000000000000000000Q+01
0.25833333333333333333333333Q+01   0.33333333333333333333333333Q+01
0.23712121212121212121212121Q+01   0.43087121212121212121212121Q+01
0.91429924242424242424242424Q+00   0.27083333333333333333333333Q+01
0.13198390151515151515151515Q+00   0.84635416666666666666666666Q+00
0.39338428030303030303030303Q-02   0.11848958333333333333333333Q+00
-0.29592803030303030303030845Q-04   0.52897135416666666666666666Q-02

173

XIV PADE APPROXIMATIONS FOR $_2F_1(1,B;C;-Z)$

Let

$$E(z) = {}_2F_1(1,b;c;-z) . \qquad (1)$$

In 1.5(1-6), put $p = 2$, $q = 1$, $\alpha_1 = 1$, $\alpha_2 = b$, $\rho_1 = c$, $\alpha = 0$, $\beta = c - 1$, $f = g = 1$, $c_1 = 2 - a$, $d_1 = c - a$, $a = 0$ or $a = 1$. Then these rational approximations occupy the $(n,n-a)$ positions of the Padé table. The rational approximations which follow are the case $a = 0$. Thus

$$E(z) = \{A_n(z)/B_n(z)\} + R_n(z) , \qquad (2)$$

$$B_n(z) = L_n z^n {}_2F_1(-n,n+c;b+1;-1/z) , \qquad (3)$$

$$A_n(z) = L_n z^n \sum_{k=0}^{n} \frac{(-n)_k (n+c)_k (b)_k}{(c)_k (b+1)_k k!} \; {}_3F_2\left(\begin{array}{c} -n+k,n+c+k,1 \\ b+1+k,1+k \end{array} \middle| -1/z \right), \qquad (4)$$

$$L_n = \{(b+1)_n/(n+c)_n\} , \qquad (5)$$

where $R_n(z)$ is the remainder which is discussed later. We also defer discussion of the case $c = 1$, in which event (1) reduces to the binomial function.

For the polynomials $A_n(z)$ and $B_n(z)$, we have

$$B_0(z) = 1, \quad B_1(z) = 1 + \frac{(b+1)z}{c+1} , \quad B_2(z) = 1 + \frac{2(b+2)z}{c+3} + \frac{(b+1)_2 z^2}{(c+2)_2} ,$$

$$B_3(z) = 1 + \frac{3(b+3)z}{c+5} + \frac{3(b+2)_2 z^2}{(c+4)_2} + \frac{(b+1)_3 z^3}{(c+3)_3} , \qquad (6)$$

$$A_0(z) = 1, \quad A_1(z) = B_1(z) - bz/c ,$$

$$A_2(z) = B_2(z) - \frac{bz}{c}\left[1 + \frac{2(b+2)z}{c+3}\right] + \frac{(b)_2 z^2}{(c)_2} ,$$

$$A_3(z) = B_3(z) - \frac{bz}{c}\left[1 + \frac{3(b+3)z}{c+5} + \frac{3(b+2)_2 z^2}{(c+4)_2}\right]$$

$$+ \frac{(b)_2 z^2}{(c)_2}\left[1 + \frac{3(b+3)z}{c+5}\right] - \frac{(b)_3 z^3}{(c)_3} .$$

Both $A_n(z)$ and $B_n(z)$ satisfy the same recursion formula

$$B_n(z) = (1 + F_1 z)B_{n-1}(z) + F_2 z^2 B_{n-2}(z) , n \geq 2 ,$$

$$F_1 = \frac{2n^2 + 2n(c-2) + (b-1)(c-1)}{(2n+c-1)(2n+c-3)} , \tag{7}$$

$$F_2 = - \frac{(n-1)(n-2+c-b)(n+b-1)(n+c-2)}{(2n+c-2)(2n+c-3)^2(2n+c-4)} .$$

This recurrence formula is stable in the forward direction.

To discuss the remainder, put

$$e^{-\zeta} = \{z +2 \mp 2(z + 1)^{\frac{1}{2}}\}/z \tag{8}$$

where the sign is chosen so that $|e^{-\zeta}| < 1$. This is possible for all z except $z \leq -1$. On the latter segment $|e^{-\zeta}| = 1$. Also set

$$(1 - e^{-\zeta}) = (e^{\zeta} - 1)e^{\pm i\pi} \tag{9}$$

where upper(lower) sign is taken when $I(z) >(<) 0$. Then

$$R_n(z) = - \frac{b\pi z^{-b}(c-b)_n(b+1)_n(1+e^{-\zeta})^{2(c-b-1)}}{2^{2c-2b-3}(c)_n n!} G_n(\zeta) , \tag{10}$$

$$G_n(\zeta) = \frac{\exp\{-(2n+b+1)\zeta\}\{1+O(n^{-1})\}}{1+\exp\{\mp i\pi(b+\frac{1}{2})-(2n+c)\zeta\}} ,$$

or

$$R_n(z) = - \frac{b\pi z^{-b}\Gamma(c)(1+e^{-\zeta})^{2(c-b-1)}}{2^{2c-2b-3}\Gamma(c-b)\Gamma(b+1)} G_n(\zeta) . \tag{11}$$

In (10) and (11), take the upper(lower) sign when $I(z)>(<) 0$. It follows that for z fixed,

$$\lim_{n\to\infty} R_n(z) = 0, \ |\arg(1+z)| < \pi . \tag{12}$$

In some considerations, we desire rational approximations for $E(1/z)$, see (1). The appropriate forms for this situation are easily recovered in view of the remarks following (1) and 1.5(13-21). For the error, in (10) and (11), z must be replaced by its reciprocal, whence $e^{-\zeta}$ must be replaced by

$$e^{-\xi} = 2z + 1 \mp 2(z^2 + z)^{\frac{1}{2}} \tag{13}$$

where the sign is chosen so that $|e^{-\xi}| < 1$ which is possible for all z except $-1 \leq z \leq 0$. On the latter segment $|e^{-\xi}| = 1$.

Also

$$(1 - e^{\xi}) = (e^{\xi} - 1)e^{\mp i\pi} \tag{14}$$

where the upper(lower) sign is taken when $I(z)>(<) 0$. Then
in place of (10) and (11), we have

$$R_n(1/z) = - \frac{b\pi z^b(c-b)_n(b+1)_n(1+e^{-\xi})^{2(c-b-1)}}{2^{2c-2b-3}(c)_n n!} H_n(\xi) , \tag{15}$$

$$H_n(\xi) = \frac{\exp\{-(2n+b+1)\xi\}\{1+O(n^{-1})\}}{1+\exp\{\pm i\pi(b+\frac{1}{2})-(2n+c)\xi\}} ,$$

$$R_n(1/z) = - \frac{b\pi z^b\Gamma(c)(1+e^{-\xi})^{2(c-b-1)}}{2^{2c-2b-3}\Gamma(c-b)\Gamma(b+1)} H_n(\xi) . \tag{16}$$

In (15) and (16), take the upper(lower) sign when $I(z)>(<) 0$.
For z fixed,

$$\lim_{n\to\infty} R_n(1/z) = 0, \; |\arg(1+1/z)| < \pi . \tag{17}$$

A measure of the rate of convergence of the approximants
for $E(z)$ is the ratio

$$R_{n+1}(z)/R_n(z) = e^{-2\zeta}\{1+O(n^{-1})\} . \tag{18}$$

Similarly,

$$R_{n+1}(1/z)/R_n(1/z) = e^{-2\xi}\{1+O(n^{-1})\} . \tag{19}$$

To facilitate evaluation of the error, values of $|e^{-\zeta}|$ and
$|e^{-\xi}|$ for complex z are tabulated in 1.8.

If $c = 1$, then (1) reduces to the binomial function
$(1+z)^{-b}$. In this event, if $b = \pm\frac{1}{2}$, it can be shown that (10)
is exact without the order term. See Luke (1969,1975a).

For a numerical example, let $b = 3/2$, $c = 3$, $z = \frac{1}{2}$. Then
$e^{-\zeta} = 1.02051(-2)$. Let t_n be the right hand side of (11) with
$O(n^{-1})$ neglected and the denominator of $G_n(\zeta)$ replaced by 1.
Then $t_n = -(0.80883)(1.02051)^n \cdot 10^{-2n-1}$. If $n = 5$, the true
error is $-1.08(-11)$ whereas $t_5 = -0.895(-11)$. The corresponding values for $n = 7$ are $-1.08(-15)$ and $-0.932(-15)$.

```
C     ...............................................................
C     .             THIS MAINLINE PROGRAM ILLUSTRATES THE SYNTAX OF THE SUB-  .
C     . ROUTINE 'R2F1P' FOR GENERATING VALUES OF THE NUMERATOR AND            .
C     . DENOMINATOR POLYNOMIALS IN THE PADE APPROXIMATION OF                  .
C     . _2F1( 1 , BP ; CP ; -Z ) .                                           .
C     ...............................................................
      IMPLICIT REAL*16(A-H,O-Z)
      DIMENSION A(26),B(26),R(26),D(26),E(26)
      DATA ZERO/0.Q0/
 10   READ(5,1,END=999) N,M
      READ(5,2) BP,CP
      N1=N+1
      D(N1)=ZERO
      E(N1)=ZERO
      DO 100 I=1,M
        READ(5,2) Z
C-------------------------------------------------------------------
      CALL R2F1P(BP,CP,Z,A,B,N)
C-------------------------------------------------------------------
C     IN THE ABOVE :
C
C  BP         IS THE NUMERATOR PARAMETER OF THE 2F1
C  CP         IS THE DENOMINATOR PARAMETER OF THE 2F1
C  Z          IS THE VALUE OF THE ARGUMENT
C  A AND B    WILL CONTAIN THE VALUES OF THE NUMERATOR AND DENOMINATOR
C             POLYNOMIALS, RESPECTIVELY, FOR ALL DEGREES FROM 0 TO
C             N INCLUSIVE
C  N          IS THE MAXIMUM DEGREE FOR WHICH VALUES OF THE POLYNOMIALS
C             ARE TO BE CALCULATED
C
C     NOTE : VALUES OF THE K-TH DEGREE POLYNOMIALS WILL BE PLACED IN
C A(K+1) AND B(K+1) RESPECTIVELY.
C-------------------------------------------------------------------
      R(N1)=A(N1)/B(N1)
      DO 50 J=1,N
        J1=N1-J
        R(J1)=A(J1)/B(J1)
        D(J1)=R(J1+1)-R(J1)
 50     E(J1)=R(N1)-R(J1)
      WRITE(6,3) N,BP,CP,Z
      DO 60 J=1,N1
        J1=J-1
 60     WRITE(6,4) J1,A(J),B(J)
      WRITE(6,5)
      DO 70 J=1,N1
        J1=J-1
 70     WRITE(6,6) J1,R(J),D(J),E(J)
100     CONTINUE
      GOTO 10
999   STOP
1     FORMAT(2I2)
2     FORMAT(Q39.32)
3     FORMAT('1','VALUES OF THE POLYNOMIALS IN THE PADE APPROXIMATION OF
     : 2F1(1,BP;CP;-Z)'//' ','N = ',I2,T20,'BP = ',Q39.32//' ',T20,
     :'CP = ',Q39.32//' ','    Z = ',Q39.32//' ',' I',T24,'A(I)',T65,
     :'B(I)'/)
4     FORMAT(' ',I2,2X,Q39.32,2X,Q39.32)
5     FORMAT('0','VALUES OF THE APPROXIMATION, 1ST DIFFERENCES AND APPRO
     :XIMATE ERRORS'//' ',' I',T12,'I-TH APPROXIMATION -- F(I)',
     :T47,'1ST DIFF''S.',T60,'F(N)-F(I)'/)
6     FORMAT(' ',I2,2X,Q39.32,2X,Q10.3,2X,Q10.3)
      END
```

```
      SUBROUTINE R2F1P(BP,CP,Z,A,B,N)
C     ****************************************************************
C     *         THIS SUBROUTINE RETURNS VALUES A(I) AND B(I),  I=1,...,N+1 *
C     *   OF THE NUMERATOR AND DENOMINATOR POLYNOMIALS RESPECTIVELY IN  *
C     *   THE PADE APPROXIMATION OF 2F1( 1 , BP ; CP ; -Z ) .          *
C     *                                                                *
C     *        NO OTHER SUBROUTINES ARE CALLED BY THIS ONE.           *
C     ****************************************************************
      IMPLICIT REAL*16(A-H,O-Z)
      DIMENSION A(1),B(1)
      DATA ONE/1.Q0/
C
C   INITIALIZATION :
C
      B(1)=ONE
      A(1)=ONE
      XI1=ONE
      CT1=CP
      B1C1=(CP-ONE)*(BP-ONE)
      ZZ=Z*Z
      B(2)=ONE+Z/(CP+ONE)*(BP+ONE)
      A(2)=B(2)-BP/CP*Z
C
C   FOR I=2,...,N , THE VALUES A(I+1) AND B(I+1) ARE CALCULATED
C   USING THE RECURRENCE RELATIONS BELOW
C
      DO 100 I=2,N
C
C   CALCULATION OF THE MULTIPLIERS FOR THE RECURSION
C
          CT2=CT1+XI1
          CT3=CT2*CT2
          G2=CT1/CT3*(BP-CT1)/(CT3-ONE)*XI1*(BP+XI1)*ZZ
          XI1=XI1+ONE
          G1=ONE+((XI1+XI1)*CT1+B1C1)/(CT3+CT2+CT2)*Z
C-------------------------------------------------------------------
C     THE RECURRENCE RELATIONS FOR A(I+1) AND B(I+1) ARE AS FOLLOWS
C-------------------------------------------------------------------
C
          B(I+1)=G1*B(I)+G2*B(I-1)
          A(I+1)=G1*A(I)+G2*A(I-1)
C
  100     CT1=CT1+ONE
      RETURN
      END
```

VALUES OF THE POLYNOMIALS IN THE PADE APPROXIMATION OF 2F1(1,BP;CP;-Z)

N = 12   BP = 0.15000000000000000000000000000Q+01

              CP = 0.30000000000000000000000000000Q+01

Z = 0.50000000000000000000000000000Q+00

I	A(I)	B(I)
0	0.10000000000000000000000000000Q+01	0.10000000000000000000000000000Q+01
1	0.10625000000000000000000000000Q+01	0.13125000000000000000000000000Q+01
2	0.13385416666666666666666666667Q+01	0.16562500000000000000000000000Q+01
3	0.16723632812500000000000000000Q+01	0.20693593750000000000000000000Q+01
4	0.20812988281250000000000000000Q+01	0.25753417968750000000000000000Q+01
5	0.25852101643880208333333333333Q+01	0.31988677779785156250000000000Q+01
6	0.32075898306710379464285714150Q+01	0.39689832414899535714285710Q+01
7	0.39770775437355041503906250000Q+01	0.49211261272430419218749999990Q+01
8	0.49289160900645785586154513888Q+01	0.60989049077033996582031249999990Q+01
9	0.61066616175917711944580078125Q+01	0.75561585314571857452392578124999Q+01
10	0.75639653856950727376070889596580Q+01	0.93594422728161920069469105113Q+01
11	0.93674926135240557293097178141226Q+01	0.11591076715791132301092147827148Q+02
12	0.11599505113725451967464404354389Q+02	0.14352907895436286568068541013277Q+02

VALUES OF THE APPROXIMATION, 1ST DIFFERENCES AND APPROXIMATE ERRORS

I	I-TH APPROXIMATION -- F(I)	1ST DIFF'S.	F(N)-F(I)
0	0.10000000000000000000000000000Q+01	-0.190Q+00	-0.192Q+00
1	0.80952380952380952380952381Q+00	-0.135Q-02	-0.136Q-02
2	0.80817610062893080817610062893Q+00	-0.119Q-04	-0.120Q-04
3	0.80816422840962718263331760264276Q+00	-0.112Q-06	-0.113Q-06
4	0.80816411656523140511537076010087Q+00	-0.109Q-08	-0.110Q-08
5	0.80816411547995415845818963529720Q+00	-0.107Q-10	-0.108Q-10
6	0.80816411546925779384977528439158Q+00	-0.106Q-12	-0.108Q-12
7	0.80816411546915150499505854913150Q+00	-0.107Q-14	-0.108Q-14
8	0.80816411546915150439659655874698010Q+00	-0.107Q-16	-0.108Q-16
9	0.80816411546915150428952483642731890Q+00	-0.108Q-18	-0.109Q-18
10	0.80816411546915150428844556194642256Q+00	-0.109Q-20	-0.110Q-20
11	0.80816411546915150428843465948961830Q+00	-0.110Q-22	-0.110Q-22
12	0.80816411546915150428834549176102200Q+00	0.0	0.0

179

```
C     .............................................................
C     .          THIS MAINLINE PROGRAM ILLUSTRATES THE SYNTAX OF THE SUB-  .
C     .   ROUTINE 'C2F1P' WHEN USED TO GENERATE COEFFICIENTS IN THE        .
C     .   POLYNOMIALS FOR THE PADE APPROXIMATION OF 2F1(1,BP;CP;-Z).       .
C     .............................................................
      IMPLICIT REAL*16(A-H,O-Z)
      DIMENSION CA(26),CB(26),NO(25)
   10 READ(5,2,END=999) BP,CP
      WRITE(6,3) BP,CP
      READ(5,1) M,(NO(J),J=1,M)
      DO 100 I=1,M
      N=NO(I)
C-----------------------------------------------------------------
      CALL C2F1P(BP,CP,CA,CB,N)
C-----------------------------------------------------------------
C     IN THE ABOVE:
C
C     BP           IS THE NUMERATOR PARAMETER OF THE 2F1
C     CP           IS THE DENOMINATOR PARAMETER OF THE 2F1
C     N            IS THE DEGREE OF THE POLYNOMIALS IN THE PADE
C                  APPROXIMATION
C     CA AND CB    WILL CONTAIN THE COEFFICIENTS IN THE NUMERATOR AND
C                  DENOMINATOR POLYNOMIALS , RESPECTIVELY
C
C     NOTE : THE COEFFICIENTS OF THE K-TH POWER OF Z WILL BE PLACED
C     IN CA(K+1) OR CB(K+1) AS APPROPRIATE
C-----------------------------------------------------------------
      N1=N+1
  100 WRITE(6,4) N,(CA(J),CB(J),J=1,N1)
      GOTO 10
  999 STOP
    1 FORMAT(26I2)
    2 FORMAT(Q39.32)
    3 FORMAT('1','COEFFICIENTS FOR THE PADE APPROXIMATION OF 2F1( 1 , BP
     : ; CP ; -Z )'//' ',T20,'BP = ',Q39.32//' ',T20,'CP = ',Q39.32/)
    4 FORMAT(' ','N = ',I2,T13,'CA(I)',T58,'CB(I)'//
     :26(1X,Q39.32,2X,Q39.32/)/)
      END
      SUBROUTINE C2F1P(BP,CP,A,B,N)
C     **********************************************************************
C     *                                                                    *
C     *       THIS SUBROUTINE RETURNS COEFFICIENTS A(I) AND B(I) ,          *
C     *   I = 1,2,...,N+1 , OF THE NUMERATOR AND DENOMINATOR POLYNOMIALS    *
C     *   RESPECTIVELY, IN THE PADE APPROXIMATION OF ORDER  N  FOR          *
C     *   2F1( 1 , BP ; CP ; -Z ) .                                         *
C     *                                                                    *
C     *       NO OTHER SUBROUTINES ARE CALLED BY THIS ONE.                  *
C     **********************************************************************
      IMPLICIT REAL*16(A-H,O-Z)
      DIMENSION A(1),B(1)
      DATA ONE/1.Q0/,ZERO/0.Q0/
C
C     INITIALIZATION :
C
      B(1)=ONE
      A(1)=ONE
      XI=ONE
      XN=N
      XN1I=XN
      CPN1=CP+XN-ONE
      XIJ=ZERO
      DO 100 I=1,N
      I1=I+1
C-----------------------------------------------------------------
C     FOR I = 1,2,...,N , B(I+1) IS CALCULATED AS FOLLOWS
C-----------------------------------------------------------------
C
      B(I1)=B(I)/XI*(BP+XN1I)/(CPN1+XN1I)*XN1I
C
      AI1=ONE
      DO 50 J=1,I
C-----------------------------------------------------------------
C     TO CALCULATE A(I+1), WE EMPLOY B(J) , J = 1,2,...,I+1 , AS FOLLOWS
C-----------------------------------------------------------------
C
      AI1=B(J+1)-AI1/(CP+XIJ)*(BP+XIJ)
C
   50 XIJ=XIJ-ONE
      A(I1)=AI1
      XIJ=XI
      XN1I=XN-XI
  100 XI=XI+ONE
C
      RETURN
      END
```

COEFFICIENTS FOR THE PADE APPROXIMATION OF 2F1(1 , BP ; CP ; -Z)

BP = 0.150000000000000000000000000000000Q+01

CP = 0.300000000000000000000000000000000Q+01

N = 3   CA(I)   CB(I)

0.100000000000000000000000000000000Q+01   0.100000000000000000000000000000000Q+01
0.118750000000000000000000000000000Q+01   0.168750000000000000000000000000000Q+01
0.312500000000000000000000000000000Q+00   0.843750000000000000000000000000000Q+00
0.390625000000000000000000000000002Q-02   0.117187500000000000000000000000000Q+00

N = 4   CA(I)   CB(I)

0.100000000000000000000000000000000Q+01   0.100000000000000000000000000000000Q+01
0.170000000000000000000000000000000Q+01   0.220000000000000000000000000000000Q+01
0.862500000000000000000000000000000Q+00   0.165000000000000000000000000000000Q+01
0.125000000000000000000000000000000Q+00   0.481250000000000000000000000000000Q+00
0.781250000000000000000000000000002Q-03   0.429687500000000000000000000000000Q-01

N = 5   CA(I)   CB(I)

0.100000000000000000000000000000000Q+01   0.100000000000000000000000000000000Q+01
0.220833333333333333333333333333330Q+01   0.270833333333333333333333333333330Q+01
0.166666666666666666666666666666670Q+00   0.270833333333333333333333333333330Q+01
0.492187500000000000000000000000000Q+00   0.121875000000000000000000000000000Q+01
0.455729166666666666666666666666670Q-01   0.236979166666666666666666666666670Q+00
0.162760416666666666666666666666601Q-03   0.148111979166666666666666666666670Q-01

N = 6   CA(I)   CB(I)

0.100000000000000000000000000000000Q+01   0.100000000000000000000000000000000Q+01
0.271428571428571428571428571428570Q+01   0.321428571428571428571428571428570Q+01
0.272321428571428571428571428571430Q+01   0.401785714285714285714285714285710Q+01
0.123214285714285714285714285714290Q+01   0.245535714285714285714285714285710Q+01
0.242187500000000000000000000000000Q+00   0.753348214285714285714285714285710Q+00
0.156250000000000000000000000000000Q-01   0.105468750000000000000000000000000Q+00
0.348772321428571428571428571426820Q-04   0.488281250000000000000000000000000Q-02

XV RATIONAL APPROXIMATIONS FOR $_1F_1(a;c;-z)$

Let

$$E(z) = {}_1F_1(a;c;-z) = e^{-z}{}_1F_1(c-a;c;z) . \tag{1}$$

The rational approximations given below are not of the Padé class. If $a = 1$, Padé approximations for $E(z)$ are given in Chapter 16. The rational approximations given here follow from those given in Chapter 13. That is, in the latter results replace z by z/b and let $b \to \infty$. This principle holds for the polynomials in the rational approximations, the recurrence formula, and the general expressions in the error analyses, but not for the asymptotic developments of the latter. If $c = 2a$, $E(z)$ is simply related to the Bessel function of the first kind. In this event, we suggest use of the rational approximations in Chapter 17. If $c = a$, $E(z) = e^{-z}$ in which event one should use the developments in Chapter 16 with $c = 1$. In this connection, if in 1.5(1-6), we put $p = q = 0$, $f = g = 0$, $a = 0$, $\beta = 0$ and $\lambda = 1$, we then get 16(1-5) with $c = 1$. Further, if in (2)-(5) of this chapter we put $a = c$ and let $c \to \infty$, then 16(1-5) emerges with $c = 1$.

For the present analysis, we write

$$E(z) = \{A_n(z)/B_n(z)\} + R_n(z) , \tag{2}$$

$$B_n(z) = L_n z^n {}_3F_1(-n,n+1,c;a+1;-1/z) , \tag{3}$$

$$A_n(z) = L_n z^n \sum_{k=0}^{n} \frac{(-n)_k(n+1)_k(a)_k}{(a+1)_k(k!)^2}$$

$$\times {}_4F_2\left(\begin{array}{c} -n+k,n+1+k,c+k,1 \\ 1+k,a+1+k \end{array} \middle| -1/z\right) , \tag{4}$$

$$L_n = \frac{(a+1)_n}{(n+1)_n(c)_n} . \tag{5}$$

Here $R_n(z)$ is the remainder which we discuss later.

For the polynomials $B_n(z)$ and $A_n(z)$, we have

$$B_0(z) = 1, \quad B_1(z) = 1 + \frac{(a+1)z}{2c}, \quad B_2(z) = 1 + \frac{(a+2)z}{2(c+1)} + \frac{(a+1)_2 z^2}{12(c)_2},$$

$$B_3(z) = 1 + \frac{(a+3)z}{2(c+2)} + \frac{(a+2)_2 z^2}{10(c+1)_2} + \frac{(a+1)_3 z^3}{120(c)_3},$$

$$A_0(z) = 1, \quad A_1(z) = B_1(z) - az/c,$$

$$A_2(z) = B_2(z) - \frac{az}{c}\left[1 + \frac{(a+2)z}{2(c+1)}\right] + \frac{(a)_2 z^2}{2(c)_2},$$

$$A_3(z) = B_3(z) - \frac{az}{c}\left[1 + \frac{(a+3)z}{2(c+2)} + \frac{(a+2)_2 z^2}{10(c+1)_2}\right]$$

$$+ \frac{(a)_2 z^2}{2(c)_2}\left|1 + \frac{(a+3)z}{2(c+2)}\right| - \frac{(a)_3 z^3}{6(c)_3}.$$

(6)

Both $A_n(z)$ and $B_n(z)$ satisfy the same recurrence formula

$$B_n(z) = (1 + F_1 z)B_{n-1}(z) + (E + F_2 z)zB_{n-2}(z)$$

$$+ F_3 z^3 B_{n-3}(z), \quad n \geq 3,$$

$$F_1 = \frac{(n-a-2)}{2(2n-3)(n+c-1)}, \quad F_2 = \frac{(n+a)(n+a-1)}{4(2n-1)(2n-3)(n+c-2)_2},$$

(7)

$$F_3 = -\frac{(n+a-2)_2(n-a-2)}{8(2n-3)^2(2n-5)(n+c-3)_3}, \quad E = -\frac{(n+a-1)(n-c-1)}{2(2n-3)(n+c-2)_2}.$$

The recurrence formula is stable in the forward direction.

As in the discussion following 13(7), we make certain assumptions on the parameters a and c. We suppose that neither a nor c - a is a negative integer, for otherwise E(z) is a polynomial or e^{-z} times a polynomial, see (1). We also exclude the case c = a, in which event $E(z) = e^{-z}$, for the reason cited in the remarks following 1.5(7) and in the opening remarks of this chapter.

We next consider the error which is written in the form

$$R_n(z) = S_n(z)/B_n^*(z), \quad B_n^*(z) = z^{-n}L_n^{-1}B_n(z).$$

(8)

Now with a, c and z fixed,

$$B_n^*(z) = \frac{(c)_n(2n)!}{(a+1)_n z^n n!}\left[\exp\{\frac{(n+a)z}{2(n+c-1)}\}\right]\left[1-uz^2+O(n^{-2})\right] ,$$

$$u = \frac{(n+a)}{8(2n-1)(n+c-1)(n+c-2)^2}$$
$$\times\left[n^2+n\{(3-2c)(2-a-c)+2(c-a-1)\}+(ac+2-2c)\right] . \tag{9}$$

The general form for $S_n(z)$ follows from that for the $_2F_1$ case in Chapter 13 by confluence. That is, in 13(14-19), replace z by z/b and let $b \to \infty$. Then

$$S_n(z) = F(z)M_n(z) + H(z)G_n(z) , \tag{10}$$

where now

$$F(z) = - \frac{az\Gamma(c)}{\Gamma(2-c)} \, _1F_1(a+1-c;2-c;-z) \tag{11}$$

or

$$F(z) = - \frac{az\Gamma(c)}{\Gamma(2-c)} e^{-z} \, _1F_1(1-a;2-c;z) , \tag{12}$$

$$M_n(z) = \frac{\Gamma(n+1-c)}{\Gamma(n+1+c)} \, _2F_2(c,c-a;c-n,n+1+c;z) , \tag{13}$$

$$H(z) = \frac{\Gamma(1-c)}{\Gamma(-a)} \, _1F_1(a;c;-z) , \tag{14}$$

$$G_n(z) = \frac{(-)^{n+1}n!\,\Gamma(n+1-a)z^{n+1}}{\Gamma(n+2-c)(2n+1)!} \, _2F_2\left(\begin{matrix}n+1,n+1-a\\2n+2,n+2-c\end{matrix}\middle| z\right) . \tag{15}$$

Next, we give asymptotic forms for $M_n(z)$ and $G_n(z)$ for n large with all parameters and z fixed. Thus

$$M_n(z) = \frac{\Gamma(n+1-c)}{\Gamma(n+1+c)}\left[_2F_2^r\left(\begin{matrix}c,c-a\\c-n,n+c+1\end{matrix}\middle| -z\right) + O(n^{-2r-2})\right] , \tag{16}$$

$$\frac{\Gamma(n+1-c)}{\Gamma(n+1+c)} = (n+\tfrac{1}{2})^{-2c}\{1+O(n^{-2})\} , \tag{17}$$

$$G_n(z) = \frac{(-)^{n+1}n!\,\Gamma(n+1-a)z^{n+1}}{\Gamma(n+2-c)(2n+1)!}\left[\exp\{\frac{(n+1-a)z}{2(n+2-c)}\}\right]\{1+vz^2+O(n^{-2})\} ,$$

$$v = \frac{(n+1-a)}{8(2n+3)(n+3-c)(n+2-c)^2}\{n^2+n(a-3c+6)+(a-5c+ac+7)\} , \tag{18}$$

$$\frac{n!\,\Gamma(n+1-a)}{\Gamma(n+2-c)(2n+1)!} = \frac{n!\,n^{c-a-1}}{(2n+1)!}\{1+O(n^{-1})\} . \tag{19}$$

The error analyses are similar to that for the $_2F_1$ in Chapter 13, see 13(23-40). We keep discussion to a minimum and state the key results.

If c is not a positive integer, then

$$S_n(z) = \frac{F(z)\Gamma(n+1-c)}{\Gamma(n+1+c)}\left[_2F_2^r(c,c-a;c-n,n+1+c;z) + O(n^{-2r-2})\right] . \quad (20)$$

If a is not a positive integer, but c is a positive integer, then

$$S_n(z) = \frac{(-)^n\Gamma(a+1)\Gamma(n+1-c)E(z)}{\Gamma(a+1-c)\Gamma(n+1+c)}$$

$$\times\left[_2F_2^r(c,c-a;c-n,n+1+c;z) + O(n^{-2r-2})\right] , \quad (21)$$

where E(z) is given by (1). If both c and a are positive integers, c > a, then

$$S_n(z) = -\frac{az^{1-c}\Gamma(c-1)e^{-z}}{\Gamma(c-a)} {_1F_1^{a-1}}(1-a;2-c;z)G_n(z) . \quad (22)$$

If the numbers c and a are arbitrary except as previously noted with the further proviso that if these numbers are positive integers, they are not so simultaneously, then the forms for the error readily follow from (8), (9) and (20) or (21) as appropriate. In these situations we have

$$R_n(z) = \frac{W\Gamma(n+1-c)\Gamma(n+a+1)n!z^n}{\Gamma(n+1+c)\Gamma(n+c)(2n)!}\{1 + O(n^{-1})\} \quad (23)$$

where W is free of n except that it might contain the factor $(-)^n$. Clearly

$$\lim_{n\to\infty} R_n(z) = 0 , \quad (24)$$

and

$$|R_{n+1}(z)/R_n(z)| = \left|\frac{(n+1-c)(n+a+1)z}{2(n+c)(n+c+1)(2n+1)}\right|\{1 + O(n^{-1})\} . \quad (25)$$

If both c and a are positive integers with c > a, then from (8), (9), (19) and (22), we have

$$R_n(z) = \frac{(-)^n z^{2n+2-c}e^{-z}\Gamma(c)\Gamma(c-1)(n!)^2\Gamma(n+1-a)\Gamma(n+1+a)}{\Gamma(a)\Gamma(c-a)\Gamma(n+c)\Gamma(n+2-c)(2n)!(2n+1)!}$$

$$\times {_1F_1^{a-1}}(1-a;2-c;z)\{\exp\frac{z(c-a-1)(2n+1)}{2(n+c-1)(n+2-c)}\}\{1 + (u+v)z^2 + O(n^{-2})\} . \quad (26)$$

Again

$$\lim_{n\to\infty} R_n(z) = 0 , \qquad\qquad (27)$$

and

$$R_{n+1}(z)/R_n(z) = -\frac{z^2(n+1-a)(n+1+a)}{4(2n+1)(2n+3)(n+c)(n+2-c)}\{1 + O(n^{-1})\} . \qquad (28)$$

It is of interest to compare the error $R_n(z)$ when $a = 1$ with the corresponding error, call it $R_{n,p}(z)$, for the Padé approximations in the next chapter. If c is not a positive integer,

$$\frac{R_{n,p}(z)}{R_n(z)} = \frac{(-)^{n+1}z^n e^{z/2}(\pi/n)^{3/2}}{\Gamma(c)\Gamma(c-1)(\sin\pi c)2^{2n+2c-1}}\{1 + O(n^{-1})\} , \qquad (29)$$

whence the Padé approximation is superior. Now consider (26). If z is small,

$$R_n(z) = O(z^{2n+2-c}) . \qquad\qquad (30)$$

This would be the situation for a Padé approximation if the numerator and denominator polynomials were of degree $n+1-c$ and n respectively. Thus the rational approximants under the conditions leading to (26), though not of the Padé class, are very much akin to this class. Indeed when $a = 1$ and c is a positive integer, $c > 1$, we find that

$$R_{n,p}(z)/R_n(z) = -(z/4n)^{c-1}\{1 + O(n^{-1})\} , \qquad (31)$$

and again the Padé approximation is superior.

Numerical Examples

Let $a = 2/3$, $c = 4/3$, $z = 3/4$ and $n = 5$. Then from (9), (20) and (8) without order terms, $B_n(z) = 1.47509$, $B_n^*(z) = 1.04541(5)$ and $R_n(z) = -0.23921(-7)$. The true values of $B_n(z)$ and $R_n(z)$ as found from the machine run are 1.47779 and $-0.23914(-7)$, respectively.

Suppose $a = 1$, $c = 2$, $z = 5/4$ and $n = 5$. Then from (26) without the order term, $R_n(z) = -0.2762(-9)$ which is in agreement with the true value.

```
C     .................................................................
C     .          THIS MAINLINE PROGRAM ILLUSTRATES THE SYNTAX OF THE SUB-  .
C     .    ROUTINE 'R1F1' FOR GENERATING VALUES OF THE NUMERATOR AND       .
C     .    DENOMINATOR POLYNOMIALS IN THE RATIONAL APPROXIMATION OF        .
C     .    1F1( AP ; CP ; -Z ) .                                          .
C     .................................................................
      IMPLICIT REAL*16(A-H,O-Z)
      DIMENSION A(26),B(26),R(26),D(26),E(26)
      DATA ZERO/0.Q0/
   10 READ(5,1,END=999) N,M
      READ(5,2) AP,CP
      N1=N+1
      D(N1)=ZERO
      E(N1)=ZERO
      DO 100 I=1,M
         READ(5,2) Z
C-------------------------------------------------------------------------
         CALL R1F1(AP,CP,Z,A,B,N)
C-------------------------------------------------------------------------
C     IN THE ABOVE :
C
C     AP          IS THE NUMERATOR PARAMETER OF THE 1F1
C     CP          IS THE DENOMINATOR PARAMETER OF THE 1F1
C     Z           IS THE VALUE OF THE ARGUMENT
C     A AND B     WILL CONTAIN THE VALUES OF THE NUMERATOR AND DENOMINATOR
C                     POLYNOMIALS, RESPECTIVELY, FOR ALL DEGREES FROM 0 TO
C                     N INCLUSIVE
C     N           IS THE MAXIMUM DEGREE FOR WHICH VALUES OF THE POLYNOMIALS
C                     ARE TO BE CALCULATED
C
C     NOTE : VALUES OF THE K-TH DEGREE POLYNOMIALS WILL BE PLACED IN
C     A(K+1) AND B(K+1) RESPECTIVELY.
C-------------------------------------------------------------------------
         R(N1)=A(N1)/B(N1)
         DO 50 J=1,N
            J1=N1-J
            R(J1)=A(J1)/B(J1)
            D(J1)=R(J1+1)-R(J1)
   50       E(J1)=R(N1)-R(J1)
         WRITE(6,3) N,AP,CP,Z
         DO 60 J=1,N1
            J1=J-1
   60       WRITE(6,4) J1,A(J),B(J)
         WRITE(6,5)
         DO 70 J=1,N1
            J1=J-1
   70       WRITE(6,6) J1,R(J),D(J),E(J)
  100    CONTINUE
      GOTO 10
  999 STOP
    1 FORMAT(2I2)
    2 FORMAT(Q39.32)
    3 FORMAT('1','VALUES OF THE POLYNOMIALS IN THE RATIONAL APPROXIMATIO
     :N OF 1F1(AP;CP;-Z)'//' ','N = ',I2,T20,'AP = ',Q39.32//' ',T20,'CP
     : = ',Q39.32//' ','        Z = ',Q39.32//' ',' I',T24,'A(I)',T65,
     :'B(I)'/)
    4 FORMAT(' ',I2,2X,Q39.32,2X,Q39.32)
    5 FORMAT('0','VALUES OF THE APPROXIMATION, 1ST DIFFERENCES AND APPRO
     :XIMATE ERRORS'//' ',' I',T12,'I-TH APPROXIMATION -- F(I)',T47,
     :'1ST DIFF''S.',T60,'F(N)-F(I)'/)
    6 FORMAT(' ',I2,2X,Q39.32,2X,Q10.3,2X,Q10.3)
      END
```

```
      SUBROUTINE R1F1(AP,CP,Z,A,B,N)
C     ********************************************************************
C     *         THIS SUBROUTINE RETURNS VALUES A(I) AND B(I),  I=1,...,N+1 *
C     *  OF THE NUMERATOR AND DENOMINATOR POLYNOMIALS IN THE RATIONAL     *
C     *  APPROXIMATION OF 1F1( AP ; CP ; -Z ) .                           *
C     *                                                                   *
C     *         NO OTHER SUBROUTINES ARE CALLED BY THIS ONE.              *
C     ********************************************************************
      IMPLICIT REAL*16(A-H,O-Z)
      DIMENSION A(1),B(1)
      DATA ZERO/0.Q0/,ONE/1.Q0/,TWO/2.Q0/,THREE/3.Q0/
C
C  INITIALIZATION :
C
      CT1=AP*Z/CP
      XN3=ZERO
      XN1=TWO
      Z2=Z/TWO
      CT2=Z2/(ONE+CP)
      XN2=ONE
      A(1)=ONE
      B(1)=ONE
      B(2)=ONE+(ONE+AP)*Z2/CP
      A(2)=B(2)-CT1
      B(3)=ONE+(TWO+B(2))*(TWO+AP)/THREE*CT2
      A(3)=B(3)-(ONE+CT2)*CT1
      CT1=THREE
      XN0=THREE
C
C  FOR I=3,...,N , THE VALUES A(I+1) AND B(I+1) ARE CALCULATED
C  USING THE RECURRENCE RELATIONS BELOW.
C
      DO 100 I=3,N
C
C  CALCULATION OF THE MULTIPLIERS FOR THE RECURSION
C
      CT2=Z2/CT1/(CP+XN1)
      G1=ONE+CT2*(XN2-AP)
      CT2=CT2*(AP+XN1)/(CP+XN2)
      G2=CT2*((CP-XN1)+(AP+XN0)/(CT1+TWO)*Z2)
      G3=CT2*Z2*Z2/CT1/(CT1-TWO)*(AP+XN2)/(CP+XN3)*(AP-XN2)
C-------------------------------------------------------------------------
C  THE RECURRENCE RELATIONS FOR A(I+1) AND B(I+1) ARE AS FOLLOWS
C-------------------------------------------------------------------------
C
      B(I+1)=G1*B(I)+G2*B(I-1)+G3*B(I-2)
      A(I+1)=G1*A(I)+G2*A(I-1)+G3*A(I-2)
C
      XN3=XN2
      XN2=XN1
      XN1=XN0
      XN0=XN0+ONE
  100 CT1=CT1+TWO
      RETURN
      END
```

VALUES OF THE POLYNOMIALS IN THE RATIONAL APPROXIMATION OF 1F1(AP;CP;-Z)

N = 12   AP = 0.66666666666666666666667Q+00

CP = 0.13333333333333333333333Q+01

Z = 0.75000000000000000000000Q+00

I	A(I)	B(I)
0	0.10000000000000000000000Q+01	0.10000000000000000000000Q+01
1	0.10937500000000000000000Q+01	0.14687500000000000000000Q+01
2	0.10602678571428571429Q+01	0.14955357142857142857Q+01
3	0.10543247767857142857Q+01	0.14887388923571428571Q+01
4	0.10497612251030219780Q+01	0.14823588594093406593Q+01
5	0.10465263492453469265Q+01	0.14777924841568801510Q+01
6	0.10441792686314379493Q+01	0.14744782836803704651Q+01
7	0.10424143361305815297Q+01	0.14719659890943789052Q+01
8	0.10410437543466516752Q+01	0.14700506001719704398Q+01
9	0.10399505127119028791Q+01	0.14685068413061030663Q+01
10	0.10390589966739233429Q+01	0.14672479377523064477Q+01
11	0.10383184973928179838Q+01	0.14662022838995438148Q+01
12	0.10376938567505824346Q+01	0.14653202032119958562Q+01

VALUES OF THE APPROXIMATION, 1ST DIFFERENCES AND APPROXIMATE ERRORS

I	I-TH APPROXIMATION -- F(I)	1ST DIFF'S. - F(K)-F(I)	
0	0.10000000000000000000000Q+01	-0.255Q+00	-0.292Q+00
1	0.74468035106382978723Q+00	-0.357Q-01	-0.365Q-01
2	0.70895522388059701492Q+00	-0.755Q-03	-0.787Q-03
3	0.70819995352010255564Q+00	-0.304Q-04	-0.313Q-04
4	0.70816957101209639178Q+00	-0.893Q-06	-0.917Q-06
5	0.70816667758151164976Q+00	-0.234Q-07	-0.239Q-07
6	0.70816665421829102586Q+00	-0.540Q-09	-0.551Q-09
7	0.70816653637850040859Q+00	-0.112Q-10	-0.114Q-10
8	0.70816653667341650041Q+00	-0.208Q-12	-0.212Q-12
9	0.70816653667133222568Q+00	-0.355Q-14	-0.360Q-14
10	0.70816653667129673898Q+00	-0.555Q-16	-0.563Q-16
11	0.70816653667129617617Q+00	-0.802Q-18	-0.802Q-18
12	0.70816653667129617947Q+00	0.0	0.0

```
C     ..............................................................
C     .          THIS MAINLINE PROGRAM ILLUSTRATES THE SYNTAX OF THE SUB- .
C     .   ROUTINE 'C1F1' WHEN USED TO GENERATE COEFFICIENTS IN THE POLY-  .
C     .   NOMIALS FOR THE RATIONAL APPROXIMATION OF 1F1(AP;CP;-Z).        .
C     ..............................................................
      IMPLICIT REAL*16(A-H,O-Z)
      DIMENSION CA(26),CB(26),NO(25)
   10 READ(5,2,END=999) AP,CP
      WRITE(6,3) AP,CP
      READ(5,1) M,(NO(J),J=1,M)
      DO 100 I=1,M
      N=NO(I)
C-----------------------------------------------------------------------
      CALL C1F1(AP,CP,CA,CB,N)
C-----------------------------------------------------------------------
C     IN THE ABOVE:
C
C   AP          IS THE NUMERATOR PARAMETER OF THE 1F1
C   CP          IS THE DENOMINATOR PARAMETER OF THE 1F1
C   N           IS THE DEGREE OF THE POLYNOMIALS IN THE RATIONAL
C                 APPROXIMATION
C   CA AND CB   WILL CONTAIN THE COEFFICIENTS IN THE NUMERATOR AND
C                 DENOMINATOR POLYNOMIALS RESPECTIVELY
C
C     NOTE : THE COEFFICIENTS OF THE K-TH POWER OF Z WILL BE PLACED
C IN CA(K+1) OR CB(K+1) AS APPROPRIATE
C-----------------------------------------------------------------------
      N1=N+1
  100 WRITE(6,4) N,(CA(J),CB(J),J=1,N1)
      GOTO 10
  999 STOP
    1 FORMAT(26I2)
    2 FORMAT(Q39.32)
    3 FORMAT('1','COEFFICIENTS FOR THE RATIONAL APPROXIMATION OF 1F1( AP
     : ; CP ; -Z )'//' ',T20,'AP = ',Q39.32/' ',T20,'CP = ',Q39.32/)
    4 FORMAT(' ','N = ',I2,T18,'CA(I)',T58,'CB(I)'//
     :26(1X,Q39.32,2X,Q39.32/)/)
      END
      SUBROUTINE C1F1(AP,CP,A,B,N)
C     ***********************************************************************
C     *        THIS SUBROUTINE RETURNS COEFFICIENTS A(I) AND B(I) ,        *
C     *  I = 1,2,...,N+1 , OF THE NUMERATOR AND DENOMINATOR POLYNOMIALS    *
C     *  RESPECTIVELY, IN THE RATIONAL APPROXIMATION OF ORDER   N   FOR    *
C     *  1F1( AP ; CP ; -Z ).                                             *
C     *                                                                   *
C     *        NO OTHER SUBROUTINES ARE CALLED BY THIS ONE.               *
C     ***********************************************************************
      IMPLICIT REAL*16(A-H,O-Z)
      DIMENSION A(1),B(1)
      DATA ONE/1.Q0/,ZERO/0.Q0/
C
C   INITIALIZATION :
C
      XN=N
      XN1I=XN
      CP1=CP-ONE
      B(1)=ONE
      A(1)=ONE
      XI=ONE
      XIJ=ZERO
      DO 100 I=1,N
      I1=I+1
C-----------------------------------------------------------------------
C FOR I = 1,2,...,N , B(I+1) IS COMPUTED AS FOLLOWS
C-----------------------------------------------------------------------
C
      B(I1)=(AP+XN1I)/(CP1+XN1I)*XN1I/(XN+XN1I)*B(I)/XI
C
      A(I1)=ONE
      DO 50 J=1,I
C-----------------------------------------------------------------------
C TO CALCULATE A(I+1) , WE EMPLOY B(J) , J = 1,2,...,I+1 AS FOLLOWS
C-----------------------------------------------------------------------
C
      A(I1)=B(J+1)-(AP+XIJ)/(CP+XIJ)*A(I1)/(ONE+XIJ)
C
   50 XIJ=XIJ-ONE
      XIJ=XI
      XN1I=XN-XI
  100 XI=XI+ONE
      RETURN
      END
```

COEFFICIENTS FOR THE RATIONAL APPROXIMATION OF 1F1(AP ; CP ; -Z)

$$AP = 0.66666666666666666666666666666667Q+00$$
$$CP = 0.13333333333333333333333333333333Q+01$$

N = 3   CA(I)   CB(I)

```
0.10000000000000000000000000000000Q+01    0.10000000000000000000000000000000Q+01
0.49999999999999999999999999999991Q-01    0.55000000000000000000000000000001Q+00
0.29285714285714285714285714285714Q-01    0.12571428571428571428571428571429Q+00
0.83333333333333333333333333333331Q-03    0.13095238095238095238095238095239Q-01
```

N = 4   CA(I)   CB(I)

```
0.10000000000000000000000000000000Q+01    0.10000000000000000000000000000000Q+01
0.38461538461538461538461538461527Q-01    0.53846153846153846153846153846154Q+00
0.36263736263736263736263736263737Q-01    0.12692307692307692307692307692308Q+00
0.11904761904761904761904761904760Q-02    0.16117216117216117216117216117217Q-01
0.45787545787545787545787545787567Q-04    0.10073260073260073260073260073261Q-02
```

N = 5   CA(I)   CB(I)

```
0.10000000000000000000000000000000Q+01    0.10000000000000000000000000000000Q+01
0.31249999999999999999999999999989Q-01    0.53125000000000000000000000000000Q+00
0.40083180708180708180708180708181Q-01    0.12713675213675213675213675213675Q+00
0.11599511599511599511599511599511Q-02    0.17481304188034188034188034188804Q-01
0.16502594627594627594627594627593Q-03    0.14270451770451770451770451770452Q-02
0.20667989417989417989417989418095Q-05    0.59460215710215710215710215710219Q-04
```

N = 6   CA(I)   CB(I)

```
0.10000000000000000000000000000000Q+01    0.10000000000000000000000000000000Q+01
0.26315789473684210526315789473671Q-01    0.52631578947368421052631578947369Q+00
0.42506835269993164730006835269994Q-01    0.12709330143540669856459330143541Q+00
0.10685586343481080323185586343478Q-02    0.18249294565084038768249294565084Q-01
0.25404055009318167212904055009322Q-03    0.16728520017993702204228520017994Q-02
0.48014850646429593798014850646370Q-05    0.95591542959964012595591542959967Q-04
0.79111921217184375079111921220009Q-07    0.28449863976179765653449863976181Q-05
```

XVI PADE APPROXIMATIONS FOR $_1F_1(1;c;-z)$

Let

$$E(z) = {}_1F_1(1;c;-z) \tag{1}$$

which is a form of the incomplete gamma function. See 2.2(21). The rational approximations of this chapter lie on the main diagonal of the Padé table. They follow from the rational approximations for 14(1) by confluence. That is, in the latter results replace z by z/b and let b → ∞. This principle does not hold for the asymptotic estimate of the error. If c = 1, we get the main diagonal Padé approximants for the exponential function. For some related comments, see the discussion following 15(1).

We write

$$E(z) = \{A_n(z)/B_n(z)\} + R_n(z) , \tag{2}$$

$$B_n(z) = L_n z^n {}_2F_0(-n,n+c;-1/z) , \tag{3}$$

$$A_n(z) = L_n z^n \sum_{k=0}^{n} \frac{(-n)_k(n+c)_k}{(c)_k k!} \, {}_3F_1(-n+k,n+c+k,1;1+k;-1/z), \tag{4}$$

$$L_n = \Gamma(n+c)/\Gamma(2n+c) , \tag{5}$$

where $R_n(z)$ is the remainder which is discussed later.

For the polynomials $A_n(z)$ and $B_n(z)$, we have

$$B_0(z) = 1, \quad B_1(z) = 1 + \frac{z}{c+1} , \quad B_2(z) = 1 + \frac{2z}{c+3} + \frac{z^2}{(c+2)_2} ,$$

$$B_3(z) = 1 + \frac{3z}{c+5} + \frac{3z^2}{(c+4)_2} + \frac{z^3}{(c+3)_3} , \tag{6}$$

$$A_0(z) = 1, \quad A_1(z) = B_1(z) - \frac{z}{c}, \quad A_2(z) = B_2(z) - \frac{z}{c}\left(1 + \frac{2z}{c+3}\right) + \frac{z^2}{(c)_2} ,$$

$$A_3(z) = B_3(z) - \frac{z}{c}\left(1 + \frac{3z}{c+5} + \frac{3z^2}{(c+4)_2}\right) + \frac{z^2}{(c)_2}\left(1 + \frac{3z}{c+5}\right) - \frac{z^3}{(c)_3} .$$

Both $A_n(z)$ and $B_n(z)$ satisfy the same recurrence formula

$$B_n(z) = (1 + F_1 z)B_{n-1}(z) + F_2 z^2 B_{n-2}(z), \ n \geq 2,$$

$$F_1 = \frac{(c-1)}{(2n+c-1)(2n+c-3)}, \quad F_2 = \frac{(n-1)(n+c-2)}{(2n+c-2)(2n+c-3)^2(2n+c-4)}. \tag{7}$$

The recursion formula is stable in the forward direction.

For the remainder, we have

$$R_n(z) = \frac{(-)^{n+1}\pi\Gamma(c)n!\Gamma(n+c)z^{2n+1}e^{-\theta}\{1 + O(n^{-3})\}}{2^{4n+2c-2}(2n+c)\{\Gamma(n+\frac{1}{2}c)\Gamma(n+\frac{1}{2}c+\frac{1}{2})\}^2},$$

$$\theta = z - z(z+4c-4)/4(2n+c), \tag{8}$$

or

$$R_n(z) = \frac{(-)^{n+1}\pi\Gamma(c)n^{1-c}z^{2n+1}e^{-\theta}\{1 + O(n^{-1})\}}{2^{4n+2c-1}(n!)^2}. \tag{9}$$

It follows that for z and c fixed,

$$\lim_{n\to\infty} R_n(z) = 0. \tag{10}$$

To facilitate computation of a priori error estimates, we have

$$\frac{R_{n+1}(z)}{R_n(z)} = -\frac{(n+1)(n+c)z^2}{(2n+c)(2n+c+1)^2(2n+c+2)}\exp\left(\frac{-z(z+4c-4)}{2(2n+c)(2n+c+2)}\right)$$

$$\times \{1 + O(n^{-3})\} = -\{z^2/4(2n+c+1)^2\}\{1+O(n^{-2})\}, \tag{11}$$

which is a measure of the rate of convergence. With $R_n(z,c)$ in place of $R_n(z)$,

$$\frac{R_n(z,c+h)}{R_n(z,c)} = \frac{\Gamma(c+h)}{\Gamma(c)}\left[\frac{n+c}{(2n+c)(2n+c+1)}\right]^h$$

$$\times \left[\exp\{\frac{-zh(8n+4-z)}{4(2n+c)(2n+c+h)}\}\right]\{1 + O(n^{-1})\}. \tag{12}$$

The latter shows that for a given z, c and n, the error changes slightly for small values of h, that is, for small changes in c.

Next we consider some results for the exponential function which is the case c = 1. We have

$$e^{-z} = \{G_n(-z)/G_n(z)\} + S_n(z), \tag{13}$$

$$G_n(z) = \{(2n)!/n!\}B_n(z), \tag{14}$$

$$G_n(z) = z^n {_2}F_0(-n,n+1;-1/z) = \{(2n)!/n!\}_1F_1^n(-n;-2n;z) , \tag{15}$$

and $S_n(z) = R_n(z)$ with $c = 1$. It is convenient to write

$$G_n(z) = M_n(z^2) + zN_n(z^2) . \tag{16}$$

Then by computing $M_n(z^2)$ and $N_n(z^2)$, evaluation of the main diagonal Padé approximation of order n only necessitates the evaluation of essentially (n+1) terms. The polynomials $G_n(z)$, $M_n(z^2)$ and $N_n(z^2)$ satisfy the same recurrence formula,

$$G_{n+1}(z) = 2(2n+1)G_n(z) + z^2 G_{n-1}(z) , \quad n \geq 1 , \tag{17}$$

$$G_0(z) = 1, \; G_1(z) = z + 2, \; G_2(z) = z^2 + 6z + 12,$$

$$G_3(z) = z^3 + 12z^2 + 60z + 120, \tag{18}$$

$$G_4(z) = z^4 + 20z^3 + 180z^2 + 840z + 1680 .$$

We also have the explicit representations,

$$M_n(z^2) = \{(2n)!/n!\}_2F_3^{[n/2]}(-\tfrac{1}{2}n,\tfrac{1}{2}-\tfrac{1}{2}n;-n,\tfrac{1}{2}-n,\tfrac{1}{2};z^2/4), \tag{19}$$

$$N_n(z^2) = \{2(2n)!/n!\}_2F_3^{[\frac{1}{2}n-\frac{1}{2}]}(\tfrac{1}{2}-\tfrac{1}{2}n,1-\tfrac{1}{2}n;1-n,\tfrac{1}{2}-n,3/2;z^2/4), \tag{20}$$

where [y] stands for the largest integer in y.

As previously noted, forms for the error $S_n(z)$ follow from (8)-(11) with $c = 1$. In closed form,

$$S_n(z) = \frac{(-)^{n+1}\pi e^{-z}I_{n+\frac{1}{2}}(z/2)}{K_{n+\frac{1}{2}}(z/2)} , \tag{21}$$

and we have the asymptotic representation

$$S_n(z) = (-)^{n+1}e^{-z}\{\exp(2\nu\zeta + 2U/\nu)\}\{1 + O(\nu^{-3})\} \tag{22}$$

uniformly in $z, z \neq 0$, $|\arg z| < \pi/2$, where

$$\nu = n + \tfrac{1}{2}, \; z = 2\nu x, \; \zeta = u^{-1} + \ln(\frac{ux}{1+u}), \; u = (1+x^2)^{-\frac{1}{2}} ,$$

$$U = (3u - 5u^3)/24. \tag{23}$$

In particular, for z large, we find

$$S_n(z) = (-)^{n+1}\left[\exp\{-\tfrac{\nu}{x}(1-\frac{1}{12x^2}+O(x^{-3}))\}\right]\{\exp \tfrac{2U}{\nu}\} \{1+O(\nu^{-3})\}. \tag{24}$$

In illustration, let $n = 4$, $z = 9$ whence $\nu = 9/2$. Neglecting order terms, from (22) $S_4(z) = -0.01474$ whereas the true value is -0.01503.

Uniform asymptotic representations for the first and second subdiagonal Padé approximations for e^{-z} are also given in Luke(1975b). For further comments on Padé approximants for (1) and other remarks on the exponential function, see Luke (1969,1970,1975a,1975b,1976). In connection with the reference (1975b), Dr. M. G. de Bruin has kindly informed me that in quoting the results of H. van Rossum, I overlooked the restriction $\mu \leq \nu + 1$. Consequently, the general results of this reference are not valid for the lower part of the Padé table unless $c = 0$, which in that notation is the case for the exponential function. Also, the concept of 'normal' employed by H. van Rossum is different from that usually used for the Padé table.

We conclude with a numerical example. Let $c = 2$. Then $E(z) = (1-e^{-z})/z$. We take $z = \frac{1}{2}$. Let $V_1(n)$ and $V_2(n)$ be the right hand sides of (8) and (9), respectively, each with the order term neglected. Then

$$V_1(n) = (-)^{n+1}\left[\exp\{-(16n+7)/32(n+1)\}\right]/2^{6n+2}\{(3/2)_n\}^2 , \qquad (25)$$

$$V_2(n) = (-)^{n+1}\left[\exp\{-(16n+7)/32(n+1)\}\right]\pi/2^{6n+4}n(n!)^2 . \qquad (26)$$

Values of $V_1(n)$, $V_2(n)$ and $V(n)$ true for $n = 1(1)5$ are recorded in the following table.

n	$(-)^{n+1}V_1(n)$	$(-)^{n+1}V_2(n)$	$(-)^{n+1}V(n)$
1	0.121(-2)	0.682(-3)	0.122(-2)
2	0.289(-5)	0.127(-5)	0.290(-5)
3	0.360(-8)	0.451(-8)	0.361(-8)
4	0.274(-11)	0.326(-11)	0.274(-11)
5	0.140(-14)	0.161(-14)	0.140(-14)

```
C     ...............................................................
C     .        THIS MAINLINE PROGRAM ILLUSTRATES THE SYNTAX OF THE SUB-  .
C     .  ROUTINE 'R1F1P' FOR GENERATING VALUES OF THE NUMERATOR AND      .
C     .  DENOMINATOR POLYNOMIALS IN THE PADE APPROXIMATION OF            .
C     .  1F1( 1 ; CP ; -Z ) .                                           .
C     ...............................................................
      IMPLICIT REAL*16(A-H,O-Z)
      DIMENSION A(26),B(26),R(26),D(26),E(26)
      DATA ZERO/0.Q0/
   10 READ(5,1,END=999) N,M
      READ(5,2) CP
      N1=N+1
      D(N1)=ZERO
      E(N1)=ZERO
      DO 100 I=1,M
         READ(5,2) Z
C------------------------------------------------------------------------
         CALL R1F1P(CP,Z,A,B,N)
C------------------------------------------------------------------------
C     IN THE ABOVE :
C
C  CP          IS THE DENOMINATOR PARAMETER OF THE 1F1
C  Z           IS THE VALUE OF THE ARGUMENT
C  A AND B     WILL CONTAIN THE VALUES OF THE NUMERATOR AND DENOMINATOR
C                POLYNOMIALS, RESPECTIVELY, FOR ALL DEGREES FROM 0 TO
C                N INCLUSIVE
C  N           IS THE MAXIMUM DEGREE FOR WHICH VALUES OF THE POLYNOMIALS
C                ARE TO BE CALCULATED
C
C     NOTE : VALUES OF THE K-TH DEGREE POLYNOMIALS WILL BE PLACED IN
C A(K+1) AND B(K+1) RESPECTIVELY.
C------------------------------------------------------------------------
         R(N1)=A(N1)/B(N1)
         DO 50 J=1,N
            J1=N1-J
            R(J1)=A(J1)/B(J1)
            D(J1)=R(J1+1)-R(J1)
   50       E(J1)=R(N1)-R(J1)
         WRITE(6,3) N,CP,Z
         DO 60 J=1,N1
            J1=J-1
   60       WRITE(6,4) J1,A(J),B(J)
         WRITE(6,5)
         DO 70 J=1,N1
            J1=J-1
   70       WRITE(6,6) J1,R(J),D(J),E(J)
  100    CONTINUE
      GOTO 10
  999 STOP
    1 FORMAT(2I2)
    2 FORMAT(Q39.32)
    3 FORMAT('1','VALUES OF THE POLYNOMIALS IN THE PADE APPROXIMATION OF
     : 1F1(1;CP;-Z)'//' ',' N = ',I2,T20,'CP = ',Q39.32//' ',' Z = ',
     :Q39.32//' ',' I',T24,'A(I)',T65,'B(I)'/)
    4 FORMAT(' ',I2,2X,Q39.32,2X,Q39.32)
    5 FORMAT('0','VALUES OF THE APPROXIMATION, 1ST DIFFERENCES AND APPRO
     :XIMATE ERRORS'//' ',' I',T12,'I-TH APPROXIMATION -- F(I)',
     :T47,'1ST DIFF''S.',T60,'F(N)-F(I)'/)
    6 FORMAT(' ',I2,2X,Q39.32,2X,Q10.3,2X,Q10.3)
      END
```

```fortran
      SUBROUTINE P1F1P(CP,Z,A,B,N)
C     ****************************************************************
C     *       THIS SUBROUTINE RETURNS VALUES A(I) AND B(I), I=1,...,N+1 *
C     *   OF THE NUMERATOR AND DENOMINATOR POLYNOMIALS IN THE PADE      *
C     *   APPROXIMATION OF 1F1( 1 ; CP ; -Z ) .                         *
C     *                                                                 *
C     *        NO OTHER SUBROUTINES ARE CALLED BY THIS ONE.             *
C     ****************************************************************
      IMPLICIT REAL*16(A-H,O-Z)
      DIMENSION A(1),B(1)
      DATA ONE/1.Q0/,TWO/2.Q0/
C
C     INITIALIZATION :
C
      XI1=ONE
      B(1)=ONE
      A(1)=ONE
      CT1=CP+ONE
      CP1=CP-ONE
      ZZ=Z*Z
      B(2)=ONE+Z/CT1
      A(2)=B(2)-Z/CP
C
C     FOR I=2,...,N , THE VALUES A(I+1) AND B(I+1) ARE CALCULATED
C     USING THE RECURRENCE RELATIONS BELOW.
C
      DO 100 I=2,N
C
C     CALCULATION OF THE MULTIPLIERS FOR THE RECURSION
C
      CT2=CT1*CT1
      G1=ONE+CP1/(CT2+CT1+CT1)*Z
      G2=XI1/(CT2-ONE)*(XI1+CP1)/CT2*ZZ
C---------------------------------------------------------------------
C     THE RECURRENCE RELATIONS FOR A(I+1) AND B(I+1) ARE AS FOLLOWS
C---------------------------------------------------------------------
C
      A(I+1)=G1*A(I)+G2*A(I-1)
      B(I+1)=G1*B(I)+G2*B(I-1)
C
      CT1=CT1+TWO
  100 XI1=XI1+ONE
      RETURN
      END
```

VALUES OF THE POLYNOMIALS IN THE PADE APPROXIMATION OF 1F1(1;CP;-Z)

N = 12   CP =   0.2000000000000000000000000000Q+01

Z =   0.5000000000000000000000000000Q+00

I	A(I)	B(I)
0	0.1000000000000000000000000000Q+01	0.1000000000000000000000000000Q+01
1	0.9166666666666666666666666666Q+00	0.1166666666666666666666666667Q+01
2	0.9541666666666666666666666666Q+00	0.1212500000000000000000000000Q+01
3	0.9700892857142857142857142857Q+00	0.1232738095238095238095238100Q+01
4	0.9790054563492063492063492000Q+00	0.1244068287037037037037037037Q+01
5	0.9846969454043915343915343800Q+00	0.1251302647005772005772005700Q+01
6	0.9886471402584683834683834600Q+00	0.1256320428341912716912716900Q+01
7	0.9915462965552483781650448371Q+00	0.1260004522822054488721155389Q+01
8	0.9937651291854938645686194705Q+00	0.1262824097681747661227661227Q+01
9	0.9955179000514993812496699002Q+00	0.1265051426020172681820482388Q+01
10	0.9969374696646901314780119097Q+00	0.1266855334727974155819452889Q+01
11	0.9981105814675718801400521114Q+00	0.1268346068258984278952231472Q+01
12	0.9990962893298790421400674225976Q+00	0.1269598653608180252584597971745Q+01

VALUES OF THE APPROXIMATION, 1ST DIFFERENCES AND APPROXIMATE ERRORS

I	I-TH APPROXIMATION -- F(I)	1ST DIFF'S.	F(N)-F(I)
0	0.1000000000000000000000000000Q+01	-0.214Q+00	-0.213Q+00
1	0.7857142857142857142857142900Q+00	0.123Q-02	0.122Q-02
2	0.7869415807560137457044673353Q+00	-0.290Q-05	-0.290Q-05
3	0.7869386769676484789956542732Q+00	0.361Q-08	0.361Q-08
4	0.7869386680577475786220989115Q+00	-0.274Q-11	-0.274Q-11
5	0.7869386680574731749756334075Q+00	0.140Q-14	0.140Q-14
6	0.7869386680574733153307738598Q+00	-0.515Q-18	-0.515Q-18
7	0.7869386680574733152792258509Q+00	0.142Q-21	0.142Q-21
8	0.7869386680574733152792409609Q+00	-0.307Q-25	-0.307Q-25
9	0.7869386680574733152792400930Q+00	0.530Q-29	0.530Q-29
10	0.7869386680574733152792400930Q+00	0.193Q-32	0.270Q-32
11	0.7869386680574733152792400930Q+00	0.770Q-33	0.770Q-33
12	0.7869386680574733152792400930Q+00	0.0	0.0

```
C     ..................................................................
C     .       THIS MAINLINE PROGRAM ILLUSTRATES THE SYNTAX OF THE SUB-  .
C     .     ROUTINE 'C1F1P' WHEN USED TO GENERATE COEFFICIENTS IN THE    .
C     .     POLYNOMIALS FOR THE PADE APPROXIMATION OF 1F1(1;CP;-Z).      .
C     ..................................................................
      IMPLICIT REAL*16(A-H,O-Z)
      DIMENSION CA(26),CB(26),NO(25)
  10  READ(5,2,END=999) CP
      WRITE(6,3) CP
      READ(5,1) M,(NO(J),J=1,M)
      DO 100 I=1,M
        N=NO(I)
C-----------------------------------------------------------------------
      CALL C1F1P(CP,CA,CB,N)
C-----------------------------------------------------------------------
C     IN THE ABOVE:
C
C CP          IS THE DENOMINATOR PARAMETER OF THE 1F1 IN THE PADE
C                 APPROXIMATION
C N           IS THE DEGREE OF THE POLYNOMIALS IN THE PADE
C                 APPROXIMATION
C CA AND CB   WILL CONTAIN THE COEFFICIENTS IN THE NUMERATOR AND
C                 DENOMINATOR POLYNOMIALS , RESPECTIVELY
C
C     NOTE : THE COEFFICIENTS OF THE K-TH POWER OF Z WILL BE PLACED
C IN CA(K+1) OR CB(K+1) AS APPROPRIATE
C-----------------------------------------------------------------------
        N1=N+1
 100    WRITE(6,4) N,(CA(J),CB(J),J=1,N1)
      GOTO 10
 999  STOP
 1    FORMAT(26I2)
 2    FORMAT(Q39.32)
 3    FORMAT('1','COEFFICIENTS FOR THE PADE APPROXIMATION OF 1F1( 1 ; CP
     : ; -Z )'//' ',T20,'CP = ',Q39.32/)
 4    FORMAT(' ','N = ',I2,T18,'CA(I)',T58,'CB(I)'//
     :26(1X,Q39.32,2X,Q39.32)/)
      END
      SUBROUTINE C1F1P(CP,A,B,N)
C     ****************************************************************
C     *       THIS SUBROUTINE RETURNS COEFFICIENTS A(I) AND B(I) ,    *
C     *   I = 1,2,...,N+1 , OF THE NUMERATOR AND DENOMINATOR POLYNOMIALS *
C     *   RESPECTIVELY, IN THE PADE APPROXIMATION OF ORDER  N  FOR    *
C     *   1F1( 1 ; CP ; -Z ).                                         *
C     *                                                               *
C     *       NO OTHER SUBROUTINES ARE CALLED BY THIS ONE.            *
C     ****************************************************************
      IMPLICIT REAL*16(A-H,O-Z)
      DIMENSION A(1),B(1)
      DATA ONE/1.Q0/,ZERO/0.Q0/
C
C     INITIALIZATION :
C
      XN=N
      XN1I=XN
      B(1)=ONE
      A(1)=ONE
      XI=ONE
      CP2NI=CP+XN+XN-ONE
      XIJ=ZERO
      DO 100 I=1,N
        I1=I+1
C-----------------------------------------------------------------------
C FOR I = 1,2,...,N , B(I+1) IS COMPUTED AS FOLLOWS
C-----------------------------------------------------------------------
C
        B(I1)=XN1I/CP2NI*B(I)/XI
C
        A(I1)=ONE
        DO 50 J=1,I
C-----------------------------------------------------------------------
C TO CALCULATE A(I+1), WE EMPLOY B(J), J = 1,2,...,I+1 AS FOLLOWS
C-----------------------------------------------------------------------
C
          A(I1)=B(J+1)-A(I1)/(CP+XIJ)
 50       XIJ=XIJ-ONE
        XIJ=XI
        XN1I=XN-XI
        CP2NI=CP2NI-ONE
 100    XI=XI+ONE
      RETURN
      END
```

COEFFICIENTS FOR THE PADE APPROXIMATION OF 1F1(1 ; CP ; -Z)

CP =   0.20000000000000000000000000000000000Q+01

N = 3   CA(I)   CB(I)

  0.10000000000000000000000000000000000Q+01   0.10000000000000000000000000000000000Q+01
 -0.71428571428571428571428571428572Q-01   0.42857142857142857142857142857143Q+00
  0.23809523809523809523809523809524Q-01   0.71428571428571428571428571428571Q-01
 -0.11904761904761904761904761904761Q-02   0.47619047619047619047619047619047Q-02

N = 4   CA(I)   CB(I)

  0.10000000000000000000000000000000000Q+01   0.10000000000000000000000000000000000Q+01
 -0.55555555555555555555555555555556Q-01   0.44444444444444444444444444444444Q+00
  0.27777777777777777777777777777778Q-01   0.83333333333333333333333333333333Q-01
 -0.13227513227513227513227513227513Q-02   0.79365079365079365079365079365079Q-02
  0.66137566137566137566137566137560Q-04   0.33068783068783068783068783068783Q-03

N = 5   CA(I)   CB(I)

  0.10000000000000000000000000000000000Q+01   0.10000000000000000000000000000000000Q+01
 -0.45454545454545454545454545454546Q-01   0.45454545454545454545454545454545Q+00
  0.30303030303030303030303030303030Q-01   0.90909090909090909090909090909091Q-01
 -0.12626262626262626262626262626263Q-01   0.10101010101010101010101010101010Q-01
  0.12626262626262626262626262626264Q-03   0.63131313131313131313131313131313Q-03
 -0.30062530062530062530062530062529Q-05   0.18037518037518037518037518037518Q-04

N = 6   CA(I)   CB(I)

  0.10000000000000000000000000000000000Q+01   0.10000000000000000000000000000000000Q+01
 -0.38461538461538461538461538461539Q-01   0.46153846153846153846153846153846Q+00
  0.32051282051282051282051282051282Q-01   0.96153846153846153846153846153846Q-01
 -0.11655011655011655011655011655011Q-01   0.11655011655011655011655011655012Q-01
  0.17482517482517482517482517482517Q-03   0.87412587412587412587412587412587Q-03
 -0.48562548562548562548562548562521Q-05   0.38850038850038850038850038850039Q-04
  0.11562511562511562511562511562441Q-06   0.80937580937580937580937580937581Q-06

```
C
C        ----------- IBM S/370 ----------- MULTIPLE PRECISION --------------
C
C        ***********************************************************************
C        *THIS SUBROUTINE, EXP, COMPUTES MULTIPLE PRECISION VALUES OF        *
C        *THE COEFFICIENTS IN THE MAIN DIAGONAL PADE APPROXIMATION OF EXP    *
C        *(-X).   THE OUTPUT POLYNOMIALS CALLED ODD AND EVEN CAN BE COM-     *
C        *BINED TO GIVE AN APPROXIMATION OF EXP(-X) BY (EVEN(X)-X*ODD(X))/   *
C        *(EVEN(X)+X*ODD(X)).   SEE TEXT FOR DETAILS.                        *
C        *                                                                   *
C        *DESCRIPTION OF VARIABLES.                                          *
C        *                                                                   *
C        *N       -INPUT  - THE ORDER OF THE DESIRED PADE APPROXIMATION.  N  *
C        *                  MUST BE LESS THAN OR EQUAL TO 100.               *
C        *                                                                   *
C        *EVEN    -OUTPUT - AN ARRAY CONTAINING THE COEFFICIENTS OF THE EVEN *
C        *                  POLYNOMIALS ARRANGED IN ASCENDING ORDER.         *
C        *                                                                   *
C        *ODD     -OUTPUT - AN ARRAY CONTAINING THE COEFFICIENTS OF THE ODD  *
C        *                  POLYNOMIAL ARRANGED IN ASCENDING ORDER.          *
C        *                                                                   *
C        *ALL OTHER VARIABLES ARE FOR INTERNAL USE.                          *
C        ***********************************************************************
         SUBROUTINE EXP(N,EVEN,ODD)
         IMPLICIT REAL*16 (A-H,O-Z)
         DIMENSION G(101),EVEN(1),ODD(1)
         DATA ZERO/0.Q0/,ONE/1.Q0/
         IF(N.EQ.0) GO TO 3
         G(1)=ONE
         COUNT=ZERO
         F1=-(N+1)
         F2=N
         DO 1 I=1,N
         F1=F1+ONE
         F2=F2+ONE
         COUNT=COUNT+ONE
       1 G(I+1)=-(F1*F2*G(I)/COUNT)
         INDEX=N+3
         NN=(N+1)/2
         DO 2 I=1,NN
         INDEX=INDEX-2
         EVEN(I)=G(INDEX)
       2 ODD(I)=G(INDEX-1)
         IF((N-2*(N/2)).EQ.0) RETURN
         EVEN(NN+1)=ONE
         ODD(NN+1)=ZERO
         RETURN
       3 EVEN(1)=ONE
         ODD(1)=ZERO
         RETURN
         END
```

COEFFICIENTS IN THE MAIN DIAGONAL PADE APPROXIMATION FOR EXP(-X)

(See (16), (19) and (20) of text for details.)

In the machine output which is based on the subroutine given on the previous page, the index n is changed to i to avoid confusion with the index in the polynomial $N_n(x^2)$. Thus dropping subscripts and simplifying the notation, we write

G = M + xN.

I	M
3	0.1200000000000000000000000000000000Q+03
	0.1200000000000000000000000000000000Q+02
4	0.1680000000000000000000000000000000Q+04
	0.1800000000000000000000000000000000Q+03
	0.1000000000000000000000000000000000Q+01
5	0.3024000000000000000000000000000000Q+05
	0.3360000000000000000000000000000000Q+04
	0.3000000000000000000000000000000000Q+02
6	0.6652800000000000000000000000000000Q+06
	0.7560000000000000000000000000000000Q+05
	0.8400000000000000000000000000000000Q+01
	0.1000000000000000000000000000000000Q+01

I	N
3	0.6000000000000000000000000000000000Q+02
	0.1000000000000000000000000000000000Q+01
4	0.8400000000000000000000000000000000Q+03
	0.2000000000000000000000000000000000Q+02
	0.0
5	0.1512000000000000000000000000000000Q+05
	0.4200000000000000000000000000000000Q+03
	0.1000000000000000000000000000000000Q+01
6	0.3326400000000000000000000000000000Q+06
	0.1008000000000000000000000000000000Q+05
	0.4200000000000000000000000000000000Q+02
	0.0

XVII RATIONAL APPROXIMATIONS FOR BESSEL FUNCTIONS
OF THE FIRST KIND

The modified Bessel functions can be represented in terms of hypergeometric functions in two different ways. Thus

$$I_\nu(z) = \frac{(z/2)^\nu}{\Gamma(\nu+1)} \; {}_0F_1(\nu+1;z^2/4) \; , \quad R(\nu) > -1, \tag{1}$$

$$I_\nu(z) = \frac{(z/2)^\nu e^z}{\Gamma(\nu+1)} \; {}_1F_1(\nu+\tfrac{1}{2};2\nu+1;-2z) \; , \quad R(\nu) > -1. \tag{2}$$

In the sequel, we give rational approximations to each of the above hypergeometric forms. Results for the Bessel function $J_\nu(z)$ are easily achieved since

$$J_\nu(z) = e^{i\nu\pi/2}I_\nu(ze^{-i\pi/2}) \; . \tag{3}$$

Thus in the approximations for ${}_0F_1$ we need only replace z^2 by $-z^2$, but in the approximations for ${}_1F_1$ we must replace z by $-iz$. If z and ν are real, $J_\nu(z)$ is real. In this event, the $J_\nu(z)$ representation in terms of the ${}_1F_1$ is not convenient as complex arithmetic is required.

Rational approximations for (2) can be found from the results in Chapter 15. These data can also be used to derive approximants for (1) by means of the confluence principle. That is, in Chapter 15, replace z by z/a and let $a \to \infty$. We omit these equations. The rational approximations we present are equivalent to forms deduced by use of the recurrence formula for $I_\nu(z)$ employed in the backward direction together with certain normalization relations. See Chapter 19.

If in 1.5(1-6), we put $p = 0$, $q = 1$, $\rho_1 = \nu + 1$, $f = g = a = 0$, $\alpha = 1 - \delta$, $\beta = \nu$, $\lambda = \nu + 2 - \delta$, $\delta = 0$ or $\delta = 1$, we get the following rational approximations.

$$_0F_1(\nu+1;z^2/4) = \{A_n(z)/B_n(z)\} + R_n(z), \quad R(\nu) > -1, \quad (4)$$

$$B_n(z) = {}_0F_1^n(-2n+1-\lambda;z^2/4), \quad (5)$$

$$A_n(z) = {}_2F_3^{[n+\frac{1}{2}-\frac{1}{2}\delta]}(-n+\frac{1}{2}\delta-\frac{1}{2}, -n+\frac{1}{2}\delta; -2n-\lambda+1, \nu+1, -2n-1+\delta; z^2), \quad (6)$$

$$\lambda = \nu + 2 - \delta, \quad \delta = 0 \text{ or } \delta = 1. \quad (7)$$

For the polynomials $B_n(z)$ and $A_n(z)$, we have

$$B_0(z) = 1, \quad B_1(z) = 1 - \frac{z^2}{4(\lambda+1)}, \quad B_2(z) = 1 - \frac{z^2}{4(\lambda+3)} + \frac{z^4}{32(\lambda+2)_2},$$

$$B_3(z) = 1 - \frac{z^2}{4(\lambda+5)} + \frac{z^4}{32(\lambda+4)_2} - \frac{z^6}{384(\lambda+3)_3}, \quad (8)$$

$$A_0(z) = 1, \quad A_1(z) = 1 + \frac{(2-\delta)z^2}{4(\lambda+1)(\nu+1)},$$

$$A_2(z) = 1 + \frac{(4-\delta)z^2}{4(\lambda+3)(\nu+1)} + \frac{(2-\delta)_2 z^4}{32(\lambda+2)_2(\nu+1)_2},$$

$$A_3(z) = 1 + \frac{(6-\delta)z^2}{4(\lambda+5)(\nu+1)} + \frac{(4-\delta)_2 z^4}{32(\lambda+4)_2(\nu+1)_2} \quad (9)$$

$$+ \frac{(2-\delta)_3 z^6}{384(\lambda+3)_3(\nu+1)_3}.$$

Both $A_n(z)$ and $B_n(z)$ satisfy the recurrence formula

$$B_n(z) = (1+F_1 y)B_{n-1}(z) + (E+F_2 y)yB_{n-2}(z) + F_3 y^3 B_{n-3}(z),$$

$$n \geq 3, \quad y = z^2/4,$$

$$F_1 = \frac{(n-\lambda+1)}{n(2n+\lambda-1)(2n+\lambda-4)}, \quad E = \frac{(n+\lambda-3)}{n(2n+\lambda-4)(2n+\lambda-3)}, \quad (10)$$

$$F_2 = \frac{E(n+2\lambda-4)}{(n+\lambda-3)(2n+\lambda-5)(2n+\lambda-2)}, \quad F_3 = \frac{-E}{(2n+\lambda-6)(2n+\lambda-5)^2(2n+\lambda-4)}.$$

This formula is stable when used in the forward direction.

For the error, we have from 19(18) with $m = 0$,

$$R_n(z) = \frac{(-)^n(z/2)^{2n+2}\Gamma(n+1-\delta+\nu)F_\nu(z)\{1 + O(n^{-1})\}}{(n+1)!\,\Gamma(2n+2-\delta+\nu)},$$

$$F_\nu(z) = \Gamma(\nu+1)(z/2)^{-\nu}I_\nu(z). \quad (11)$$

A more convenient form is

$$R_n(z) = \frac{(-)^n (z/2)^{2n+2}(2-\delta+\nu)_n F_\nu(z)\{1 + O(n^{-1})\}}{(n+1-\delta+\nu)(n+1)!(2-\delta+\nu)_{2n}},\qquad (12)$$

or if ν is an integer or zero,

$$R_n(z) = \frac{(-)^n(z/2)^{2n+2}(\pi/n)^{\frac{1}{2}}F_\nu(z)\{1 + O(n^{-1})\}}{2^{2n+1-\delta+\nu}\{(n+1)!\}^2}.\qquad (13)$$

We also have

$$\frac{R_{n+1}(z)}{R_n(z)} = -\frac{(n+1-\delta+\nu)z^2\{1 + O(n^{-1})\}}{4(n+2)(2n+2-\delta+\nu)(2n+3-\delta+\nu)} = -\frac{z^2\{1 + O(n^{-1})\}}{16n^2}.$$
$$(14)$$

If $F_\nu(z)$ is omitted in (11)-(13), then the expressions represent the relative error. Further comments on the error are
given in Chapter 19.

Next we turn to approximations for (2). In 1.5(1-6),
put $p = q = 1$, $\alpha_1 = \nu + \frac{1}{2}$, $\rho_1 = 2\nu + 1$, $f = g = a = 0$, $\alpha = 1$,
$\beta = 2\nu$, $\lambda = 2\nu + 2$ and replace z by $2z$. We then obtain

$$_1F_1(\nu+\tfrac{1}{2};2\nu+1;-2z) = \{G_n(z)/H_n(z)\} + S_n(z),\qquad (15)$$

$$H_n(z) = {}_1F_1^n(-n-\nu-\tfrac{1}{2};-2n-2\nu-1;2z),\qquad (16)$$

$$G_n(z) = {}_2F_3^{[n/2]}(-\tfrac{1}{2}n,\tfrac{1}{2}-\tfrac{1}{2}n;-n-\nu,\nu+1,-n;z^2),\qquad (17)$$

$$H_0(z) = 1,\ H_1(z) = 1 + z,\ H_2(z) = 1 + z + (2\nu+3)z^2/2(2\nu+4),$$

$$H_3(z) = 1 + z + \frac{(2\nu+5)z^2}{2(2\nu+6)} + \frac{(2\nu+3)z^3}{6(2\nu+6)},\qquad (18)$$

$$G_0(z) = 1,\ G_1(z) = 1,\ G_2(z) = 1 + \frac{z^2}{4(\nu+1)_2},\qquad (19)$$

$$G_3(z) = 1 + \frac{z^2}{2(\nu+1)(\nu+3)},\ G_4(z) = 1 + \frac{3z^2}{4(\nu+1)(\nu+4)} + \frac{z^4}{16(\nu+1)_4}.$$

Note that if in $A_n(z)$, see (6), we put $\delta = 1$ and replace n by
n/2, we get $G_n(z)$. Both $G_n(z)$ and $H_n(z)$ satisfy the same recurrence formula

$$H_n(z) = (1+Q_1z)H_{n-1}(z) + (P+Q_2z)zH_{n-2}(z) + Q_3z^3H_{n-3}(z),$$

$$n \geq 3,\ Q_1 = \frac{(n+2\nu-1)}{2n(n+\nu-1)},\ P = -Q_1,$$

$$Q_2 = \{4(n+\nu)(n+\nu-1)\}^{-1}, \quad Q_3 = P\{4(n+\nu-1)(n+\nu-2)\}^{-1}. \quad (20)$$

This formula is stable in the forward direction.

For the error, we have from 19(33) with m = 0, δ = 1 and n replaced by n/2,

$$S_n(z) = -\frac{(2z)^{n+1}(\nu+\frac{1}{2})_{n+1}e^{-z}V_\nu(z)\{1 + O(n^{-1})\}}{(n+1)!(n+1+2\nu)_{n+1}},$$

$$V_\nu(z) = {}_1F_1(\nu +\tfrac{1}{2}; 2\nu+1; -2z), \quad (21)$$

or

$$S_n(z) = -\frac{(2z)^{n+1}\pi^{\frac{1}{2}}(n/4)^\nu e^{-z}V_\nu(z)\{1 + O(n^{-1})\}}{\Gamma(\nu+\frac{1}{2})(n+1)!2^{2n+1}}. \quad (22)$$

Further,

$$\frac{S_{n+1}(z)}{S_n(z)} = \frac{(n+1+2\nu)z\{1 + O(n^{-1})\}}{(n+2)(2n+2+\nu)} = \frac{z\{1 + O(n^{-1})\}}{2n}. \quad (23)$$

If $V_\nu(z)$ is omitted in (21-22), then the expressions represent the relative error. Further discussion of the error follows from the remarks in Chapter 19.

This chapter concludes with four sets of programs. The first set gives programs to get rational approximations for ${}_0F_1(\nu+1; \text{VAL } z^2/4)$ where VAL is either 1.0 or - 1.0 which must be specified by the user. We have

$$I_\nu(z) = \frac{(z/2)^\nu}{\Gamma(\nu+1)} {}_0F_1(\nu+1; \text{VAL } z^2/4) \text{ if VAL = 1.0}, \quad (24)$$

$$J_\nu(z) = \frac{(z/2)^\nu}{\Gamma(\nu+1)} {}_0F_1(\nu+1; \text{VAL } z^2/4) \text{ if VAL = - 1.0}. \quad (25)$$

The second program generates coefficients in the numerator and denominator polynomials which define the rational approximations for the ${}_0F_1$ in (1,4). If in (4), we write

$$A_n(z) = \sum_{k=0}^n a_k z^{2k}, \quad B_n(z) = \sum_{k=0}^n b_k z^{2k}, \quad (26)$$

then this program generates the a_k's and b_k's. If we desire to evaluate the numerator and denominator polynomials in the

rational approximation for the $_0F_1$ related to $J_\nu(z)$, see
(1,3,4), then in (26) we must replace z^2 by $-z^2$ or equivalently, in (26) we must replace a_k and b_k by $(-)^k a_k$ and $(-)^k b_k$
respectively. Finally, for $_1F_1(\nu+\frac{1}{2};2\nu+1;-2z)$, the third and
fourth set of programs are designed to get rational approximations as a number, and to get the coefficients which define
the numerator and denominator polynomials in the rational
approximations, respectively.

Numerical Examples

1.   Consider $_0F_1(\nu+1;z^2/4)$ with $\nu = 1/3$, $z = \frac{1}{2}$ and $\delta = 0$.
Here VAL = 1.0. Then from (11) without the order term, we
get for n = 3 and 4, the values $- 0.619(-9)$ and $0.431(-12)$,
respectively, whereas the true values are $- 0.623(-9)$ and
$0.434(-12)$, respectively.

2.   Consider $_1F_1(\nu+\frac{1}{2};2\nu+1;-2z)$ with $\nu = 1/3$ and $z = \frac{1}{2}$.
Let n = 4 and use (21) without the order term. We find the
estimate $- 0.106(-4)$, whereas the true value is $- 0.107(-4)$.

```
C
C        ...............THIS MAINLINE PROGRAM ILLUSTRATES THE SYNTAX OF THE SUB-...
C        .   ROUTINE 'RBOF1' FOR GENERATING VALUES OF THE NUMERATOR AND        .
C        .   DENOMINATOR POLYNOMIALS IN THE RATIONAL APPROXIMATIONS OF         .
C        .   OF1( V+1 ; VAL*(Z**2)/4 ) WHICH EXCEPT FOR AN AUXILIARY           .
C        .   MULTIPLIER INCLUDES THE BESSEL FUNCTIONS I( V ; Z ) AND           .
C        .   J( V ; Z ) .                                                      .
C        ...............................................................
         IMPLICIT REAL*16(A-H,O-Z)
         DIMENSION A(26),B(26),R(26),D(26),E(26)
         DATA ZERO/0.Q0/
  10     READ(5,1,END=999) N,IDELTA,M
         READ(5,2) V,VAL
         N1=N+1
         D(N1)=ZERO
         E(N1)=ZERO
         DO 100 I=1,M
         READ(5,2) Z
C--------------------------------------------------------------------
         CALL RBOF1(V,VAL,IDELTA,Z,A,B,N)
C--------------------------------------------------------------------
C        IN THE ABOVE :
C
C   V           IS THE PARAMETER IN THE ABOVE OF1
C   VAL         IS THE PARAMETER IN THE ABOVE OF1.  EXCEPT FOR AN
C               AUXILIARY MULTIPLIER, VAL=1 GIVES I( V ; Z ) AND VAL=-1
C               GIVES J( V ; Z ).
C   DELTA       IS A PARAMETER WHICH MUST BE 0 OR 1
C   Z           IS THE VARIABLE IN THE ARGUMENT OF THE OF1
C   A AND B     WILL CONTAIN THE VALUES OF THE NUMERATOR AND DENOMINATOR
C                  POLYNOMIALS RESPECTIVELY, FOR ALL DEGREES FROM 0 TO
C               N INCLUSIVE
C   N           IS THE MAXIMUM DEGREE FOR WHICH VALUES OF THE POLYNOMIALS
C               ARE TO BE CALCULATED
C
C   NOTE : VALUES OF THE 2K-TH DEGREE POLYNOMIALS WILL BE PLACED IN
C   A(K+1) OR B(K+1) AS APPROPRIATE.
C--------------------------------------------------------------------
         R(N1)=A(N1)/B(N1)
         DO 50 J=1,N
         J1=N1-J
         R(J1)=A(J1)/B(J1)
         D(J1)=R(J1+1)-R(J1)
  50     E(J1)=R(N1)-R(J1)
         WRITE(6,3) VAL,N,IDELTA,V,Z
         DO 60 J=1,N1
         J1=J-1
  60     WRITE(6,4) J1,A(J),B(J)
         WRITE(6,5)
         DO 70 J=1,N1
         J1=J-1
  70     WRITE(6,6) J1,R(J),D(J),E(J)
 100     CONTINUE
         GOTO 10
 999     STOP
  1      FORMAT(3I2)
  2      FORMAT(Q39.32)
  3      FORMAT('1','VALUES OF THE POLYNOMIALS IN THE RATIONAL APPROXIMATIO
        :NS OF OF1( V+1 ; VAL*(Z**2)/4 )'//' ','VAL = ',Q39.32//' ','N = ',
        :I2,'   DELTA = ',I2,'   V = ',Q39.32//' ','   Z = ',Q39.32//' ',
        :'  I',T24,'A(I)',T65,'B(I)'/)
  4      FORMAT(' ',I2,2X,Q39.32,2X,Q39.32)
  5      FORMAT('0','VALUES OF THE APPROXIMATION, 1ST DIFFERENCES AND APPRO
        :XIMATE ERRORS'//' ','  I',T12,'I-TH APPROXIMATION -- F(I)',T47,
        :'1ST DIFF''S',T60,'F(N)-F(I)'/)
  6      FORMAT(' ',I2,2X,Q39.32,2X,Q10.3,2X,Q10.3)
         END
C
         SUBROUTINE RBOF1(V,VAL,IDELTA,Z,A,B,N)
C   ************************************************************
C   *       THIS SUBROUTINE RETURNS VALUES A(I) AND B(I), I=1,....,N+1 *
C   *   OF THE NUMERATOR AND DENOMINATOR POLYNOMIALS IN THE RATIONAL   *
C   *   APPROXIMATIONS OF OF1( V+1 ; VAL*(Z**2)/4 ) WHICH EXCEPT FOR   *
C   *   AN AUXILIARY MULTIPLIER INCLUDES THE BESSEL FUNCTIONS          *
C   *   I( V ; Z ) AND J( V ; Z ).                                     *
C   *       THE NATURE OF THE APPROXIMATION OBTAINED DEPENDS UPON THE  *
C   *   PARAMETER  IDELTA , WHICH MUST BE 0 OR 1 , AS DISCUSSED IN THE *
C   *   MAIN TEXT.                                                     *
C   *                                                                  *
C   *       NO OTHER SUBROUTINES ARE CALLED BY THIS ONE .              *
C   ************************************************************
         IMPLICIT REAL*16(A-H,O-Z)
         DIMENSION A(1),B(1)
         DATA ONE/1.Q0/,TWO/2.Q0/,THREE/3.Q0/,FOUR/4.Q0/,EIGHT/8.Q0/
C
```

```
C     INITIALIZATION :
C
      Y=VAL*Z*Z/FOUR
      YY=Y*Y
      DEL=IDELTA
      CT1=TWO-DEL
      CT3=V-DEL
      XIL3=ONE+CT3
      CT3=CT3+CT3
      CT4=Y/(CT3+EIGHT)
      B(1)=ONE
      A(1)=ONE
      V1=V+ONE
      CT2=XIL3+TWO
      B(2)=ONE-Y/CT2
      A(2)=ONE+Y/CT2*CT1/V1
      CT2=Y/(CT2+TWO)
      B(3)=ONE-CT2*(ONE-CT4)
      A(3)=ONE+CT2/V1*(TWO+CT1+CT4/(V+TWO)*CT1*(CT1+ONE))
      XI=THREE
      DO 100 I=3,N
C
C     FOR I=3,...,N , THE VALUES A(I+1) AND B(I+1) ARE CALCULATED USING
C     THE RECURRENCE RELATIONS BELOW
C
      X2IL5=XIL3-ONE+XI
      CT1=X2IL5*X2IL5
      XIL3=XIL3+ONE
      CT2=Y/(X2IL5+ONE)/XI
      EY=CT2*XIL3/(X2IL5+TWO)
      G1=ONE+CT2*(XIL3-CT3)/(X2IL5+FOUR)
      G2=EY*(ONE+(CT3+XI)/XIL3*Y/(CT1+X2IL5+X2IL5+X2IL5))
      G3=EY/(ONE-CT1)*YY/CT1
C--------------------------------------------------------------------
C     THE RECURRENCE RELATIONS FOR A(I+1) AND B(I+1) ARE AS FOLLOWS
C--------------------------------------------------------------------
C
      B(I+1)=G1*B(I)+G2*B(I-1)+G3*B(I-2)
      A(I+1)=G1*A(I)+G2*A(I-1)+G3*A(I-2)
C
  100 XI=XI+ONE
      RETURN
      END
```

VALUES OF THE POLYNOMIALS IN THE RATIONAL APPROXIMATIONS OF 0F1(V+1 ; VAL*(Z**2)/4 ;

VAL = 0.1000000000000000000000000000000Q+01

N = 12   DELTA = 0   V = 0.33333333333333333333333333333Q+00

Z = 0.5000000000000000000000000000000Q+00

I	A(I)	B(I)
0	0.10000000000000000000000000000000Q+01	0.10000000000000000000000000000000Q+01
1	0.10125000000000000000000000000000Q+01	0.98125000000000000000000000000000Q+00
2	0.10359319339886758241782624Q+01	0.98835760213446153446153870Q+00
3	0.10386229316155349367735256523Q+01	0.99151961386344494617224800382477Q+00
4	0.10405180490981346677210256657Q+01	0.99332611872063184980514498647721469538770Q+00
5	0.10417447465833206318808931Q+01	0.99452013449805305829681815651566448Q+00
6	0.10426085877914549250866694797Q+01	0.99539327850924968818229634Q+00
7	0.10437361268762435900973680Q+01	0.99647121544564512290682666Q+00
8	0.10441262168762435900973680Q+01	0.99673748025584586342Q+00
9	0.10444704028682860279596931Q+01	0.99707481725498609397956Q+00
10	0.10444925734615580324965Q+01	0.99732517481819521852375Q+00
11	0.10449253461580324965987037510Q+01	0.99753660281022308961608143027Q+00
12		

VALUES OF THE APPROXIMATION, 1ST DIFFERENCES AND APPROXIMATE ERRORS

I	I-TH APPROXIMATION -- F(I)	1ST DIFF'S.	F(N)-F(I)
0	0.10000000000000000000000000000000Q+01	0.4780-01	0.4750-01
1	0.10477070063942675592356878980Q+01	0.2650-03	0.2640-03
2	0.10477196420997081802965455Q+01	-0.5570-06	-0.5560-06
3	0.10477542864067243804499Q+01	0.6240-08	0.6230-09
4	0.10477506284830706880704Q+01	-0.4340-11	-0.4340-12
5	0.10477506280483706880757Q+01	0.2060-15	0.2060-15
6	0.10477506280483169644670Q+01	-0.7070-22	-0.1840-22
7	0.10477506280483169644670Q+01	-0.1840-26	-0.1840-26
8	0.10477506280483169644670Q+01	0.3770-26	-0.3770-30
9	0.10477506280483169644670Q+01	-0.6290-32	-0.1620-30
10	0.10477506280483169644670Q+01	0.3080-32	-0.3080-32
11	0.10477506280483169644670Q+01	-0.0	-0.0
12	0.10477506280483169644670Q+01	-0.0	0.0

210

VALUES OF THE POLYNOMIALS IN THE RATIONAL APPROXIMATIONS OF 0F1(V+1 ; VAL*(Z**2)/4)

VAL = -0.1000000000000000000000000000000Q+01

N = 12   DELTA = 0   V = 0.3333333333333333333333333333333Q+00

Z = 0.5000000000000000000000000000000Q+00

I	A(I)	B(I)
0	0.1000000000000000000000000000000Q+01	0.1000000000000000000000000000000Q+01
1	0.9651806733986675882457582417990Q+00	0.1018075900052163462163464634Q+01
2	0.9601976871500345695719903031Q+00	0.1018064442472695245315110478Q+01
3	0.9601596104523358124730375080Q+00	0.1006721616168695734811676704Q+01
4	0.9580244662160350664895483750Q+00	0.1005531404893097944813135615Q+01
5	0.9582309271904331910432662080Q+00	0.1004089876774307475302603661Q+01
6	0.9576444931292544326620489462Q+00	0.1036126747527301336637810Q+01
7	0.9579194446381882055187280609Q+00	0.1003238756340610938857286Q+01
8	0.9568373826538979757130305550Q+00	0.1002934140923465050571903Q+01
9	0.9562977463375389777717204550Q+00	0.1002693409269258176433986Q+01
10	0.9562104904018546529164253439Q+00	0.1002470264655633249548558700Q+01

VALUES OF THE APPROXIMATION, 1ST DIFFERENCES AND APPROXIMATE ERRORS

I	I-TH APPROXIMATION -- F(I)	1ST DIFF'S.	F(N)-F(I)
0	0.1000000000000000000000000000000Q+01	-0.460Q-01	-0.460Q-01
1	0.9539877300613496931496931190Q+00	-0.230Q-03	-0.239Q-03
2	0.9537493185347168025153371790Q+00	-0.560Q-09	-0.560Q-09
3	0.9537488811570159582037915382Q+00	-0.360Q-12	-0.360Q-12
4	0.9537488860615807251170614713Q+00	-0.630Q-15	-0.630Q-15
5	0.9537488860604766533557304040Q+00	-0.160Q-19	-0.160Q-19
6	0.9537488860604766334491607070Q+00	-0.170Q-22	-0.170Q-22
7	0.9537488860604766334491299972Q+00	-0.310Q-26	-0.310Q-26
8	0.9537488860604766334491772835Q+00	-0.550Q-30	-0.540Q-30
9	0.9537488860604766334491772835Q+00	-0.650Q-32	-0.100Q-31
10	0.9537488860604766334491772835Q+00	0.360Q-32	0.360Q-32
11	0.9537488860604766334491772835Q+00	0.0	0.0
12	0.9537488860604766334491772836Q+00		

```
C     ...........THIS MAINLINE PROGRAM ILLUSTRATES THE SYNTAX OF THE SUB-...
C     .     ROUTINE 'CBOF1' FOR GENERATING COEFFICIENTS IN THE POLYNOMIALS  .
C     .     FOR THE RATIONAL APPROXIMATION OF OF1(V+1;(Z**2)/4) WHICH        .
C     .     EXCEPT FOR AN AUXILIARY MULTIPLIER IS THE BESSEL FUNCTION        .
C     .     I(V;Z) .                                                         .
C     ...........................................................
      IMPLICIT REAL*16(A-H,O-Z)
      DIMENSION CA(26),CB(26),NO(25)
   10 READ(5,2,END=999) V
      READ(5,3) IDELTA
      WRITE(6,4) IDELTA,V
      READ(5,1) M,(NO(J),J=1,M)
      DO 100 I=1,M
      N=NO(I)
C-----------------------------------------------------------------
      CALL CBOF1(V,IDELTA,CA,CB,N)
C-----------------------------------------------------------------
C     IN THE ABOVE:
C
C     V           IS THE PARAMETER IN THE RELATED BESSEL FUNCTION
C     IDELTA      IS A PARAMETER WHICH MUST BE 0 OR 1 AS DISCUSSED IN
C                 THE MAIN TEXT
C     N           IS THE DEGREE OF THE POLYNOMIALS IN THE RATIONAL
C                 APPROXIMATION
C     CA AND CB   WILL CONTAIN THE COEFFICIENTS IN THE NUMERATOR AND
C                 DENOMINATOR POLYNOMIALS RESPECTIVELY
C
C     NOTE : THE COEFFICIENTS OF THE 2K-TH POWER OF Z WILL BE PLACED
C     IN CA(K+1) OR CB(K+1) AS APPROPRIATE
C-----------------------------------------------------------------
      N1=N+1
  100 WRITE(6,5) N,(CA(J),CB(J),J=1,N1)
      GOTO 10
  999 STOP
    1 FORMAT(26I2)
    2 FORMAT(Q39.32)
    3 FORMAT(I1)
    4 FORMAT('1','COEFFICIENTS FOR THE RATIONAL APPROXIMATIONS OF OF1( V
     :+1 ; (Z**2)/4 )'//' ','IDELTA = ',I1,T20,'V = ',Q39.32/)
    5 FORMAT(' ','N = ',I2,T18,'CA(I)',T58,'CB(I)'//,Q39.32/)
     :26(1X,Q39.32,2X,Q39.32/))
      END
      SUBROUTINE CBOF1(V,IDELTA,A,B,N)
C     ****************************************************************
C     *      THIS SUBROUTINE RETURNS COEFFICIENTS A(I) AND B(I) ,      *
C     *   I = 1,2,...,N+1 , OF THE NUMERATOR AND DENOMINATOR POLYNOMIALS *
C     *   RESPECTIVELY, IN THE RATIONAL APPROXIMATION OF ORDER   N   FOR *
C     *   OF1(V+1;(Z**2)/4) WHICH EXCEPT FOR AN AUXILIARY MULTIPLIER    *
C     *   IS THE BESSEL FUNCTION I(V;Z) .                              *
C     *                                                                *
C     *      NO OTHER SUBROUTINES ARE CALLED BY THIS ONE.              *
C     ****************************************************************
      IMPLICIT REAL*16(A-H,O-Z)
      DIMENSION A(1),B(1)
      DATA ONE/1.Q0/
C
C     INITIALIZATION :
C
      B(1)=ONE
      A(1)=ONE
      XI=ONE
      CT1=IDELTA-2-(N+N)
      CT2=CT1-ONE
      CT3=CT1-V
C
C     FOR EACH I = 1,2,...,N , A(I+1) AND B(I+1) ARE CALCULATED
C     USING THE RECURRENCE RELATIONS BELOW
C
      DO 100 I=1,N
      I1=I+1
      XII=XI+XI
      X4I=XII+XII
C-----------------------------------------------------------------
C     THE RECURRENCE RELATIONS FOR A(I+1) AND B(I+1) ARE AS FOLLOWS
C-----------------------------------------------------------------
      B(I1)=B(I)/(CT3+XI)/X4I
      A(I1)=A(I)/(CT3+XI)*(CT1+XII)/(CT1+XI)*(CT2+XII)/(V+XI)/X4I
C
  100 XI=XI+ONE
      RETURN
      END
```

COEFFICIENTS FOR THE RATIONAL APPROXIMATIONS OF 0F1(V+1 ; (Z**2)/4)

DELTA = 0   V =   0.333333333333333333333333333333333333Q+00

N = 3   CA(I)   CB(I)

0.10000000000000000000000000000000000Q+01   0.10000000000000000000000000000000000Q+01
0.15340909090909090909090909090909090Q+00   -0.34090909090909090909090909090909090Q-01
0.43254442925495557074504442925495Q-02   0.67284688995215311004784688995215Q-03
0.24330624145591250854408749145591Q-04   -0.10513232655502392344497607655502Q-04

N = 4   CA(I)   CB(I)

0.10000000000000000000000000000000000Q+01   0.10000000000000000000000000000000000Q+01
0.16071428571428571428571428571429Q+00   -0.26785714285714285714285714285714Q-01
0.54241071428571428571428571428571Q-02   0.40178571428571428571428571428571Q-03
0.52832212430426716141001855287569Q-04   -0.45657467532467532467532467532467Q-05
0.12031627324742926246685645181886Q-06   0.45056711380724538619275461380724Q-07

N = 5   CA(I)   CB(I)

0.10000000000000000000000000000000000Q+01   0.10000000000000000000000000000000000Q+01
0.16544117647058823529411764705882Q+00   -0.22058823529411764705882352941176Q-01
0.61754540525887774466462455950122Q-02   0.26684060721062618595825426944971Q-03
0.77193175657359718080780699376524Q-04   -0.23825054215234480889129854865582Q-05
0.33400893313280647246491648768688Q-06   0.17868790661425860666847384114936Q-07
0.36600004848156553395100426816339Q-09   -0.12183266360063086818305034623820Q-09

N = 6   CA(I)   CB(I)

0.10000000000000000000000000000000000Q+01   0.10000000000000000000000000000000000Q+01
0.16875000000000000000000000000000Q+00   -0.18750000000000000000000000000000Q-01
0.67190516409266640926640926640926Q-02   0.19003378378378378378378378378378Q-03
0.97013045082898024074494662729956Q-04   -0.13973072337042925278219395866455Q-05
0.56871791317207588554837131687226Q-06   0.84514550425662854505359249192266Q-08
0.12694596276162408159561859751613Q-08   -0.45275652013747957770728169210143Q-10
0.75165372687803732523721538002971Q-12   0.22637826006873978885364084605071Q-12

```
C       ..............................................................
C       .        THIS MAINLINE PROGRAM ILLUSTRATES THE SYNTAX OF THE SUB- .
C       . ROUTINE 'RB1F1' FOR GENERATING VALUES OF THE NUMERATOR AND      .
C       . DENOMINATOR POLYNOMIALS IN THE RATIONAL APPROXIMATION OF        .
C       . 1F1( V+1/2 ; 2V+1 ; -2Z ) WHICH EXCEPT FOR AN AUXILLIARY        .
C       . MULTIPLIER IS THE BESSEL FUNCTION I( V ; Z ) .                  .
C       ..............................................................
        IMPLICIT REAL*16(A-H,O-Z)
        DIMENSION A(26),B(26),R(26),D(26),E(26)
        DATA ZERO/0.Q0/
   10   READ(5,1,END=999) N,M
        READ(5,2) V
        N1=N+1
        D(N1)=ZERO
        E(N1)=ZERO
        DO 100 I=1,M
          READ(5,2) Z
C------------------------------------------------------------------------
          CALL RB1F1(V,Z,A,B,N)
C------------------------------------------------------------------------
C       IN THE ABOVE :
C
C    V            IS THE VARIABLE IN THE PARAMETERS OF THE 1F1
C    Z            IS THE VARIABLE IN THE ARGUMENT OF THE 1F1
C    A AND B      WILL CONTAIN THE VALUES OF THE NUMERATOR AND DENOMINATOR
C                 POLYNOMIALS RESPECTIVELY FOR ALL DEGREES FROM 0 TO
C                 N INCLUSIVE
C    N            IS THE MAXIMUM DEGREE FOR WHICH VALUES OF THE POLYNOMIALS
C                 ARE TO BE CALCULATED
C
C    NOTE : VALUES OF THE K-TH DEGREE POLYNOMIALS WILL BE PLACED IN
C A(K+1) OR B(K+1) AS APPROPRIATE.
C------------------------------------------------------------------------
          R(N1)=A(N1)/B(N1)
          DO 50 J=1,N
            J1=N1-J
            R(J1)=A(J1)/B(J1)
            D(J1)=R(J1+1)-R(J1)
   50       E(J1)=R(N1)-R(J1)
          WRITE(6,3) N,V,Z
          DO 60 J=1,N1
            J1=J-1
   60       WRITE(6,4) J1,A(J),B(J)
          WRITE(6,5)
          DO 70 J=1,N1
            J1=J-1
   70       WRITE(6,6) J1,R(J),D(J),E(J)
  100   CONTINUE
        GOTO 10
  999   STOP
    1   FORMAT(2I2)
    2   FORMAT(Q39.32)
    3   FORMAT('1','VALUES OF THE POLYNOMIALS IN THE RATIONAL APPROXIMATIO
       :N OF 1F1(V+1/2;2V+1;-2Z)'//' ','N = ',I2,'     V = ',Q39.32//
       :' ','     Z = ',Q39.32//' ',' I',T24,'A(I)',T65,'B(I)'/)
    4   FORMAT(' ',I2,2X,Q39.32,2X,Q39.32)
    5   FORMAT('0','VALUES OF THE APPROXIMATION, 1ST DIFFERENCES AND APPRO
       :XIMATE ERRORS'//' ',' I',T12,'I-TH APPROXIMATION -- F(I)',T47,
       :'1ST DIFF''S.',T60,'F(N)-F(I)'/)
    6   FORMAT(' ',I2,2X,Q39.32,2X,Q10.3,2X,Q10.3)
        END
```

```
      SUBROUTINE RB1F1(V,Z,A,B,N)
C     ******************************************************************
C     *         THIS SUBROUTINE RETURNS VALUES A(I) AND B(I), I=1,...,N+1 *
C     *   OF THE NUMERATOR AND DENOMINATOR POLYNOMIALS IN THE RATIONAL  *
C     *   APPROXIMATION OF 1F1( V+1/2 ; 2V+1 ; -2Z ) WHICH EXCEPT FOR   *
C     *   AN AUXILLIARY MULTIPLIER IS THE BESSEL FUNCTION I( V ; Z ) .  *
C     *                                                                *
C     *         NO OTHER SUBROUTINES ARE CALLED BY THIS ONE.           *
C     ******************************************************************
      IMPLICIT REAL*16(A-H,O-Z)
      DIMENSION A(1),B(1)
      DATA ONE/1.Q0/,TWO/2.Q0/,THREE/3.Q0/
C
C  INITIALIZATION :
C
      Z2=Z/TWO
      Y=Z2*Z2
      V2I=V+ONE
      V1I=V2I+ONE
      B(1)=ONE
      A(1)=ONE
      A(2)=ONE
      B(2)=ONE+Z
      B(3)=B(2)+Y*(ONE+V2I/V1I)
      A(3)=ONE+Y/V1I/V2I
      XI=THREE
      DO 100 I=3,N
C
C  FOR I=3,...,N , THE VALUES OF A(I+1) AND B(I+1) ARE CALCULATED
C  USING THE RECURRENCE RELATIONS BELOW
C
      E=Z2/V1I
      G1=E/XI*(V+V1I)
      G3=-G1/V1I*Y/V2I
      V2I=V1I
      V1I=V1I+ONE
      G2=E/V1I*Z2-G1
      G1=ONE+G1
C-----------------------------------------------------------------------
C     THE RECURRENCE RELATIONS FOR A(I+1) AND B(I+1) ARE AS FOLLOWS.
C-----------------------------------------------------------------------
C
      B(I+1)=G1*B(I)+G2*B(I-1)+G3*B(I-2)
      A(I+1)=G1*A(I)+G2*A(I-1)+G3*A(I-2)
C
  100 XI=XI+ONE
      RETURN
      END
```

VALUES OF THE POLYNOMIALS IN THE RATIONAL APPROXIMATION OF 1F1(V+1/2;2V+1;-2Z)

N = 12   V = 0.3333333333333333333333333333333333333Q+00

Z = 0.500000000000000000000000000000000Q+00

I	A(I)	B(I)
0	0.1000000000000000000000000000000000000Q+01	0.1000000000000000000000000000000000000Q+01
1	0.1020008928571428571428571428571430Q+01	0.1500000000000000000000000000000000Q+01
2	0.1020008928571428571428571428571430Q+01	0.1598214285714285714285714285714143Q+01
3	0.1028125000000000000000000000000000Q+01	0.1617708333333333333333333333333333Q+01
4	0.1032538848478708791208791208791209Q+01	0.1625135216346153846153846153846153846Q+01
5	0.1035319239886675824175824175824Q+01	0.1629537572616185897435897435897435Q+01
6	0.1037229770508165508422498554075Q+01	0.1632545826749334022154745458395630Q+01
7	0.1038622993161559349367375256322Q+01	0.1634738731166989962477215766689Q+01
8	0.1039683796335942700455896899153Q+01	0.1636640838180566447323245673369028Q+01
9	0.1040518404098304667790210259665Q+01	0.1637720111864759788088312034897Q+01
10	0.1041192162318564739757015529850Q+01	0.1638782471688478395271553759025Q+01
11	0.1041747467444383320631898893678Q+01	0.1639656493160319092080041704071Q+01
12	0.1042213024353210313292162269439Q+01	0.1640389255593010251202323769468Q+01

VALUES OF THE APPROXIMATION, 1ST DIFFERENCES AND APPROXIMATE ERRORS

I	I-TH APPROXIMATION -- F(I)	1ST DIFF'S.	F(N)-F(I)
0	0.1000000000000000000000000000000Q+01	-0.333Q+00	-0.365Q+00
1	0.6666666666666666666666666666670Q+00	-0.284Q-01	-0.313Q-01
2	0.6382681564245810055865921877090Q+00	-0.272Q-02	-0.292Q-02
3	0.6355441081777205408886027044430Q+00	-0.188Q-03	-0.199Q-03
4	0.6353556537851483723221466937319Q+00	-0.102Q-04	-0.107Q-04
5	0.6353454203123931986396708591132Q+00	-0.456Q-06	-0.473Q-06
6	0.6353449646025926236278312802330646Q+00	-0.172Q-07	-0.178Q-07
7	0.6353449473972627864787403681021Q+00	-0.564Q-09	-0.581Q-09
8	0.6353449468333343154221184240222000Q+00	-0.163Q-10	-0.168Q-10
9	0.6353449468170017145548739399462820Q+00	-0.424Q-12	-0.434Q-12
10	0.6353449468165779540992382056815Q+00	-0.996Q-14	-0.102Q-13
11	0.6353449468165679965996538308956700Q+00	-0.214Q-15	-0.214Q-15
12	0.6353449468165677827366300159544Q+00	0.0	0.0

```
C      . . . . . . . . . . . . . . . . . . . . . . . . . . . . . . . . . . . . . . . . .
C      .       THIS MAINLINE PROGRAM ILLUSTRATES THE SYNTAX OF THE SUB-  .
C      .  ROUTINE 'CB1F1' FOR GENERATING COEFFICIENTS IN THE POLYNOMIALS .
C      .  FOR THE RATIONAL APPROXIMATION OF 1F1(V+1/2;2V+1;-2Z) WHICH     .
C      .  EXCEPT FOR AN AUXILLIARY MULTIPLIER IS THE BESSEL FUNCTION      .
C      .  I(V;Z) .                                                        .
C      . . . . . . . . . . . . . . . . . . . . . . . . . . . . . . . . . . . . . . . . .
       IMPLICIT REAL*16(A-H,O-Z)
       DIMENSION CA(26),CB(26),NO(25)
  10   READ(5,2,END=999) V
       WRITE(6,3) V
       READ(5,1) M,(NO(J),J=1,M)
       DO 100 I=1,M
          N=NO(I)
C-------------------------------------------------------------------------
       CALL CB1F1(V,CA,CB,N)
C-------------------------------------------------------------------------
C      IN THE ABOVE:
C
C   V            IS THE VARIABLE IN THE PARAMETERS OF THE 1F1
C   N            IS THE DEGREE OF THE POLYNOMIALS IN THE RATIONAL
C                APPROXIMATION
C   CA AND CB    WILL CONTAIN THE COEFFICIENTS OF THE NUMERATOR AND
C                DENOMINATOR POLYNOMIALS RESPECTIVELY
C
C      NOTE : THE COEFFICIENTS OF THE K-TH POWER OF Z WILL BE PLACED
C  IN CA(K+1) OR CB(K+1) AS APPROPRIATE
C-------------------------------------------------------------------------
          N1=N+1
 100      WRITE(6,4) N,(CA(J),CB(J),J=1,N1)
       GOTO 10
 999   STOP
   1   FORMAT(26I2)
   2   FORMAT(Q39.32)
   3   FORMAT('1','COEFFICIENTS FOR THE RATIONAL APPROXIMATION OF 1F1( V+
      :1/2 ; 2V+1 ; -2Z )'//' ',T20,'V = ',Q39.32/)
   4   FORMAT(' ','N = ',I2,T18,'CA(I)',T58,'CB(I)'//
      :26(1X,Q39.32,2X,Q39.32/)/)
       END
```

```
      SUBROUTINE CB1F1(V,A,B,N)
C     ***************************************************************
C     *         THIS SUBROUTINE RETURNS COEFFICIENTS A(I) AND B(I) ,  *
C     *   I = 1,2,...,N+1 , OF THE NUMERATOR AND DENOMINATOR POLYNOMIALS *
C     *   RESPECTIVELY IN THE RATIONAL APPROXIMATION OF ORDER  N  FOR  *
C     *   1F1( V+1/2 ; 2V+1 ; -2Z ) WHICH EXCEPT FOR AN AUXILLIARY     *
C     *   MULTIPLIER IS THE BESSEL FUNCTION I(V;Z)                     *
C     *                                                               *
C     *         NO OTHER SUBROUTINES ARE CALLED BY THIS ONE.          *
C     ***************************************************************
      IMPLICIT REAL*16(A-H,O-Z)
      LOGICAL*1 SWITCH
      DIMENSION A(1),B(1)
      DATA ZERO/0.Q0/,ONE/1.Q0/,TWO/2.Q0/,FOUR/4.Q0/
C
C   INITIALIZATION :
C
      B(1)=ONE
      B(2)=ONE
      A(1)=ONE
      XI1=ONE
      A(2)=ZERO
      XI=TWO
      SWITCH=.FALSE.
      V2=V+V
      CT1=-(N+1)
      CT2=CT1+CT1
      CT3=CT2-V2+TWO
C
C   FOR EACH I = 2,3,...,N , A(I+1) AND B(I+1) ARE CALCULATED
C   USING THE RECURRENCE RELATIONS BELOW
C
      DO 100 I=2,N
      I1=I+1
C---------------------------------------------------------------
C   THE RECURRENCE RELATIONS FOR A(I+1) AND B(I+1) ARE AS FOLLOWS
C---------------------------------------------------------------
C
      B(I1)=B(I)/XI*(CT3+XI1)/CT3
C
      A(I1)=ZERO
      IF(SWITCH) GOTO 50
      I2=I-1
C
C   A(I+1) IS CALCULATED FOR EVEN I ONLY, SINCE THE NUMERATOR POLYNOMIAL
C   IS A POLYNOMIAL IN Z**2
C
      A(I1)=A(I2)/XI*(CT1+XI1)/(CT2+XI)*(CT1+XI)/CT3*FOUR/(V2+XI)
C
  50  CT3=CT3+ONE
      XI1=XI
      XI=XI+ONE
 100  SWITCH=.NOT.SWITCH
      RETURN
      END
```

COEFFICIENTS FOR THE RATIONAL APPROXIMATION OF 1F1(V+1/2 ; 2V+1 ; -2Z)

V =   0.33333333333333333333333333333333333Q+00

N = 3	CA(I)	CB(I)

	CA(I)	CB(I)
0.10000000000000000000000000000000000Q+01		0.10000000000000000000000000000000000Q+01
0.0		0.10000000000000000000000000000000000Q+01
0.11250000000000000000000000000000000Q+00		0.42500000000000000000000000000000000Q+00
0.0		0.91666666666666666666666666666666660Q-01

N = 4   CA(I)   CB(I)

0.10000000000000000000000000000000000Q+01   0.10000000000000000000000000000000000Q+01
0.0   0.10000000000000000000000000000000000Q+01
0.12980769230769230769230769230769Q+00   0.44230769230769230769230769230769Q+00
0.0   0.10897435897435897435897435897436Q+00
0.13907967032967032967032967032967Q-02   0.14983974358974358974358974358974Q-01

N = 5   CA(I)   CB(I)

0.10000000000000000000000000000000000Q+01   0.10000000000000000000000000000000000Q+01
0.0   0.10000000000000000000000000000000000Q+01
0.14062500000000000000000000000000000Q+00   0.45312500000000000000000000000000000Q+00
0.0   0.11979166666666666666666666666666670Q+00
0.26077438186813186813186813186813Q-02   0.19581330128205128205128205128205Q-01
0.0   0.18729967948717948717948717948718Q-02

N = 6   CA(I)   CB(I)

0.10000000000000000000000000000000000Q+01   0.10000000000000000000000000000000000Q+01
0.0   0.10000000000000000000000000000000000Q+01
0.14802631578947368421052631578947Q+00   0.46052631578947368421052631578947Q+00
0.35684915413533834586466165413534Q-02   0.12719298245614035087719298245614Q+00
0.0   0.22854989035087719292456140350880Q-01
0.10293725600057836899942163100058Q-04   0.26795504385964912280701754385965Q-02
                                                    0.18894265913180386864597390913180Q-03

XVIII PADÉ APPROXIMATIONS FOR $I_{\nu+1}(z)/I_\nu(z)$

Let

$$S_\nu(z) = \frac{2(\nu+1)I_{\nu+1}(z)}{zI_\nu(z)} .$$ (1)

Then

$$S_\nu(z) = \{C_n(z,a)/D_n(z,a)\} + R_n(z) ,$$ (2)

$$C_n(z,a) = {}_2F_3^{n-a}(a-n,\tfrac{1}{2}-n;-2n-\nu-1+a,\nu+2,-2n+a;z^2) , \quad C_0(z,1)=0,$$ (3)

$$D_n(z,a) = {}_2F_3^n(-n,-n-\tfrac{1}{2}+a;-2n-\nu-1+a,\nu+1,-2n+a-1:z^2) , \quad a=0 \text{ or } a=1 .$$ (4)

This approximation to $S_\nu(z)$ is the same as $2(\nu+1)/z$ times the ratio $W_{1,\nu}(z)/W_{0,\nu}(z)$, where $W_{m,\nu}(z)$ is found by the backward recursion process described by 19.2(1-7). Indeed, in 17(7), let

$$A_n(z) \equiv A_n(z,\delta,\nu) .$$ (5)

Then

$$C_n(z,0) = A_n(z,1,\nu+1) , \quad C_n(z,1) = A_{n-1}(z,0,\nu+1) ,$$
$$D_n(z,0) = A_n(z,0,\nu) , \quad D_n(z,1) = A_n(z,1,\nu) .$$ (6)

Both $C_n(z,a)$ and $D_n(z,a)$ satisfy the same recurrence formula

$$D_{n+1}(z,a) = (1+Ez^2)D_n(z,a) - Fz^4D_{n-1}(z,a) , \quad n \geq 1,$$

$$E = \{2(2n+\nu+1-a)(2n+\nu+3-a)\}^{-1} ,$$ (7)

$$F = \{16(2n+\nu-a)(2n+\nu+1-a)^2(2n+\nu+2-a)\}^{-1} .$$

This formula is stable when used in the forward direction.

For the error, we have

$$R_n(z) = \frac{(-)^{1-a}(\nu+1)(z/2)^{2n-a}I_{2n+2+\nu-a}(z)}{(\nu+1)_{2n+1-a}D_n(z,a)I_\nu(z)} .$$ (8)

Thus for z and ν fixed,

$$\lim_{n\to\infty} R_n(z) = 0 ,$$ (9)

unless z is a zero of $I_\nu(z)$ in which case if ν is real, we know that z is pure imaginary. Also for z and ν fixed,

$$D_n(z,a) = (z/2)^{-\nu}\Gamma(\nu+1)I_\nu(z)\{1 - \frac{z^2}{4(2n+\nu+1-a)} + O(n^{-2})\},$$

(10)

so that approximations for zeros of $I_\nu(z)$ follow by find-

ing zeros of $D_n(z,a)$. Evaluation of (8) can be simplified by

use of (10) and the defining series for $I_\mu(z)$, $\mu = 2n+2+\nu-a$,

see 17(1). Thus

$$R_n(z) = \frac{(-)^{1-a}(\nu+1)(z/2)^{4n+2-2a}\{1+\frac{(z^2/2)(2n+\nu+2-a)}{(2n+\nu+1-a)(2n+\nu+3-a)}+O(n^{-2})\}}{(\nu+1)_{2n+1-a}(\nu+1)_{2n+2-a}\{(z/2)^{-\nu}\Gamma(\nu+1)I_\nu(z)\}^2}.$$

(11)

Further simplification follows by use of 1.3(14). We find

$$R_n(z) = \frac{(-)^{1-a}(2n+2-a)^{-2\nu}\Gamma(\nu+1)\Gamma(\nu+2)(z/2)^{4n+2-2a}}{(2n+1-a)!(2n+2-a)!\{(z/2)^{-\nu}\Gamma(\nu+1)I_\nu(z)\}^2}$$

$$\times\{1 + \frac{(z^2/2)}{(2n+2+\nu-a)} - \frac{\nu^2}{n} + O(n^{-2})\}.$$

(12)

To facilitate estimation of the error, we record the result

$$\frac{R_{n+1}(z)}{R_n(z)} = \frac{(z/2)^4\{1 - z^2\{(2n+\nu+1-a)(2n+\nu+5-a)\}^{-1} + O(n^{-3})\}}{(2n+2+\nu-a)(2n+3+\nu-a)^2(2n+4+\nu-a)}.$$

(13)

The approximations to $S_\nu(z)$ occupy the (n,n-a) positions

of the Padé table. In view of (8), we have the inequality

$$\frac{C_n(z,1)}{D_n(z,1)} < S_\nu(z) < \frac{C_n(z,0)}{D_n(z,0)}, \quad z > 0, \nu > -1,$$

(14)

with equality if z = 0 provided n > 0.

Notice that

$$S_{-\frac{1}{2}}(z) = z^{-1}\tanh z.$$

(15)

To clarify the programs at the end of this chapter, it

is helpful to introduce some new notation, namely

S(V;SQRT(VAL)*Z). If VAL = 1.0, the first mainline program and

and subroutine give values of the polynomials $C_n(z,a)$,

$D_n(z,a)$ and the quotient of these polynomials, see (1,2).

Thus VAL = 1.0 relates to the rational approximations for

$S_\nu(z)$. If in (1-4), we replace z by iz, we have

$$S_\nu(iz) = \frac{2(\nu+1)J_{\nu+1}(z)}{zJ_\nu(z)} = \{C_n(iz,a)/D_n(iz,a)\} + R_n(iz) . \qquad (16)$$

Then upon setting VAL = - 1.0, we get data for (16) corresponding to that for (1). The second mainline program and subroutine are designed to get the coefficients in the polynomials (3,4). If for the latter equations, we write

$$C_n(z,a) = \sum_{k=0}^{n-a} a_k z^{2k} , \quad D_n(z,a) = \sum_{k=0}^{n} b_k z^{2k} , \qquad (17)$$

then this program generates the a_k's and the b_k's. If we desire to evaluate the numerator and denominator polynomials in the rational approximation for $S_\nu(iz)$, see (16), then in (17) we must replace z^2 by $-z^2$, or equivalently, we must replace a_k and b_k by $(-)^k a_k$ and $(-)^k b_k$, respectively.

Numerical Example

Let

z = 2.5, ν = - ½, VAL = 1.0 and a = 0.

Then $S_\nu(z)$ is given by (15) and to facilitate computation of the error, we note that

$$\{(z/2)^{-\nu}\Gamma(\nu+1)I_\nu(z)\}^2 = \sinh^2 z , \quad \text{if } \nu = -\tfrac{1}{2} . \qquad (18)$$

Let n = 4 and use (12) without the order term. Then $R_4(z) = - 2.26004(-11)$. The true value is - 2.38457(-11). From (13) without the order term, $R_5(z)/R_4(z) = 0.191(-3)$ while the true value is 0.192(-3).

```
C     ....................................................................
C     .         THIS MAINLINE PROGRAM ILLUSTRATES THE SYNTAX OF THE SUB-  .
C     .  ROUTINE 'RSVZ' FOR GENERATING VALUES OF THE NUMERATOR AND        .
C     .  DENOMINATOR POLYNOMIALS IN THE MAIN DIAGONAL AND FIRST SUB-      .
C     .  DIAGONAL PADE APPROXIMATIONS OF S( V ; SQRT(VAL)*Z )  .
C     ....................................................................
      IMPLICIT REAL*16(A-H,O-Z)
      DIMENSION A(26),B(26),R(26),D(26),E(26)
      DATA ZERO/0.Q0/
   10 READ(5,1,END=999) N,IA,M
      READ(5,2) V,VAL
      N1=N+1
      D(N1)=ZERO
      E(N1)=ZERO
      DO 100 I=1,M
      READ(5,2) Z
C-------------------------------------------------------------------------
      CALL RSVZ(V,VAL,IA,Z,A,B,N)
C-------------------------------------------------------------------------
C     IN THE ABOVE :
C
C     V               IS THE PARAMETER IN S(V;Z)
C     VAL             IS THE PARAMETER IN S(V;SQRT(VAL)*Z)
C     IA              IS A PARAMETER WHICH MUST EQUAL 0 OR 1 .
C     Z               IS THE ARGUMENT OF THE FUNCTION ABOVE
C     A AND B         WILL CONTAIN THE VALUES OF THE NUMERATOR AND DENOMINATOR
C                     POLYNOMIALS RESPECTIVELY FOR ALL DEGREES FROM 0 TO
C                     N INCLUSIVE
C     N               IS THE MAXIMUM DEGREE FOR WHICH VALUES OF THE POLYNOMIALS
C                     ARE TO BE CALCULATED
C
C     NOTE : VALUES OF THE K-TH DEGREE POLYNOMIALS WILL BE PLACED IN
C     A(K+1) AND B(K+1) AS APPROPRIATE.
C-------------------------------------------------------------------------
      R(N1)=A(N1)/B(N1)
      DO 50 J=1,N
      J1=N1-J
      R(J1)=A(J1)/B(J1)
      D(J1)=R(J1+1)-R(J1)
   50 E(J1)=R(N1)-R(J1)
      WRITE(6,3) VAL,N,IA,V,Z
      DO 60 J=1,N1
      J1=J-1
   60 WRITE(6,4) J1,A(J),B(J)
      WRITE(6,5)
      DO 70 J=1,N1
      J1=J-1
   70 WRITE(6,6) J1,R(J),D(J),E(J)
  100 CONTINUE
      GOTO 10
  999 STOP
    1 FORMAT(3I2)
    2 FORMAT(Q39.32)
    3 FORMAT('1',5'VALUES OF THE POLYNOMIALS IN THE PADE APPROXIMATION OF
     : S( V ; SQRT(VAL)*Z )'//' ','VAL = ',Q39.32//' ','N = ',I2,'      IA
     : = ',I2,'     V = ',Q39.32//' ','     Z = ',Q39.32//' ','I',T24,'A
     :(I)',T65,'B(I)'/)
    4 FORMAT(' ',I2,2X,Q39.32,2X,Q39.32)
    5 FORMAT('0',5'VALUES OF THE APPROXIMATION, 1ST DIFFERENCES AND APPRO
     :XIMATE ERRORS'//' ',' I',T12,'I-TH APPROXIMATION -- F(I)',T47,
     :'1ST DIFF''S.',T60,'F(N)-F(I)'/)
    6 FORMAT(' ',I2,2X,Q39.32,2X,Q10.3,2X,Q10.3)
      END
```

```
      SUBROUTINE RSVZ(V,VAL,IA,Z,A,B,N)
C     ************************************************************
C     *        THIS SUBROUTINE RETURNS VALUES A(I) AND B(I), I=1,...,N+1 *
C     *  OF THE NUMERATOR AND DENOMINATOR POLYNOMIALS IN THE PADE        *
C     *  APPROXIMATIONS OF S( V ; SQRT(VAL)*Z ) .                        *
C     *        IF  IA = 0 , THE MAIN DIAGONAL PADE APPROXIMATION WILL BE *
C     *  OBTAINED, BUT IF IA = 1, THE 1ST SUBDIAGONAL APPROXIMATION      *
C     *  WILL ENSUE.                                                     *
C     *                                                                  *
C     *        NO OTHER SUBROUTINES ARE CALLED BY THIS ONE.             *
C     ************************************************************
      IMPLICIT REAL*16(A-H,O-Z)
      DIMENSION A(1),B(1)
      DATA ZERO/0.Q0/,ONE/1.Q0/,TWO/2.Q0/,THREE/3.Q0/
C
C     INITIALIZATION :
C
      XA=IA
      CT2=THREE-XA
      CT1=CT2+V
      Z2=Z/TWO
      Y=VAL*Z2*Z2
      A(1)=ZERO
      B(1)=ONE
      A(2)=ONE
      B(2)=ONE+Z/CT1*Z2/(V+ONE)*(ONE-XA/CT2)
      IF(IA.EQ.1) GOTO 10
      A(1)=ONE
      A(2)=ONE+Y/(V+TWO)/(V+THREE)
  10  DO 100 I=2,N
C
C     FOR I=2,...,N , THE VALUES A(I+1) AND B(I+1) ARE CALCULATED
C     USING THE RECURRENCE RELATIONS BELOW
C
          CT2=CT1*CT1
          G2=Y/CT2*Y/(ONE-CT2)
          G1=Y/CT1
          CT1=CT1+TWO
          G1=ONE+(G1+G1)/CT1
C-------------------------------------------------------------------
C     THE RECURRENCE RELATIONS FOR A(I+1) AND B(I+1) ARE AS FOLLOWS
C-------------------------------------------------------------------
C
          B(I+1)=G1*B(I)+G2*B(I-1)
 100      A(I+1)=G1*A(I)+G2*A(I-1)
C
      RETURN
      END
```

VALUES OF THE POLYNOMIALS IN THE PADE APPROXIMATION OF S(V ; SQRT(VAL)*Z)

VAL = 0.1000000000000000000000000000000000Q+01

N = 12   IA = 0   V = -0.5000000000000000000000000000000Q+00

Z = 0.2500000000000000000000000000000000Q+01

I	A (I)	B(I)
0	0.10000000000000000000000000000000Q+01	0.10000000000000000000000000000000Q+01
1	0.14166788042332800420Q+01	0.35978174603174603174603174603Q+01
2	0.17357804233280423Q+01	0.48574917582491582491582491582Q+01
3	0.19123564275774630729504537945Q+01	0.51136692329980305342180698435Q+01
4	0.20878842454058204054730290497Q+01	0.54421239887697453756198425250Q+01
5	0.21737801105164228843528449044Q+01	0.55778282913761692757682153204Q+01
6	0.22055310685524986322170290135Q+01	0.56778696861183152027587694747Q+01
7	0.22452809341227781804264307095Q+01	0.56809996864118159234350527408Q+01
8	0.22579753507441227872860109518Q+01	0.57152498994940502716762331244Q+01
9	0.27072054241494945263330808248Q+01	0.57539056692889875252197685063Q+01

VALUES OF THE APPROXIMATION, 1ST DIFFERENCES AND APPROXIMATE ERRORS

I	I-TH APPROXIMATION -- F(I)	1ST DIFF'S.	F(N)-F(I)
0	0.10000000000000000Q+01	0.5950+01	0.6050+00
1	0.40476190476190476Q+01	-0.1010Q+04	-0.4560Q+04
2	0.39469133176763980Q+01	0.5490Q+07	-0.5490Q+07
3	0.39464577928445771Q+01	-0.3680Q+14	-0.3680Q+14
4	0.39464577660732317Q+01	0.4560Q+18	-0.4560Q+18
5	0.39464577199260577Q+01	-0.2380Q+22	-0.2380Q+22
6	0.39464577192605772Q+01	0.1760Q+27	-0.1760Q+27
7	0.39464577192605772Q+01	-0.1620Q+31	-0.1620Q+31
8	0.39464577192605772Q+01	0.3850Q+33	-0.3850Q+33
9	0.39464577192605772Q+01	0.0	0.0

VALUES OF THE POLYNOMIALS IN THE PADE APPROXIMATION OF S(V : SQRT(VAL)*Z)

VAL = -0.1000000000000000000000000000000000Q+01

N = 12   IA = 0   V = -0.500000000000000000000000000000000Q+00

Z = 0.250000000000000000000000000000000Q+01

I	A (I)	B(I)
0	0.100000000000000000000000000000000Q+01	0.100000000000000000000000000000000Q+01
1	0.583333333333333333333333333333333Q+00	0.350000000000000000000000000000000Q+01
2	0.346891534391534391534391534391540Q+00	0.245337301587301587301587301587300Q+01
3	0.306177040552040552040552040552040Q+00	0.216933991933991933991933991933990Q+01
4	0.288373394148516697536305379442630Q+00	0.204320290786283433342256871668640Q+01
5	0.278131078707840388389801137356300Q+00	0.197063335667227059310208359301680Q+01
6	0.271450458729530991421064082346710Q+00	0.192329937072043912217529337909140Q+01
7	0.266742286148283958528301097956420Q+00	0.188994070407779907231769850315610Q+01
8	0.263243127888237647574363671114778Q+00	0.186514822845958921863119841233480Q+01
9	0.260539355180329341953589288377740Q+00	0.184599127299880473422534490876370Q+01
10	0.258387021868214918809943686355180Q+00	0.183074141369060985289417793401660Q+01
11	0.256632822078099921010426459418740Q+00	0.181831243726435243881081114448510Q+01
12	0.255175521222806355817979652343100Q+00	0.180798706949354618744452764026840Q+01

VALUES OF THE APPROXIMATION, 1ST DIFFERENCES AND APPROXIMATE ERRORS

I	I-TH APPROXIMATION -- F(I)	1ST DIFF'S.	F(N)-F(I)
0	0.100000000000000000000000000000000Q+01	-0.833Q+00	-0.859Q+00
1	0.166666666666666666666666666666670Q+00	-0.253Q-01	-0.255Q-01
2	0.141393718830031001482679606415960Q+00	-0.255Q-03	-0.256Q-03
3	0.141138342507984291519381894313140Q+00	-0.430Q-06	-0.430Q-06
4	0.141137912949699058203081963046710Q+00	-0.224Q-09	-0.224Q-09
5	0.141137912725434208113892988850420Q+00	-0.483Q-13	-0.483Q-13
6	0.141137912725385917155234030696260Q+00	-0.507Q-17	-0.507Q-17
7	0.141137912725385912087522386558820Q+00	-0.290Q-21	-0.290Q-21
8	0.141137912725385912087232395896680Q+00	-0.982Q-26	-0.982Q-26
9	0.141137912725385912087232386081340Q+00	-0.209Q-30	-0.209Q-30
10	0.141137912725385912087232386081140Q+00	0.385Q-33	0.385Q-33
11	0.141137912725385912087232386081140Q+00	0.0	0.0
12	0.141137912725385912087232386081140Q+00	0.0	0.0

```
C       ....................................................................
C       .          THIS MAINLINE PROGRAM ILLUSTRATES THE SYNTAX OF THE SUB-     .
C       .     ROUTINE 'CSVZ' FOR GENERATING COEFFICIENTS IN THE POLYNOMIALS     .
C       .     FOR THE MAIN DIAGONAL AND 1ST SUBDIAGONAL PADE APPROXIMATIONS     .
C       .     OF S(V;Z) = 2(V+1)*I(V+1;Z)/(Z*I(V;Z)) .                          .
C       ....................................................................
        IMPLICIT REAL*16(A-H,O-Z)
        DIMENSION CA(26),CB(26),NO(25)
   10   READ(5,2,END=999) V
        READ(5,3) IA
        WRITE(6,4) IA,V
        READ(5,1) M,(NO(J),J=1,M)
        DO 100 I=1,M
        N=NO(I)
C------------------------------------------------------------------------
        CALL CSVZ(V,IA,CA,CB,N)
C------------------------------------------------------------------------
C       IN THE ABOVE :
C
C
C    V            IS THE PARAMETER IN S(V;Z)
C    IA           IS 0 FOR THE MAIN DIAGONAL, OR 1 FOR THE 1ST SUBDIAGONAL
C                 PADE APPROXIMATIONS
C    2(N-IA)      IS THE DEGREE OF THE NUMERATOR POLYNOMIAL IN THE PADE
C                 APPROXIMATION
C    2N           IS THE DEGREE OF THE DENOMINATOR POLYNOMIAL
C    CA AND CB    WILL CONTAIN THE COEFFICIENTS OF THE NUMERATOR AND
C                 DENOMINATOR POLYNOMIALS RESPECTIVELY
C
C
C       NOTE : BOTH THE NUMERATOR AND DENOMINATOR ARE POLYNOMIALS IN
C    Z**2, SO THE COEFFICIENTS OF THE 2K-TH POWER OF Z WILL BE PLACED
C    IN CA(K+1) OR CB(K+1) AS APPPOPRIATE.
C------------------------------------------------------------------------
        N1=N+1
  100   WRITE(6,5) N,(CA(J),CB(J),J=1,N1)
        GOTO 10
  999   STOP
    1   FORMAT(26I2)
    2   FORMAT(Q39.32)
    3   FORMAT(I1)
    4   FORMAT('1','COEFFICIENTS FOR THE PADE APPROXIMATIONS OF S( V ; Z )
       ://' ','IA = ',I1,T20,'V = ',Q39.32/)
    5   FORMAT(' ','N = ',I2,T18,'CA(I)',T58,'CB(I)'//
       :26(1X,Q39.32,2X,Q39.32/)/)
        END
```

```
      SUBROUTINE CSVZ(V,IA,A,B,N)
C     ************************************************************
C     *        THIS SUBROUTINE RETURNS COEFFICIENTS A(I) AND B(I) ,       *
C     *     I = 1,2,...,N+1 , OF THE NUMERATOR AND DENOMINATOR POLYNOMIALS *
C     *     RESPECTIVELY IN THE PADE APPROXIMATIONS OF ORDER  N  FOR       *
C     *     S(V;Z) = 2(V+1)*I(V+1;Z)/(Z*I(V;Z)).  IF IA = 0 , THE MAIN     *
C     *     DIAGONAL APPROXIMATION IS OBTAINED, BUT IF IA = 1 , THE 1ST    *
C     *     SUBDIAGONAL APPROXIMATION WILL ENSUE .                         *
C     *                                                                    *
C     *         NO THER SUBROUTINES ARE CALLED BY THIS ONE.                *
C     ************************************************************
      IMPLICIT REAL*16(A-H,O-Z)
      DIMENSION A(1),B(1)
      DATA ONE/1.Q0/,HALF/0.5Q0/
C
C  INITIALIZATION :
C
      A(1)=ONE
      B(1)=ONE
      XI=ONE
      V1=V+ONE
      XN=N
      XN1I=-XN
      XA=IA
      CT2=XA-HALF
      A2NI=XA+XN1I-(XN+ONE)
      DO 100 I=1,N
C
C  FOR EACH I = 1,2,...,N , A(I+1) AND B(I+1) ARE CALCULATED
C  USING THE RECURRENCE RELATIONS BELOW
C
      I1=I+1
      CT1=A2NI-V
C------------------------------------------------------------
C  THE RECURRENCE RELATIONS FOR A(I+1) AND B(I+1) ARE AS FOLLOWS
C------------------------------------------------------------
C
      B(I1)=B(I)/XI*XN1I/CT1*(CT2+XN1I)/A2NI/(V+XI)
      A2NI=A2NI+ONE
      A(I1)=A(I)/XI*(XA+XN1I)/CT1*(HALF+XN1I)/A2NI/(V1+XI)
C
      XN1I=XN1I+ONE
  100 XI=XI+ONE
      RETURN
      END
```

COEFFICIENTS FOR THE PADE APPROXIMATIONS OF S(V ‡ Z)

IA = 0   V = -0.500000000000000000000000000000000000Q+00

N = 3	CA(I)		CB(I)

```
0.1000000000000000000000000000000000Q+01    0.1000000000000000000000000000000000Q+01
0.12820512820512820512820512820513Q+00      0.46153846153846153846153846153846Q+00
0.27972027972027972027972027972028Q-02      0.23310023310023310023310023310023Q-01
0.74000074000074000074000074000074Q-05      0.20720020720020720020720020720021Q-03
```

N = 4	CA(I)		CB(I)

```
0.1000000000000000000000000000000000Q+01    0.1000000000000000000000000000000000Q+01
0.13725490196078431372549019607843Q+00      0.47058823529411764705882352941176Q+00
0.39215686274509803921568627450980Q-02      0.27450980392156862745098039215686Q-01
0.28729440494146376499317675788264Q-04      0.40221216691804927099044746103569Q-03
0.29019636862774117676078460392186Q-07      0.13058836586824835295423530717648Q-05
```

N = 5	CA(I)		CB(I)

```
0.1000000000000000000000000000000000Q+01    0.1000000000000000000000000000000000Q+01
0.14285714285714285714285714285714Q+00      0.47619047619047619047619047619048Q+00
0.46783625730994152046783625730994Q-02      0.30075187969924812030075187969925Q-01
0.49142464003145117696201287532557Q-04      0.55039559683522531819745442036463Q-03
0.15607822232206722845083452484831Q-06      0.32761642668763411797467525021704Q-05
0.72730919455574229764607670155854Q-10      0.48002406840678991644641062302864Q-08
```

N = 6	CA(I)		CB(I)

```
0.1000000000000000000000000000000000Q+01    0.1000000000000000000000000000000000Q+01
0.14666666666666666666666666666667Q+00      0.48000000000000000000000000000000Q+00
0.52173913043478260869565217391304Q-02      0.31884057971014492753623188405797Q-01
0.66252587991718426501035196687370Q-04      0.66252587991718426501035196687370Q-03
0.32284836253274086988808575383709Q-06      0.52304674730304020921869892121608Q-05
0.51790635079263427627945059631690Q-09      0.15193805295658393877086388415863Q-07
0.12648855557491170393844812201018Q-12      0.11510458557316965058398779102926Q-10
```

XIX EVALUATION OF BESSEL FUNCTIONS OF THE FIRST KIND BY USE OF THE BACKWARD RECURRENCE FORMULA

19.1. Introduction

In the main body of this chapter, it is convenient to deal with $I_\nu(z)$. We assume $-\pi < \arg z \le \pi$ and $-\pi < \arg \nu \le \pi$, ν not a negative integer. This is no burden for if ν is an integer n, then $I_{-n}(z) = I_n(z)$. Actually it is sufficient to have $0 \le \arg z \le \pi/2$ and $0 \le \arg \nu \le \pi$, $\nu \ne -n$, in view of known relations for the analytic continuation of $I_\nu(z)$. Even so, it is convenient to restate some of our results for the Bessel function $J_\nu(z)$. Only a skeleton set of equations are are presented in 19.2 to cover the theoretical aspects of the subject. Some numerical examples are given in 19.3. For more complete details, see Luke (1972). The material in 19.2 is quite lengthy and to facilitate understanding of the machine programs, an outline of the relevant mathematics is developed in 19.4.

19.2. Backward Recurrence Schemata for Generating $I_\nu(z)$

The difference equation satisfied by $I_{m+\nu}(z)$ is

$$Q_{m,\nu}(z) = \{2(m+\nu+1)/z\}Q_{m+1,\nu}(z) + Q_{m+2,\nu}(z) . \tag{1}$$

Let N be a large positive integer with $m \le N + 2$. Set

$$W_{N+2,\nu}(z) = 0, \quad W_{N+1,\nu}(z) = 1 \tag{2}$$

and evaluate $W_{m,\nu}(z)$ for $m = N, N-1, \ldots, 1, 0$ from (1) with $Q_{m,\nu}(z)$ replaced by $W_{m,\nu}(z)$. We have

$$W_{N+1,\nu}(z) = 1, \quad W_{N,\nu}(z) = (N+\nu+1)y ,$$

$$W_{N-1,\nu}(z) = 1 + (N+\nu)_2 y^2 , \quad W_{N-2,\nu}(z) = 2(N+\nu)y + (N+\nu-1)_3 y^3 ,$$

$$W_{N-3,\nu}(z) = 1 + 3(N+\nu-1)_2 y^2 + (N+\nu-2)_4 y^4 , \quad y = 2/z , \tag{3}$$

and in general,

$$W_{m,\nu}(z) = \{ (2/z)^{2p} \Gamma(2n+2-\delta+\nu)/\Gamma(m+\nu+1) \}$$

$$\times {}_2F_3^r(-p, \tfrac{1}{2}-p; -2n-1+\delta-\nu, m+\nu+1, -2p; z^2) , \tag{4}$$

$$p = n + \tfrac{1}{2}(1-m-\delta), \quad N = 2n-\delta, \quad \delta = 0 \text{ or } \delta = 1 \text{ and } r = [p] \tag{5}$$

is the largest integer $\leq p$.   Clearly

$$W_{0,\nu}(z) = (2/z)^{N+1}(\nu+1)_{N+1} A_n(z) , \tag{6}$$

$$W_{0,\nu}(z) = (2/z)^{N+1}(\nu+1)_{N+1} G_{N+1}(z) , \tag{7}$$

where $A_n(z)$ and $G_n(z)$ are given by 17(6) and 17(17), respectively.

We suppose that a normalization relation is known of the form

$$P(z) = \sum_{k=0}^{\infty} a_k I_{k+\nu}(z) . \tag{8}$$

Let

$$P_N(z) = \sum_{k=0}^{\infty} a_k W_{k,\nu}(z) , \tag{9}$$

and consider

$$i_{m+\nu}(z) = P(z) W_{m,\nu}(z)/P_N(z) , \quad m \leq N + 1 . \tag{10}$$

Observe that both $W_{m,\nu}(z)$ and $i_{m+\nu}(z)$ are N dependent.   We have omitted adding an N to the notation for the sake of simplicity.   We have

$$\lim_{N\to\infty} i_{m+\nu}(z) = I_{m+\nu}(z) , \quad m = 0,1,\ldots, \tag{11}$$

provided that

$$\lim_{N\to\infty} \{ I_{N+2+\nu}(z)/K_{N+2+\nu}(z) \} \sum_{k=0}^{N+1} (-)^k a_k K_{k+\nu}(z) = 0 . \tag{12}$$

We next consider three choices for the a_k's in (8).   In each situation, the condition (11) holds as will be evident from the error analysis.

CASE I. Consider the normalization relation

$$P(z) = \frac{(z/2)^\nu}{\Gamma(\nu+1)} = \sum_{k=0}^{\infty} \frac{(-)^k (2k+\nu)\Gamma(k+\nu)}{\Gamma(\nu+1)k!} I_{2k+\nu}(z) , \quad \nu \neq 0 ,$$

$$P(z) = 1 = I_0(z) + 2 \sum_{k=1}^{\infty} (-)^k I_{2k}(z) , \quad \nu = 0 . \tag{13}$$

Then

$$P_N(z) = (2/z)^{N+1}(\nu+1)_{N+1} B_n(z) , \tag{14}$$

where $B_n(z)$ is given by 17(5). It follows that

$$i_\nu(z) = P(z)A_n(z)/B_n(z) , \tag{15}$$

$$i_{m+\nu}(z) = (z/2)^\nu W_{m,\nu}(z)/\Gamma(\nu+1)P_N(z) . \tag{16}$$

That is, the result produced by use of the recurrence formula for $I_\nu(z)$ employed in the backward direction together with the normalization relation (13) and the rational approximation 17(4) are identical.

Next we turn to the error. Let

$$E_{m,\nu}(z) = I_{m+\nu}(z) - i_{m+\nu}(z) . \tag{17}$$

A closed form representation of the error has been given by Luke (1972,1975a). Here we give only asymptotic forms sufficient for the applications. Thus for z and ν fixed,

$$E_{m,\nu}(z) = \{Z_n(z) + 2(-)^m V_n(z)K_{m+\nu}(z)\}\{1 + O(n^{-1})\} , \tag{18}$$

$$Z_n(z) = \frac{(-)^n(z/2)^{2n+2}\Gamma(n+1-\delta+\nu)}{(n+1)!\,\Gamma(2n+2-\delta+\nu)} I_{m+\nu}(z) , \tag{19}$$

$$V_n(z) = (-)^\delta(z/2)^{2n+2-\delta+\nu} I_{2n+2-\delta+\nu}(z)/\Gamma(2n+2-\delta+\nu) . \tag{20}$$

Further simplification often obtains by use of 1.3(14). Thus

$$Z_n(z) = \frac{(-)^n(z/2)^{2n+2}2^{\delta+1-\nu}}{(2n+2)!} I_{m+\nu}(z)\{1 + O(n^{-1})\} , \tag{21}$$

$$V_n(z) = \frac{(-)^\delta(2n)^{2\delta-2\nu}(z/2)^{4n+4-2\delta+\nu}}{(2n+2)!\,(2n+1)!} \{1 + O(n^{-1})\} . \tag{22}$$

Here, we have used the fact that

$$I_\mu(z) = \frac{(z/2)^\mu}{\Gamma(\mu+1)} \{1 + O(\mu^{-1})\} . \tag{23}$$

We also have for z fixed,

$$K_\mu(z) = \frac{1}{2}(z/2)^{-\mu}\Gamma(\mu)\{1 + O(\mu^{-1})\} . \tag{24}$$

Clearly for z and v fixed,

$$\lim_{n\to\infty} E_{m,v}(z) = 0 , \quad m = 0,1,\dots . \tag{25}$$

Further for $n \gg m$, the $Z_n(z)$ term in (18) is dominant, and so the relative error is independent of m.   If v, n and z are fixed, so that $E_{m,v}(z)$ is conceived as function of m only, then the latter satisfies the recurrence formula for $W_{m,v}(z)$.

CASE II.   Consider the normalization relation

$$M(z) = \frac{(z/2)^v e^z}{\Gamma(v+1)} = \sum_{k=0}^\infty \frac{(2k+2v)\Gamma(k+2v)}{\Gamma(2v+1)k!} I_{k+v}(z) , \quad v \neq 0 ,$$

$$M(z) = e^z = I_0(z) + 2 \sum_{k=1}^\infty I_k(z) , \quad v = 0 . \tag{26}$$

Then

$$M_N(z) = (2/z)^q (v+1)_q H_q(z) , \quad q = N + 1 = 2n + 1 - \delta , \tag{27}$$

where $H_n(z)$ is given by 17(16).   It follows that

$$i_v(z) = M(z)G_q(z)/H_q(z) , \tag{28}$$

$$i_{m+v}(z) = (z/2)^v e^z W_{m,v}(z)/\Gamma(v+1)M_N(z) . \tag{29}$$

That is, the result produced by use of the recurrence formula for $I_v(z)$ employed in the backward direction together with the normalization relation (26) and the rational approximation 17(15) are identical.

   We now turn to the error.   Let

$$F_{m,v}(z) = I_{m+v}(z) - i_{m+v}(z) . \tag{30}$$

A complete description of the error has been given by Luke (1972,1975a).   Here we present only asymptotic forms sufficient for the applications. If $\nu = -\frac{1}{2}$,

$$F_{m,-\frac{1}{2}}(z) = (-)^{\delta+m} I_{q+\frac{1}{2}}(z) K_{m-\frac{1}{2}}(z)/K_{q+\frac{1}{2}}(z) ,$$

$$F_{m,-\frac{1}{2}}(z) = F_{0,-\frac{1}{2}}(z)\{(-)^m (2z/\pi)^{\frac{1}{2}} e^{-z} K_{m-\frac{1}{2}}(z)\} , \qquad (31)$$

$$F_{0-\frac{1}{2}}(z) = -F_{1,-\frac{1}{2}}(z) = (-)^{\delta}(\pi/2z)^{\frac{1}{2}} e^{-z} I_{q+\frac{1}{2}}(z)/K_{q+\frac{1}{2}}(z)$$

$$= \frac{(-)^{\delta}(2\pi z)^{\frac{1}{2}} e^{-z}(z/2)^{2q+1}}{(q!)^2} \{1 + O(n^{-1})\} . \qquad (32)$$

Thus $(2\pi z)^{\frac{1}{2}} e^{-z} i_{-\frac{1}{2}}(z)$ is the main diagonal Padé approximation to $1 + e^{-2z}$. The main diagonal Padé approximation for e^{-z} follows from 16(2) with $c = 1$.

We now suppose that $\nu \neq -\frac{1}{2}$. Then

$$F_{m,\nu}(z) = \{-L_n(z) + (-)^m N_n(z) K_{m+\nu}(z)\}\{1 + O(n^{-1})\} , \quad (33)$$

$$L_n(z) = \frac{(2z)^{q+1}(\nu+\frac{1}{2})_{q+1}}{(q+1)! \ (q+1+2\nu)_{q+1}} e^{-z} I_{m+\nu}(z) , \qquad (34)$$

$$N_n(z) = \frac{2(-)^{\delta}(z/2)^{q+1+\nu}}{\Gamma(q+1+\nu)} I_{q+1+\nu}(z) . \qquad (35)$$

Again, simplification often obtains by use of 1.3(14). Thus

$$L_n(z) = \frac{(z/2)^{q+1}(q/4)^{\nu}}{(q+1)!(\frac{1}{2})_{\nu}} e^{-z} I_{m+\nu}(z)\{1 + O(n^{-1})\} , \qquad (36)$$

$$N_n(z) = \frac{2(-)^{\delta}(z/2)^{2q+2+2\nu}}{q^{2\nu} q!(q+1)!} \{1 + O(n^{-1})\} , \qquad (37)$$

where again we make use of (23). Clearly for z and ν fixed,

$$\lim_{n\to\infty} F_{m,\nu}(z) = 0, \quad m = 0,1,\ldots, \qquad (38)$$

with the latter also holding for $\nu = -\frac{1}{2}$. Also if $n \gg m$, the $L_n(z)$ term in (33) is dominant whence the relative error is independent of m. With n, ν and z fixed and m variable, $F_{m,\nu}(z)$ and $W_{m,\nu}(z)$ satisfy the same recurrence formula.

We now present some comments on the relative merits of the Case I and Case II procedures. In (18) and (33), for z, m and v fixed and n sufficiently large, the term involving $K_{m+v}(z)$ is of lower order than the term involving $I_{m+v}(z)$. Neglecting the former term in each equation, we have

$$\frac{E_{m,v}(z)}{F_{m,v}(z)} = \frac{(-)^{n+1}(\tfrac{1}{2})_v(z/2)^{\delta}e^z}{n^{v+\delta}} \{1 + O(n^{-1})\} , \quad v \neq -\tfrac{1}{2} . \quad (39)$$

This shows that there is little difference in the accuracy of the two schemes. Computation wise, if the backward recursion scheme is used, Case I requires less operations since the associated normalization relation uses the sequence $W_{k,v}(z)$, k = 0,2,4..., while the Case II normalization relation uses the same sequence but for k = 0,1,2... . Also to get $I_v(z)$ by the Case II scheme, e^z must be evaluated. On the other hand, if $|z|$ is large, R(z) > 0, one often wants $e^{-z}I_v(z)$ and this is furnished by the Case II procedure. It appears that for the same n, the Case II technique might be more accurate than the Case I scheme even for moderate values of $|z|$, R(z) > 0, in view of the presence of e^z in the numerator of (39). Also, Case II is favored when $R(v+\delta)$ < 0. Improved information cannot be derived from (39) as the estimate is for fixed z, m and v. For error analyses it is suggested than one use (18) or (33) as appropriate. Further discussion is deferred to 19.3 where numerical examples are presented.

If z is pure imaginary and v is real, then $z^{-v}I_v(z)$ is real and definitely the Case I procedure is better than the Case II scheme since the former requires real arithmetic while the latter demands complex arithmetic.

If only $I_v(z)$ or only $e^{-z}I_v(z)$ is required, use of the rational approximation scheme or the equivalent backward recursion scheme demands about the same number of operations.

If one requires $I_{k+\nu}(z)$ or $e^{-z}I_{k+\nu}(z)$ for a given ν and $k = 0$, $1,\ldots,r$, then obviously the backward recursion scheme is high-ly advantageous.

CASE III. Here we assume that $I_\nu(z)$ is known. Thus in the notation of (8), $a_0 = 1$ and $a_k = 0$ for $k > 0$. Put

$$i_{m+\nu}(z) = I_\nu(z)W_{m,\nu}(z)/W_{0,\nu}(z) , \tag{40}$$

and represent the error by

$$G_{m,\nu}(z) = I_{m+\nu}(z) - i_{m+\nu}(z) . \tag{41}$$

Then

$$G_{m,\nu}(z) = (-)^m(2/z)^{m+\nu}\Gamma(m+\nu)V_n(z)\Omega_m(z)/A_n(z) ,$$

$$\Omega_m(z) = {}_2F_3^s(\tfrac{1}{2}-\tfrac{1}{2}m,1-\tfrac{1}{2}m;1-\nu-m,\nu+1,1-m;z^2) , \tag{42}$$

$$s = [\tfrac{1}{2}m-\tfrac{1}{2}] , \quad 0 < m \leq N + 2 ,$$

where $V_n(z)$ is given in (20) and $A_n(z)$ is the hypergeometric polynomial in (4) with $m = 0$. We also have

$$G_{m,\nu}(z) = \frac{(-)^{m+\delta}\Gamma(m+\nu)(z/2)^{2q+2+2\nu-m}\Omega_m(z)}{\Gamma(q+1+\nu)\Gamma(q+2+\nu)\Gamma(\nu+1)I_\nu(z)}$$

$$\times[1 + \{z^2/2(q+1+\nu)\} + O(z^4/n^2)] , \quad m > 0 , \tag{43}$$

with q as in (27). Clearly, the backward recurrence scheme is convergent. Note that with n, ν and z fixed, $(-)^m G_{m,\nu}(z)$ satisfies the same recurrence formula as does $W_{m,\nu}(z)$. Also $G_{m,\nu}(z)$ is 0 when $m = 0$.

Next we compare the errors in Cases I and II with Case III when z, m and ν are fixed and n is sufficiently large. Now

$$\frac{G_{m,\nu}(z)}{E_{m,\nu}(z)} = \frac{(z/2)^{2n}}{\Gamma(2n+3-\delta+\nu)} O(n^{\delta+1-\nu}) , \tag{44}$$

and so Case III is superior to Case I. Similarly, Case III is superior to Case II. Now suppose that m is sufficiently large so that in (18), the term involving $K_{m+\nu}(z)$ dominates the term involving $I_{m+\nu}(z)$ which is certainly the case if m = 2n + 1 - δ - d, d << n. Then

$$\frac{G_{m,\nu}(z)}{E_{m,\nu}(z)} = (2/z)^\delta \{1 + O(n^{-1})\}\{1 + O(m^{-1})\} , \qquad (45)$$

and under these conditions there is little to choose between the two cases. Overall, it appears that Case III gives better accuracy than Case I. However, for Case III, one must know $I_\nu(z)$, while for Case I no such knowledge is required. For all positive z and all ν, $0 \leq \nu \leq 1$, coefficients can be easily derived to facilitate the rapid evaluation of $J_\nu(z)$ and $I_\nu(z)$, see Luke (1971-1972). All of this can often make the Case III approach rather attractive. See the numerical examples in 19.3.

An analytical formulation of the round off error has been given by Luke (1972) who shows that if ω is the round off error in a particular value of $W_{r,\nu}(z)$, r fixed, then the effect of ω on the evaluation of $W_{m,\nu}(z)$, m < r, approaches zero as N → ∞. Unfortunately, it seems difficult to deduce a pragmatic assessment of the round off error from the analytical formulations. Heuristic evidence is abundant to indicate that if 2 or 3 extra decimals are carried beyond that required for the truncation error, then the round off error is insignificant. These statements hold for all three cases.

As previously noted, the forms for $J_\nu(z)$ follow from those for $I_\nu(z)$ in view of known connecting relations. Nonetheless, we find it convenient for the applications to restate the key results for $J_\nu(z)$. We omit discussion of Case II since it requires complex arithmetic to generate $J_\nu(z)$ which is real when z and ν are real.

It is convenient to use the following notation. Unless indicated otherwise, if Q is used to signify a function in the developments for $I_\nu(z)$, then Q* is used to signify the corresponding function in the developments for $J_\nu(z)$. For example, the difference equation satisfied by $J_{m+\nu}(z)$ is

$$Q^*_{m,\nu}(z) = \{2(m+\nu+1)/z\}Q^*_{m+1,\nu}(z) - Q^*_{m+2,\nu}(z) . \qquad (46)$$

With

$$W^*_{N+2,\nu}(z) = 0, \quad W^*_{N+1,\nu}(z) = 1 , \qquad (47)$$

we use (46) with $Q^*_{m,\nu}(z)$ replaced by $W^*_{m,\nu}(z)$ to compute the latter for m = N, N-1, ..., 1, 0.   Then

$$W^*_{m,\nu}(z) = \{(2/z)^{2p}\Gamma(2n+2-\delta+\nu)/\Gamma(m+\nu+1)\}$$
$$\times {}_2F^r_3(-p,\tfrac{1}{2}-p;-2n-1+\delta-\nu,m+\nu+1,-2p;-z^2) , \qquad (48)$$

where r and p are given in (4). Clearly

$$W^*_{0,\nu}(z) = (2/z)^{N+1}(\nu+1)_{N+1}A^*_n(z) , \qquad (49)$$

where $A^*_n(z)$ is the hypergeometric polynomial in (48) with m = 0. We consider the normalization relation

$$P^*(z) = \frac{(z/2)^\nu}{\Gamma(\nu+1)} = \sum_{k=0}^\infty \frac{(2k+\nu)\Gamma(k+\nu)}{\Gamma(\nu+1)k!} J_{2k+\nu}(z) , \quad \nu \neq 0 ,$$

$$P^*(z) = 1 = J_0(z) + 2 \sum_{k=1}^\infty J_{2k}(z) , \quad \nu = 0 . \qquad (50)$$

With

$$P^*_N(z) = (2/z)^{N+1}(\nu+1)_{N+1}B^*_n(z) , \qquad (51)$$

$$B^*_N(z) = {}_0F^n_1(-2n-1+\delta-\nu;-z^2/4) , \qquad (52)$$

the approximations to $J_\nu(z)$ and $J_{m+\nu}(z)$ are given by

$$j_\nu(z) = P^*(z)A^*_n(z)/B^*_n(z) , \qquad (53)$$

and

$$j_{m+\nu}(z) = (z/2)^\nu W^*_{m,\nu}(z)/\Gamma(\nu+1)P^*_N(z) , \qquad (54)$$

respectively. For the error, let

$$E^*_{m,\nu}(z) = J_{m+\nu}(z) - j_{m+\nu}(z) . \qquad (55)$$

Then for z and ν fixed,

$$E^*_{m,\nu}(z) = - \{Z^*_n(z) + \pi V^*_n(z) Y_{m+\nu}(z)\}\{1 + O(n^{-1})\} , \qquad (55)$$

$$Z^*_n(z) = \frac{(z/2)^{2n+2} \Gamma(n+1-\delta+\nu)}{(n+1)! \Gamma(2n+2-\delta+\nu)} J_{m+\nu}(z) , \qquad (56)$$

$$V^*_n(z) = (z/2)^{2n+2-\delta+\nu} J_{2n+2-\delta+\nu}(z)/\Gamma(2n+2-\delta+\nu) . \qquad (57)$$

Also,

$$Z^*_n(z) = \frac{(z/2)^{2n+2} 2^{\delta+1-\nu}}{(2n+2)!} J_{m+\nu}(z)\{1 + O(n^{-1})\} , \qquad (58)$$

$$V^*_n(z) = \frac{(2n)^{2\delta-2\nu}(z/2)^{4n+4-2\delta+2\nu}}{(2n+2)!(2n+1)!} \{1 + O(n^{-1})\} , \qquad (59)$$

$$J_\mu(z) = \frac{(z/2)^\mu}{\Gamma(\mu+1)} \{1 + O(\mu^{-1})\} , \qquad (60)$$

$$Y_\mu(z) = - (2/\pi)(z/2)^{-\mu}\Gamma(\mu)\{1 + O(\mu^{-1})\} . \qquad (61)$$

The backward recurrence scheme is clearly convergent and for n sufficiently large, n >> m, the relative error is essentially independent of m. If ν, n and z are fixed and $E^*_{m,\nu}(z)$ is treated as a function of m only, then $E^*_{m,\nu}(z)$ and $W^*_{m,\nu}(z)$ satisfy the same recurrence formula.

We now turn to the Case III formulation. We have

$$G^*_{m,\nu}(z) = J_{m+\nu}(z) - j_{m+\nu}(z) , \qquad (62)$$

$$j_{m+\nu}(z) = J_\nu(z) W^*_{m,\nu}(z)/W^*_{0,\nu}(z) , \qquad (63)$$

where $W^*_{m,\nu}(z)$ is given in (48). Also

$$G^*_{m,\nu}(z) = (2/z)^{m+\nu}\Gamma(m+\nu)V^*_n(z)\Omega^*_m(z)/A^*_n(z) ,$$

$$\Omega^*_m(z) = {}_2F^s_3(\tfrac{1}{2}-\tfrac{1}{2}m, 1-\tfrac{1}{2}m; 1-\nu-m, \nu+1, 1-m; -z^2) , \qquad (64)$$

s as in (42), and $A^*_n(z)$ as in (49). Also

$$G^*_{m,\nu}(z) = \frac{\Gamma(m+\nu)(z/2)^{2q+2+2\nu-m}\Omega^*_m(z)}{\Gamma(q+1+\nu)\Gamma(q+2+\nu)\Gamma(\nu+1)J_\nu(z)}$$

$$\times\left[1 - \{z^2/2(q+1+\nu)\} + O(z^4/n^2)\right] , \quad m > 0 , \qquad (65)$$

with q as in (27). Clearly, the backward recurrence scheme is convergent. Note that with n, ν and z fixed, $G^*_{m,\nu}(z)$ and $W^*_{m,\nu}(z)$ satisfy the same recurrence formula. Also $G^*_{m,0}(z)$ is zero. The discussion surrounding (44) and (45) with slight modification applies here. We omit details.

19.3. Numerical Examples

Let

$N = 5$, $n = 3$, $\delta = 1$, $z = 2/3$, $\nu = 1/3$.

Values of $W_{m,\nu}(z)$, $P_N(z)$ and $M_N(z)$ are given in the table below.

TABLE A

m	$W_{m,\nu}(z)$	
6	1	
5	19	
4	305	
3	3984	$P_N(z) = 880\ 75120/81$
2	40145	
1	2 84999	$M_N(z) = 1\ 38952\ 97360/6561$
0	11 80141	

Since $(z/2)^{1/3}/\Gamma(4/3) = 0.77645\ 82114$, $e^{-2/3} = 0.51341\ 71190$, the Case I and Case II approximations are $0.84272\ 08930$ and $0.84272\ 10326$, respectively. To 10 decimals, $I_{1/3}(2/3) = 0.84272\ 08819$. Thus the errors in the Case I and Case II approximations are $-0.111(-7)$ and $-0.151(-6)$, respectively. Using 19.2(18,33), each with $O(n^{-1})$ and the term involving $K_{m+\nu}(z)$ neglected, the approximate Case I and Case II errors are $-0.110(-7)$ and $-0.149(-6)$, respectively.

For a second example, let

$N = 5$, $n = 3$, $\delta = 1$, $z = 2$, $\nu = 0$.

Again we illustrate the Case I and Case II schemes in the following table.

TABLE B

m	$\dfrac{W_{m,\nu}(z)}{1}$	
6	1	
5	6	$P_N(z) = 611$
4	31	
3	130	$M_N(z) = 4515$
2	421	
1	972	$e^2 = 7.38905\ 6099$
0	1393	

$i_m(z)$

m	CASE I	CASE II	$I_m(z)$
0	2.27986 9067	2.27972 4285	2.27985 5302
1	1.59083 4697	1.59073 3672	1.59063 6855
2	0.68903 4370	0.68899 0613	0.68894 8448
3	0.21276 5957	0.21275 2446	0.21273 9959
4	0.50736 4975(-1)	0.50733 2755(-1)	0.50728 5700(-1)
5	0.98199 6727(-2)	0.98193 4365(-2)	0.98256 7932(-2)
6	0.16366 6121(-2)	0.16365 5728(-2)	0.16001 7336(-2)

m	ERROR		RELATIVE ERROR	
	CASE I	CASE II	CASE I	CASE II
0	-0.284(-3)	-0.139(-3)	-0.124(-3)	-0.610(-4)
1	-0.198(-3)	-0.968(-3)	-0.124(-3)	-0.609(-4)
2	-0.859(-4)	-0.422(-4)	-0.125(-3)	-0.612(-4)
3	-0.260(-4)	-0.125(-4)	-0.122(-3)	-0.587(-4)
4	-0.793(-5)	-0.471(-5)	-0.156(-3)	-0.929(-4)
5	0.571(-5)	0.634(-5)	0.581(-3)	0.645(-4)
6	-0.365(-4)	-0.364(-4)	-0.228(-1)	-0.228(-1)

Use 19.2(18,33), each with $O(n^{-1})$ and the term involving $K_{m+\nu}(z)$ neglected. Then the approximate relative errors for Cases I and II are -0.116(-3) and -0.537(-4), respectively.

In Table C below, we record the approximate errors found by use of 19.2(18) with $O(n^{-1})$ omitted for m = 6 and 5, and by use of the recursion formula for the lower values of m. In this connection, see the remarks following 19.2(25). We call this Case I-A. We also present the analogous Case II-A data based on 19.2(33) and the remarks following 19.2(38). In each instance known tabular values of $K_m(2)$ and $I_7(2)$ were used. In practice, we suggest using the approximations 19.2(23,24) as appropriate.

TABLE C

m	APPROXIMATE ERROR	
	CASE I-A	CASE II-A
0	-0.264(-3)	-0.114(-3)
1	-0.184(-3)	-0.765(-4)
2	-0.797(-4)	-0.373(-4)
3	-0.242(-4)	-0.110(-4)
4	-0.723(-5)	-0.410(-5)
5	-0.475(-5)	-0.536(-5)
6	-0.310(-4)	-0.309(-4)

For a final example, we illustrate Case III using the
data of our second example.   We get the following numbers.

TABLE D

m	$i_m(z)$	$G_m(z)$	APPROXIMATE ERROR
0	2.27958 5302	0	0
1	1.59063 6693	0.162(-6)	0.155(-6)
2	0.68894 8609	-0.162(-6)	-0.155(-6)
3	0.21273 9475	0.484(-6)	0.467(-6)
4	0.50730 1826(-1)	-0.161(-5)	-0.155(-5)
5	0.98187 4502(-2)	0.693(-5)	0.669(-5)
6	0.16364 5750(-2)	-0.363(-4)	-0.350(-4)

Here $G_m(z)$ is the true error while the approximate error
means that $G_m(z)$ is approximated for m = 1 by use of 19.2(43)
with the order term neglected and subsequent values of the error are found by use of the recurrence formula as explained in
the remark after 19.2(43).   Use of the recurrence formula in
this fashion is stable as the error is an increasing function of m.

A measure of the accuracy of the three cases obtains by
use of normalization relations.   Thus if the Case III procedure is employed, then 19.2(13,26) with $I_{k+\nu}(z)$ replaced by
$i_{k+\nu}(z)$ are available as checks.   Similarly, 19.2(13 and 26)
are available as checks for Cases II and I, respectively.   Other
useful normalization relations can be found in Luke (1969).

Analyses of the error in the backward recursion process
for solution of a general second and higher order linear difference equations have been given by a number of authors.   Some

authors have studied the case of Bessel functions directly.
We make no attempt to survey the various contributions here.
Some pertient references are given in Luke (1975a). Suffice
it to say that none of the analyses has the precision and
simplicity of those given here. We deliberately chose N and
as a consequence n small (N = 5, n = 3) in our numerical examples to put our asymptotic estimates under a severe test.
The efficiency and realism of our error formulas are manifest.

19.4. Mathematical Description of the Programs

The purpose of this section is to present a basic outline
of the mathematics relevant to the machine programs given at
the end of this chapter.

19.4.1. Evaluation of Functions Related to $I_{m+\nu}(z)$ and $J_{m+\nu}(z)$

The following developments are based on the $_0F_1$ representation for the Bessel function. Let

$$I_{m+\nu}(z) = E_\nu(z)A_{m+\nu}(z) \; , \quad J_{m+\nu}(z) = E_\nu(z)B_{m+\nu}(z) \; ,$$

$$E_\nu(z) = (z/2)^\nu/\Gamma(\nu+1) \; . \tag{1}$$

The first set of programs is designed to produce approximations for $A_{m+\nu}(z)$ and $B_{m+\nu}(z)$. The input data are z, ν, N (a
large positive integer) and VAL which for purposes of this
subsection is called L. The machine output is called A(M),
where for N sufficiently large,

$$A(M) = A_{m+\nu}(z) \text{ if } L = VAL = 1.0 \; ,$$

$$A(M) = B_{m+\nu}(z) \text{ if } L = VAL = -1.0 \; . \tag{2}$$

Thus A(M) and B(M) are approximations for $A_{m+\nu}(z)$ and $B_{m+\nu}(z)$,
respectively. Let

$$W_m = \{2(m+\nu+1)/z\}W_{m+1} + LW_{m+2} \; ,$$

$$W_{N+2} = 0 \; , \quad W_{N+1} = 1.0 \; . \tag{3}$$

The machine evaluates W_m by use of (3) for m = N,N-1,...,1,0.

Next the computer computes a sum based on the normalization relation 19.2(13). The pertinent forms are

$$P_N(z) = \sum_{m=0}^{[y]} (-L)^m \{(2m+\nu)(\nu)_m/m!\nu\}W_{2m} \text{ if } \nu \neq 0 ,$$

$$P_N(z) = W_0 + 2 \sum_{m=1}^{[y]} (-L)^m W_{2m} \text{ if } \nu = 0 , \tag{4}$$

where $y = \frac{1}{2}(N+1)$ and $[y]$ is the largest integer contained in y. The printed machine output is

$$A(M) = W_m/P_N(z) , \tag{5}$$

see (2). We do not furnish a program for the evaluation of $E_\nu(z)$, see (1).

Certain checks on the calculations are performed. The basic mathematical equations are

$$\sum_{m=0}^{\infty} \{(2m+2\nu)(2\nu)_m I_{m+\nu}(z)/m! 2\nu E_\nu(z)\} = e^z \text{ if } \nu \neq 0 ,$$

$$I_0(z) + 2 \sum_{m=1}^{\infty} I_m(z) = e^z \text{ if } \nu = 0, \tag{6}$$

$$\sum_{m=0}^{\infty} \{(z/2)^m J_{m+\nu}(z)/m! E_\nu(z)\} = 1 \text{ if } \nu \neq 0,$$

$$J_0^2(z) + 2 \sum_{m=1}^{\infty} J_m^2(z) = 1 \text{ if } \nu = 0 . \tag{7}$$

The corresponding machine calculations and output are

$$TEST = \sum_{m=0}^{N+1} \{(2m+2\nu)(2\nu)_m W_m/m! 2\nu P_N(z)\} - e^z \text{ if } \nu \neq 0 \text{ and } L = 1,$$
$$\tag{8}$$

$$TEST = \{W_0/P_0(z)\} + 2 \sum_{m=1}^{N+1} \{W_m/P_N(z)\} - e^z \text{ if } \nu = 0 \text{ and } L = 1,$$

$$TEST = \sum_{m=0}^{N+1} \{(z/2)^m W_m/m! P_N(z)\} - 1 \text{ if } \nu \neq 0 \text{ and } L = -1,$$
$$\tag{9}$$

$$TEST = \{W_0/P_0(z)\}^2 + 2 \sum_{m=1}^{N+1} \{W_m/P_N(z)\}^2 - 1 \text{ if } \nu = 0 \text{ and } L = -1 .$$

Except for truncation and round off error, the value of TEST should be zero.

19.4.2. Evaluation of Functions Related to $e^{-z}I_{m+\nu}(z)$

The following developments are based on the $_1F_1$ representation for the Bessel function. Let

$$e^{-z}I_{m+\nu}(z) = E_\nu(z)F_{m+\nu}(z) . \tag{1}$$

We describe the programs to get approximations for $F_{m+\nu}(z)$. The machine output is again called A(M) and for N sufficiently large

$$A(M) = F_{m+\nu}(z) . \tag{2}$$

To this end, the machine calculations described by 19.4.1(1-3) with L = 1 apply for the present situation. The normalization relation used in the present instance is based on 19.2(26). Thus the machine evaluates

$$M_N(z) = \sum_{m=0}^{N+1} (2m+2\nu)(2\nu)_m W_m/m!\,2\nu \quad \text{if } \nu \neq 0 ,$$

$$M_N(z) = W_0 + 2\sum_{m=1}^{N+1} W_m \quad \text{if } \nu = 0 , \tag{3}$$

and presents as output

$$A(M) = W_m/M_N(z) , \tag{4}$$

see (1,2). Machine checks are based on 19.2(13). Thus

$$\text{TEST} = \sum_{m=0}^{[y]} \{(-)^m(2m+\nu)(\nu)_m W_{2m}/m!\,\nu M_N(z)\} - e^{-z} \quad \text{if } \nu \neq 0 ,$$

$$\tag{5}$$

$$\text{TEST} = \{W_0/M_N(z)\} + 2\sum_{m=1}^{[y]} \{(-)^m W_{2m}/M_N(z)\} - e^{-z} \quad \text{if } \nu = 0 ,$$

with y as in 19.4.1(4). Again, except for truncation and round off error, the value of TEST should be zero.

```
C     ..............THIS.MAINLINE.PROGRAM.ILLUSTRATES.THE.SYNTAX.OF.THE.........
C     .     SUBROUTINE 'BRIVZ1'  THAT GENERATES VALUES OF THE FUNCTIONS    .
C     .     ((Z/2)**M)*(GAMMA(V+1))/GAMMA(M+V+1)*0F1(1+M+V ; VAL*(Z**2)/4) .
C     .     WHICH, IF MULTIPLIED BY ((Z/2)**V)/GAMMA(V+1), INCLUDES THE     .
C     .     BESSEL FUNCTIONS I( M+V ; Z ) AND J( M+V ; Z ), WHERE M=0,1,2,  .
C     .     ..,N+1   THE VALUES ARE CALCULATED BY BACKWARD RECURRENCE.      .
C     .     SEE TEXT FOR DETAILS, ESPECIALLY 19.4                           .
C     .....................................................................
      IMPLICIT REAL*16 (A-H,O-Z)
      DIMENSION A(102)
      DATA ZERO/0.Q0/,ONE/1.Q0/,TWO/2.Q0/
 10   READ(5,2,END=999) V,VAL
      READ(5,1) N,L
      N2=N+2
      DO 100 I=1,L
      READ(5,2) Z
C-------------------------------------------------------------------------------
      CALL BRIVZ1(V,VAL,Z,A,N)
C-------------------------------------------------------------------------------
C     IN THE ABOVE :
C
C     V             IS THE PARAMETER IN THE ABOVE 0F1
C     VAL           IS THE PARAMETER IN THE ABOVE 0F1. EXCEPT FOR AN
C                     AUXILIARY PARAMETER, VAL=1 GIVES I( M+V ; Z ) AND
C                     VAL=-1 GIVES J( M+V ; Z ).
C     Z             IS THE ARGUMENT OF THE ABOVE 0F1
C     A             WILL CONTAIN THE GENERATED VALUES
C     N             IS A SUITABLY LARGE INTEGER WHICH DETERMINES THE
C                     INITIAL ORDERS OF M FOR THE BACKWARD RECURRENCE
C
C     NOTE : THE VALUE OF ((Z/2)**M)*(GAMMA(V+1))/GAMMA(M+V+1)*
C            0F1( 1+M+V ; VAL*(Z**2)/4 ) WILL BE PLACED IN A(M+1)
C-------------------------------------------------------------------------------
      M=N2-1
      IF(VAL.LT.ZERO) GOTO 60
      IF(V.EQ.ZERO) GOTO 30
      XM=M
      TEST=A(N2)*(V+XM)
      V1=V-ONE
      V2=V+V1
 20   IF(M.LE.0) GOTO 25
      TEST=A(M)*(V1+XM)+TEST/XM*(V2+XM)
      M=M-1
      XM=XM-ONE
      GOTO 20
 25   TEST=TEST/V
      TEST=TEST-QEXP(Z)
      GOTO 105
 30   TEST=A(N2)
      DO 35 J=2,M
 35   TEST=TEST+A(J)
      TEST=A(1)+(TEST+TEST)
C
C     THE VALUE OF TEST SHOULD BE NEAR ZERO, AND IS COMPUTED AS A CHECK
C     ON THE ACCURACY OF THE VALUES RETURNED BY BRIVZ1
C
 40   TEST=TEST-QEXP(Z)
      GOTO 105
 60   IF(V.EQ.ZERO) GO TO 70
      Z2=Z/TWO
      XM=M
      TEST=A(N2)
 90   IF(M.LE.0) GOTO 80
      TEST=A(M)+TEST/XM*Z2
      M=M-1
      XM=XM-ONE
      GOTO 90
 70   TEST=A(N2)*A(N2)
      DO 95 J=2,M
 95   TEST=TEST+A(J)*A(J)
 80   TEST=A(1)*A(1)+TWO*TEST
      TEST=TEST-ONE
 105  WRITE(6,3) V,VAL,Z,N
      DO 50 J=1,N2
      J1=J-1
 50   WRITE(6,4) J1,A(J)
 100  WRITE(6,5) TEST
      GOTO 10
 999  STOP
 1    FORMAT(2I2)
 2    FORMAT(Q39.32)
 3    FORMAT('1',4X,'V = ',Q39.32/3X,'VAL = ',Q39.32/5X,'Z = ',Q39.32//
     :5X,'N = ',I2//5X,'M',25X,'A(M)'/)
 4    FORMAT(4X,I2,8X,Q39.32)
 5    FORMAT('0',' TEST = ',Q39.32)
      END
```

```
      SUBROUTINE BRIVZ1(V,VAL,Z,W,N)
C     *******************************************************************
C     *         THIS SUBROUTINE RETURNS VALUES W(M) , M = 0,1,...,N+1   *
C     *  WHICH EXCEPT FOR AN AUXILIARY MULTIPLIER ARE APPROXIMATIONS    *
C     *  FOR THE BESSEL FUNCTIONS I( M+V ; Z ) OR J( M+V ; Z).          *
C     *                                                                 *
C     *         NO OTHER SUBROUTINES ARE CALLED BY THIS ONE.            *
C     *******************************************************************
      IMPLICIT REAL*16 (A-H,O-Z)
      DIMENSION W(1)
      DATA ZERO/0.Q0/,ONE/1.Q0/,TWO/2.Q0/,START/1.Q-40/
C
C   INITIALIZATION :
C
      ZZ=TWO/Z
      M2=N+3
      W(M2)=ZERO
      M1=N+2
      W(M1)=START
      M=N+1
      XM=M
      CT1=ZZ*(XM+V)
 10   IF(M.LE.0) GOTO 20
C-----------------------------------------------------------------------
C  FOR M = N+1,N,...,1 , VALUES FOR W(M) ARE CALCULATED AS FOLLOWS
C-----------------------------------------------------------------------
         W(M)=W(M1)*CT1+VAL*W(M2)
C
C   NOTE THAT THESE VALUES ARE TO BE NORMALIZED USING ONE OF THE
C   RELATIONS BELOW
C
         M2=M1
         M1=M
         M=M-1
         CT1=CT1-ZZ
         GOTO 10
 20   M=1+N/2
      M2=M+M-1
      IF(V.NE.ZERO) GOTO 40
      PN=W(M2+2)
 30   IF(M2.EQ.1) GOTO 35
C-----------------------------------------------------------------------
C  FOR V = 0 , THE NORMALIZATION FACTOR PN IS COMPUTED AS A WEIGHTED
C  SUM OF VALUES OF W(M) , M = 0,2,4,... (THE GREATEST INTEGER LESS
C  THAN (N+2)/2 ) , USING THE FOLLOWING RECURRENCE RELATION
C-----------------------------------------------------------------------
         PN=W(M2)-VAL*PN
C
         M2=M2-2
         GOTO 30
 35   PN=W(1)-VAL*(PN+PN)
      GOTO 50
 40   V1=V-1
      XM=M
      CT1=V+XM+XM
      PN=CT1*W(M2+2)
 42   CT1=CT1-TWO
         IF(M2.EQ.1) GOTO 45
C-----------------------------------------------------------------------
C  FOR ALL OTHER VALUES OF V , THE RECURRENCE RELATION FOR PN
C  IS AS FOLLOWS
C-----------------------------------------------------------------------
         PN=CT1*W(M2)-VAL*PN/XM*(V1+XM)
C
         XM=XM-ONE
         M2=M2-2
         GOTO 42
 45   PN=W(1)-VAL*PN
 50   M1=N+2
      DO 100 M=1,M1
C
C   NORMALIZATION OF W(M) , M = 1,2,...,N+2
C
 100     W(M)=W(M)/PN
      RETURN
      END
```

```
 V  =   0.50000000000000000000000000000000Q+00
VAL =   0.10000000000000000000000000000000Q+01
 Z  =   0.50000000000000000000000000000000Q+00

 N  =  20

 M                            A ( M )

 0      0.10421906109874947232448512528230Q+01
 1      0.17087070843772123962747817159380Q+00
 2      0.16966360360861979469364349866716Q-01
 3      0.12071048291523292791043104922139Q-02
 4      0.66892752729369560903890975721410Q-04
 5      0.30352800236771828342809292285474Q-05
 6      0.11659220847153854971053269336715Q-06
 7      0.38826034171805418070792010013958Q-08
 8      0.11410595612229549815666332527923Q-09
 9      0.30009090224948697526479419020536Q-11
10      0.71413267490447556041533011192385Q-13
11      0.15517878960723989035558519734493Q-14
12      0.31024271117206477963810413719083Q-16
13      0.57434021207500536533128749508318Q-18
14      0.98996655156188235920888945916355Q-20
15      0.15963301608768191972638476832101Q-21
16      0.24181587519568978528809557331557Q-23
17      0.34538458526661436245689932727537Q-25
18      0.46665509059731568266044222817777Q-27
19      0.59818224600757288172078423818433Q-29
20      0.72938711408834918230522393993835Q-31
21      0.84812455126552230500607348765553Q-33

TEST =    0.0

 V  =   0.50000000000000000000000000000000Q+00
VAL =  -0.10000000000000000000000000000000Q+01
 Z  =   0.50000000000000000000000000000000Q+00

 N  =  20

 M                            A ( M )

 0      0.95885107720840600054657587043114Q+00
 1      0.16253703063606656886058857565463Q+00
 2      0.16371106607993412616955583496615Q-01
 3      0.11740354438675573089672593115209Q-02
 4      0.65389606152389708586046864677755Q-04
 5      0.29774668754574455815842526787043Q-05
 6      0.11466510767409420880669425373944Q-06
 7      0.38259240690038473897979185210909Q-08
 8      0.11261439602121288724330189328805Q-09
 9      0.29653957173907647434585070274820Q-11
10      0.70641239636618781840433416381485Q-13
11      0.15363473472123629523527852742022Q-14
12      0.30738335149913966779470623181686Q-16
13      0.56941028333543738252631664087402Q-18
14      0.98201501996508617148667903355299Q-20
15      0.15842824431259693595719858670978Q-21
16      0.24009477301483144795220404761595Q-23
17      0.34305877191819691256084716753058Q-25
18      0.46367327906390840388969655454650Q-27
19      0.59454589095306317528282833383114Q-29
20      0.72515879480872830909549336398783Q-31
21      0.84320790094038175476220158603236Q-33

TEST =    0.0
```

```
C       ..........THIS MAINLINE PROGRAM ILLUSTRATES THE SYNTAX OF THE.........
C       .    SUBROUTINE 'BREVZ1' THAT GENERATES VALUES OF THE FUNCTIONS      .
C       .    EXP(-Z)*((Z/2)**M)*(GAMMA(V+1))/GAMMA(M+V+1)*0F1 (1+M+V ;       .
C       .    (Z**2)/4 ) WHICH, IF MULTIPLIED BY ((Z/2)**V)/GAMMA(V+1), ARE   .
C       .    THE BESSEL FUNCTIONS EXP(-Z)*I( M+V ; Z ), WHERE M = 0,1,2,..., .
C       .    N+1.    THE VALUES ARE CALCULATED BY BACKWARD RECURRENCE. SEE   .
C       .    TEXT FOR DETAILS.                                               .
C       ...................................................................
        IMPLICIT REAL*16 (A-H,O-Z)
        DIMENSION A(102)
        DATA ZERO/0.Q0/,ONE/1.Q0/,TWO/2.Q0/
   10   READ(5,2,END=999) V
        READ(5,1) N,L
        N2=N+2
        DO 100 I=1,L
        READ(5,2) Z
C-----------------------------------------------------------------------
        CALL BREVZ1(V,Z,A,N)
C-----------------------------------------------------------------------
C       IN THE ABOVE :
C
C       V           IS THE PARAMETER IN THE RELATED BESSEL FUNCTIONS
C       Z           IS THE ARGUMENT OF THE RELATED BESSEL FUNCTIONS
C       A           WILL CONTAIN THE GENERATED VALUES WHICH MAY THEN BE
C                       USED TO OBTAIN EXP(-Z)*I( M+V ; Z ), M = 0,1,2,...,N+1
C       N           IS A SUITABLY LARGE INTEGER WHICH DETERMINES THE
C                       INITIAL ORDERS OF M FOR THE BACKWARD RECURRENCE
C
C       NOTE : THE VALUE OF EXP(-Z)*((Z/2)**M)*(GAMMA(V+1))/GAMMA(M+V+1)*
C              0F1( 1+M+V ; VAL*(Z**2)/4 ) WILL BE PLACED IN A(M+1)
C-----------------------------------------------------------------------
        M=1+N/2
        M2=M+M-1
        IF(V.NE.ZERO) GOTO 40
        TEST=A(M2+2)
   30   IF(M2.EQ.1) GOTO 35
        TEST=A(M2)-TEST
        M2=M2-2
        GOTO 30
   35   TEST=A(1)-(TEST+TEST)
        GOTO 50
   40   V1=V-1
        XM=M
        CT1=V+XM+XM
        TEST=CT1*A(M2+2)
   42   CT1=CT1-TWO
        IF(M2.EQ.1) GOTO 45
        TEST=CT1*A(M2)-TEST/XM*(V1+XM)
        XM=XM-ONE
        M2=M2-2
        GOTO 42
   45   TEST=A(1)-TEST
C
C   THE VALUE OF TEST SHOULD BE NEAR ZERO, AND IS COMPUTED AS A CHECK
C   ON THE ACCURACY OF THE VALUES RETURNED BY BREVZ1
C
   50   TEST=TEST-QEXP(-Z)
        WRITE(6,3) V,Z,N
        DO 60 J=1,N2
        J1=J-1
   60   WRITE(6,4) J1,A(J)
  100   WRITE(6,5) TEST
        GOTO 10
  999   STOP
    1   FORMAT(2I2)
    2   FORMAT(Q39.32)
    3   FORMAT('1',4X,'V = ',Q39.32/5X,'Z = ',Q39.32//5X,'N = ',I2//
       :5X,'M',25X,'A(M)'/)
    4   FORMAT(4X,I2,8X,Q39.32)
    5   FORMAT('0',' TEST = ',Q39.32)
        END
```

```
      SUBROUTINE BREVZ1(V,Z,W,N)
C     ***********************************************************************
C     *         THIS SUBROUTINE RETURNS VALUES W(M) , M = 0,1,...,N+1 ,     *
C     *   WHICH EXCEPT FOR AN AUXILIARY MULTIPLIER ARE APPROXIMATIONS        *
C     *   FOR THE FUNCTIONS EXP(-Z)*I( M+V , Z ).                           *
C     *                                                                     *
C     *         NO OTHER SUBROUTINES ARE CALLED BY THIS ONE.                *
C     ***********************************************************************
      IMPLICIT REAL*16 (A-H,O-Z)
      DIMENSION W(1)
      DATA ZERO/0.Q0/,UNE/1.Q0/,TWO/2.Q0/,START/1.Q-40/
C
C   INITIALIZATION :
C
      N2=N+2
      Z2=TWO/Z
      M2=N2+1
      W(M2)=ZERO
      M1=N2
      W(M1)=START
      M=N+1
      XM=M
      CT1=Z2*(XM+V)
  10  IF(M.LE.0) GOTO 20
C-----------------------------------------------------------------------
C   FOR M = N+1,N,...,1 , VALUES FOR W(M) ARE CALCULATED AS FOLLOWS
C-----------------------------------------------------------------------
      W(M)=W(M1)*CT1+W(M2)
C
C   NOTE THAT THESE VALUES ARE TO BE NORMALIZED USING ONE OF THE
C   RELATIONS BELOW
C
      M2=M1
      M1=M
      M=M-1
      CT1=CT1-Z2
      GOTO 10
  20  M=N+1
      IF(V.EQ.ZERO) GOTO 30
      XM=M
      PN=W(N2)*(V+XM)
      V1=V-ONE
      V2=V+V1
  50  IF(M.LE.0) GOTO 25
C-----------------------------------------------------------------------
C   FOR ALL NONZERO VALUES OF V , THE RECURRENCE RELATION FOR PN
C   IS AS FOLLOWS
C-----------------------------------------------------------------------
      PN=W(M)*(V1+XM)+PN/XM*(V2+XM)
C
      M=M-1
      XM=XM-ONE
      GOTO 50
  25  PN=PN/V
      GOTO 40
  30  PN=W(N2)
      DO 35 J=2,M
C-----------------------------------------------------------------------
C   FOR V = 0 , THE NORMALIZATION FACTOR PN IS COMPUTED AS A WEIGHTED
C   SUM OF VALUES OF W(M) USING THE FOLLOWING RECURRENCE RELATION
C-----------------------------------------------------------------------
  35  PN=PN+W(J)
C
      PN=W(1)+(PN+PN)
  40  DO 100 M=1,N2
C
C   NORMALIZATION OF W(M) , M = 1,2,...,N+2
C
 100  W(M)=W(M)/PN
      RETURN
      END
```

```
V =   0.50000000000000000000000000000000000Q+00
Z =   0.50000000000000000000000000000000000Q+00

N = 20

M                                    A(M)

  0          0.63212055882855767840447622983854Q+00
  1          0.10363832351432696478657131048438Q+00
  2          0.10290617742595889685048366932244Q-01
  3          0.73214608836806793608764116194742Q-03
  4          0.40572505442938579821390664979680Q-04
  5          0.18409903951734993026091923131780Q-05
  6          0.70716749121595163988434089764407Q-07
  7          0.23549180120250389099059793033923Q-08
  8          0.69208760843996691254710662637982Q-10
  9          0.18201433291514072458167737009482Q-11
 10          0.43314336243215913673262001950260Q-13
 11          0.94120693633887153976961903727854Q-15
 12          0.18817171627822843859526235447232Q-16
 13          0.34835494772934679330726491692829Q-18
 14          0.60044504381170209339299331047151Q-20
 15          0.96822318559579139328796854812748Q-22
 16          0.14666874231142955445281063247071Q-23
 17          0.20948634035633389941837382078471Q-25
 18          0.28304061995824859948957921413791Q-27
 19          0.36281587229935796085202322658476Q-29
 20          0.44239564749390025001097401795391Q-31
 21          0.51441354359755843024531862552781Q-33

TEST =   0.57777898331617075591679338277548Q-33
```

XX RATIONAL APPROXIMATIONS FOR $z^a U(a;1+a-b;z)$

For details on the confluent hypergeometric function in the title, see Luke (1969,1975a). See also 2.2(38). In Luke (1969), the notation ψ is used in place of U. This and the next chapter are based in part on a paper by Luke (1977a). It is useful to quote the asymptotic expansion

$$z^a U(a;1+a-b;z) \sim {}_2F_0(a,b;-1/z) ,$$

$$|z| \to \infty , \quad |\arg z| < 3\pi/2 . \tag{1}$$

We suppose that neither a nor b is a negative integer or zero, for otherwise the ${}_2F_0$ in (1) terminates and asymptotic equality can be replaced by equality. If in the rational approximations of Chapter 13, we replace z by c/z, and let $c \to \infty$, we get the rational approximations of this chapter except for a change in normalization. This confluence principle does not hold for asymptotic developments of the error. The rational approximations in (2)-(4) below can also be derived from 1.5 (13-16) by putting a = 0, β = 0, λ = 1, f = g = 0, p = 2, q = 0, α_1 = a, α_2 = b. Note that the a in 1.5(13-16) has nothing to do with the a in (1). The notation employed here for the polynomials in the rational approximation differs from that used in 1.5(13-16). We now write

$$z^a U(a;1+a-b;z) = \{A_n(z)/B_n(z)\} + R_n(z) , \tag{2}$$

$$B_n(z) = {}_2F_2(-n,n+1;a+1,b+1;-z) , \tag{3}$$

$$A_n(z) = \sum_{k=0}^{n} \frac{(-n)_k (n+1)_k (a)_k (b)_k}{(a+1)_k (b+1)_k (k!)^2} \, {}_3F_3\left(\begin{matrix} -n+k,n+1+k,1 \\ 1+k,a+1+k,b+1+k \end{matrix} \middle| -z\right) . \tag{4}$$

Here $R_n(z)$ is the remainder which we discuss later. We have

$$B_0(z) = 1 , \quad B_1(z) = 1 + \frac{2z}{(a+1)(b+1)} ,$$

$$B_2(z) = 1 + \frac{6z}{(a+1)(b+1)} + \frac{12z^2}{(a+1)_2(b+1)_2} , \tag{5}$$

$$B_3(z) = 1 + \frac{12z}{(a+1)(b+1)} + \frac{60z^2}{(a+1)_2(b+1)_2} + \frac{120z^3}{(a+1)_3(b+1)_3} ,$$

$$A_0(z) = 1 , \quad A_1(z) = B_1(z) - \frac{2ab}{(a+1)(b+1)} ,$$

$$A_2(z) = B_2(z) - \frac{6ab}{(a+1)(b+1)}\{1 + \frac{2z}{(a+2)(b+2)}\} + \frac{6ab}{(a+2)(b+2)} ,$$

$$A_3(z) = B_3(z) - \frac{12ab}{(a+1)(b+1)}\{1 + \frac{5z}{(a+2)(b+2)} + \frac{10z^2}{(a+2)_2(b+2)_2}\}$$

$$+ \frac{30ab}{(a+2)(b+2)}\{1 + \frac{2z}{(a+3)(b+3)}\} - \frac{20ab}{(a+3)(b+3)} . \tag{6}$$

Both $A_n(z)$ and $B_n(z)$ satisfy the same recurrence formula

$$\sum_{m=0}^{3} \{C_m - zD_m\}B_{n-m}(z) = 0 , \quad n \geq 3 ,$$

$$C_0 = 1 , \quad D_0 = D_3 = 0 ,$$

$$C_1 = - \frac{(2n-1)\{3n^2+(a+b-6)n+2-ab-2(a+b)\}}{(2n-3)(n+a)(n+b)} , \tag{7}$$

$$C_2 = \frac{\{3n^2-(a+b+6)n+2-ab\}}{(n+a)(n+b)} , \quad C_3 = - \frac{(2n-1)(n-a-2)(n-b-2)}{(2n-3)(n+a)(n+b)} ,$$

$$D_1 = D_2 = \frac{2(2n-1)}{(n+a)(n+b)} .$$

We write the error in the form

$$R_n(z) = S_n(z)/B_n(z) . \tag{8}$$

It can be shown from the work of Fields (1972,1973) and Luke (1969,1975a) that

$$S_n(z) = \{e^{-3i\pi\tau}\Gamma(a+1)\Gamma(b+1)/4\pi^2\}Q_1(z)G_n(ze^{i\pi})$$

$$+ \text{the same expression with } i \text{ replaced by } -i ,$$

$$3\tau = - (a+b+1) , \quad b = a+1-c , \tag{9}$$

$$Q_1(z) = \frac{z^a\Gamma(1-c)}{\Gamma(b)}(1-e^{-2\pi ia})\,_1F_1\left(\begin{matrix}a\\c\end{matrix}\middle| z\right) + \frac{z^b\Gamma(c-1)}{\Gamma(a)}(1-e^{-2\pi ib})\,_1F_1\left(\begin{matrix}b\\2-c\end{matrix}\middle| z\right), \tag{10}$$

$$G_n(w) = G_{2,3}^{3,1}\left(w \middle| \begin{matrix} -n,n+1 \\ 0,-a,-b \end{matrix}\right) , \qquad (11)$$

$$G_n(w) \sim \frac{2\pi}{3^{\frac{1}{2}}}(N^2 w)^\tau \left(\exp\left\{-3(N^2 w)^{1/3} - \left[\frac{w^2}{15} + \tau w - \frac{3(a^2+b^2)-3ab-1}{9}\right](N^2 w)^{-1/3}\right\}\right)$$

$$\times \left(1 + O(w^2/N^2)^{2/3}\right)\left(1 + O\left[\frac{1 + |w|^{8/3}}{(N^2 w)^{2/3}}\right]\right) ,$$

$$N^2 = n(n+1) , \quad |w| \to \infty , \quad |arg\ w| < 4\pi , \quad w = o(N^{2/3}) . \quad (12)$$

Thus omitting order terms,

$$S_n(z) \sim \frac{2b\Gamma(a+1)\Gamma(b-a)\sin\ \pi a}{3^{\frac{1}{2}}\pi} (N^2 z)^\tau e^{-z/3-q}(\cos\ \xi)\,_1F_1\left(\begin{matrix} a \\ 1+a-b \end{matrix}\middle| z\right)$$

+ the same expression with a and b interchanged,

$$\xi = \pi(-\tfrac{1}{2} + a + 2\tau) + \omega , \quad \omega = (3^{\frac{1}{2}}/2)\{3(N^2 z)^{1/3} + K(N^2 z)^{-1/3}\} ,$$

$$q = (3/2)(N^2 z)^{1/3} - (K/2)(N^2 z)^{-1/3} , \qquad (13)$$

$$K = -\frac{z^2}{15} - \frac{(a+b)z}{3} + \frac{3(a^2+b^2)-3ab-1}{9} .$$

If z in not too large, the $_1F_1$'s in (13) can be evaluated by
their series expansions or by use of the rational approximations in Chapters 15 and 16 as appropriate. If z is sufficiently large, we can employ the asymptotic expansions for
the $_1F_1$'s as given by Luke (1969,1975a). Taking the latter
approach and again omitting order terms, we have

$$S_n(z) \sim (2ab/3^{\frac{1}{2}})e^{2z/3}z^{a+b-1}E(z)(N^2 z)^\tau e^{-q}\cos\theta$$

$$+ (2/3^{\frac{1}{2}}\pi)\Gamma(a+1)\Gamma(b+1)(\sin\pi a)(\sin\ \pi b)e^{-z/3}M(z)(N^2 z)^\tau e^{-q}e^{-i\delta\phi} ,$$

$$\theta = \omega - \pi(\tau+\tfrac{1}{2}) , \quad \phi = \theta + \pi/2 , \quad q\ \text{and}\ \omega\ \text{as in (13)} , \qquad (14)$$

$$E(z) \sim\ _2F_0(1-a,1-b;1/z) , \quad M(z) \sim\ _2F_0(a,b;-1/z) ,$$

$$|z| \to \infty , \quad - (2+\delta)\pi/2 < arg\ z < (2-\delta)\pi/2 , \quad \delta = \pm 1 .$$

If z is real and positive, then replace $\exp(-i\delta\phi)$ in (14) by
$\cos\phi$. Notice that with z large and positive in (14), the

second term in the asymptotic expansion for $S_n(z)$ is subdominant to the first term.

Again from the works of Fields (1972,1973) and Luke (1969, 1975a),

$$B_n(z) \sim \frac{\Gamma(a+1)\Gamma(b+1)}{3^{\frac{1}{2}}(2\pi)} e^{-z/3}(N^2 z)^{\tau} e^{2q}\{1 + O((z^2/N^2)^{2/3})\}$$

$$\times \left(1 + O\left(\frac{1+|z|^{8/3}}{(N^2 z)^{2/3}}\right)\right) ,$$

$$|z| \to \infty , \quad |\arg z| < \pi , \quad z = o(N^{2/3}) , \tag{15}$$

with τ, N^2 and q as in (9), (12) and (13), respectively.

Combining (13) and (15) and omitting the order terms, we get

$$R_n(z) \sim 4e^{-3q} \{\Gamma(b)\}^{-1}\Gamma(b-a)(\sin\pi a)(\cos\xi_1)\,_1F_1\left(\begin{matrix}a\\1+a-b\end{matrix}\bigg| z\right)$$

$$+ \{\Gamma(a)\}^{-1}\Gamma(a-b)(\sin\pi b)(\cos\xi_2)\,_1F_1\left(\begin{matrix}a\\1+b-a\end{matrix}\bigg| z\right) ,$$

$$\xi_1 = \omega + 2\pi\tau + \pi a - \tfrac{1}{2}\pi , \quad \xi_2 = \xi_1 + \pi(b-a) . \tag{16}$$

Now combine (14) and (15) and omit the order terms. Then

$$R_n(z) \sim 4\pi\{\Gamma(a)\Gamma(b)\}^{-1}z^{a+b-1}e^z e^{-3q}E(z)\cos\theta$$

$$+ 4(\sin\pi a)(\sin\pi b)e^{-3q}M(z)e^{-i\delta\phi} . \tag{17}$$

If z is real and positive, replace $\exp(-i\delta\phi)$ by $\cos\phi$ in (17). Notice that in (17), the second term is subdominant to the first term for z large and positive. Clearly, for z fixed,

$$\lim_{n\to\infty} R_n(z) = 0 , \quad |\arg z| < \pi . \tag{18}$$

It is of interest to compare the error in the approximations of this chapter when $a = 1$, call it $R_n(z)$, with the error in the corresponding Padé approximations of the next chapter, call it $R_{n,p}(z)$. To simplify matters, we consider only the case $z > 0$. In (17), put $a = 1$, $b = 1 - \nu$, $c = 1 + \nu$ whence $E(z) = 1$ and

$$R_n(z) \sim \frac{4\pi z^{1-\nu}}{\Gamma(1-\nu)} e^z \left(\exp\{-\tfrac{9}{2}(N^2 z)^{1/3}\} \right) \cos\theta .$$ (19)

For the Padé approximation, we use the uniform error representation 21(15) in the form

$$T_n(\nu, z) \sim -\frac{2\pi z^{1-\nu}}{\Gamma(1-\nu)} e^z \left(\exp\{-2k(2\alpha + \sinh 2\alpha)\} \right) .$$ (20)

Since n is large, we approximate N by $n + \tfrac{1}{2}$ and k by $n + \tfrac{1}{2}$. Then upon neglecting $2\cos\theta$, the right hand sides of (19) and (20) would be identical if

$$4k(2\alpha + \sinh 2\alpha) = 9(k^2 z)^{1/3} ,$$ (21)

or by cubing both sides, we seek α such that

$$729 \sinh^2\alpha = 16(2\alpha + \sinh 2\alpha)^3 .$$ (22)

We find

$$n + \tfrac{1}{2} = 0.69z \text{ or } n \sim 0.69z .$$

Strictly speaking, this result is not completely valid because of the restriction $z = o(N^{2/3})$ in (15). However, further examination shows that if n >> z, the approximations of this chapter with a = 1 are superior to the corresponding Padé approximations in the next chapter. We have conducted several numerical experiments. The results show that (23) is quite realistic, that the Padé approximations and those of this chapter with a = 1 give about the same order of accuracy when n ~ 0.69z, and that the Padé approximations are superior when n << z. This and other numerical examples seem to indicate the the order restrictions noted in (12) and (15) are too strong.

Identification of $U(a; c; z)$ with several named mathematical functions is given in 2.2. In particular, from 2.2(50) we see that the results presented herein can be employed to compute the modified Bessel functions $K_\nu(z)$ and $K_{\nu+1}(z)$. Then

values of $K_{\nu+m}(z)$, $m = 2,3,\ldots$, can be found by use of the recurrence formula

$$K_{\nu+m+1}(z) = \{2(\nu+m)/z\}K_{\nu+m}(z) - K_{\nu+m-1}(z) \qquad (24)$$

which is stable in the forward direction. This formula is not
stable when used in the backward direction.

Wimp (1974) gives two recursion forumulas which are stable when used in the backward direction and which when appro-
priately combined with given normalization relations can be
used to generate values of $U(a;c;z)$. The technique for using
recurrence formulas in this manner is described in the discussion surrounding 1.3(28). See also 19.2. It is clear from
the examples of Chapter 17 and 19.2, that the general backward
recurrence schemata produces rational approximations.

One of Wimp's recursion formulas is or order 3 and generates values of

$$U_m = G_{2,3}^{3,1}\left(z \left| \begin{array}{c} 1-m, m+\gamma \\ \gamma, a, b \end{array} \right. \right)/\Gamma(a)\Gamma(b) \;,\; m = 0,1,\ldots, \qquad (25)$$

$$U_0 = z^a U(a;c;z) \;,\; c = a + 1 - b \qquad (26)$$

with an error of $O(\exp\{-3(n^2 z)^{1/3}\})$ where n is the order of
the rational approximation as in (2). Call this Method I.
In (25), γ is arbitrary, but is usually taken as unity. Let

$$V_m = \lim_{\gamma \to \infty} \gamma^m U_m/m! \;. \qquad (27)$$

Then

$$\begin{aligned} V_m &= \frac{z^m (a)_m (b)_m}{m!} U(m+a;c;z) \\ &= \frac{z^{m+1-a}}{m!} \frac{d^m}{dz^m}\left(z^{m+a-1}U(a;c;z)\right) , \end{aligned} \qquad (28)$$

and V_m satisfies a second order recurrence formula. This too
is studied by Wimp. Call it Method II. Here the error is
$O(\exp\{-2(nz)^{1/2}\})$.

From (13), (16) and (17), the error is $O(\exp\{(-9/2)(n^2 z)^{1/3}\})$, and so our rational approximations for $U(a;c;z)$ are superior to those given by Method I. If $z > 0$ and

$$2(nz)^{\frac{1}{2}} < (9/2)(n^2 z)^{1/3} , \tag{29}$$

then $z < 129.75n$, and so for virtually all z of interest, our rational approximations are superior to those given by Method II. Even so, use of Wimp's techniques might be more efficient in view of machine operations if one desires numerical values of U_m and V_m for values of m other than 0.

Numerical Examples

1. Let $a = 5/6$, $b = 1/6$, $z = 2$. We have the following table.

n	$S_n(z)$, Eq.(13)	$B_n(z)$, Eq.(15)	$R_n(z)$, Eq.(16)
9	$-0.27460(-5)$	$0.34104(5)$	$-0.80518(-10)$
10	$-0.12705(-5)$	$0.95361(5)$	$-0.13463(-10)$

n	$B_n(z)$, True	$R_n(z)$, True
9	$0.33703(5)$	$-1.01224(-10)$
10	$0.94372(5)$	$-0.19900(-10)$

From (15), $B_{10}(z)/B_9(z) = 2.796$ whereas the true value is 2.800. Again, from (16), $R_{10}(z)/R_9(z) = 0.167$ whereas the true value is 0.196. The asymptotic results are quite satisfactory in view of the fact that n is rather small. Indeed, in (12) and (15), we have the restriction $z = o(N^{2/3})$. In our example, if $n = 9$, $z/N^{2/3} = 0.446$.

2. Let $a = b = 1$, $z = 6$. We have the following table.

n	$B_n(z)$, Eq.(15)	$R_n(z)$, Eq.(17)	$B_n(z)$, True	$R_n(z)$, True
20	$4.17157(12)$	$1.53995(-23)$	$3.94976(12)$	$1.47712(-23)$
21	$1.39659(13)$	$5.53729(-24)$	$1.32658(13)$	$6.01134(-24)$
22	$4.60632(13)$	$3.49782(-25)$	$4.38811(13)$	$4.15152(-25)$

Again the asymptotic results are most satisfactory. In (12) and (15), there is the condition $z = o(N^{2/3})$. In our example, if $n = 20$, $z/N^{2/3} = 0.801$.

```
C    ..............................................................
C    .          THIS MAINLINE PROGRAM ILLUSTRATES THE SYNTAX OF THE SUB- .
C    . ROUTINE 'RU2G2' FOR GENERATING VALUES OF THE NUMERATOR AND        .
C    . DENOMINATOR POLYNOMIALS IN THE RATIONAL APPPOXIMATION OF          .
C    . (Z**AP)*U( AP ; 1+AP-BP ; Z ) .                                   .
C    ..............................................................
      IMPLICIT REAL*16(A-H,O-Z)
      DIMENSION A(26),B(26),R(26),D(26),E(26)
      DATA ZERO/0.Q0/
 10   READ(5,1,END=999) N,M
      READ(5,2) AP,BP
      N1=N+1
      D(N1)=ZERO
      E(N1)=ZERO
      DO 100 I=1,M
         READ(5,2) Z
C-------------------------------------------------------------------------
      CALL RU2G2(AP,BP,Z,A,B,N)
C-------------------------------------------------------------------------
C     IN THE ABOVE :
C
C AP AND BP    ARE THE PARAMETERS OF THE U-FUNCTION
C Z            IS THE VALUE OF THE ARGUMENT
C A AND B      WILL CONTAIN THE VALUES OF THE NUMERATOR AND DENOMINATOR
C              POLYNOMIALS, RESPECTIVELY, FOR ALL DEGREES FROM 0 TO
C              N INCLUSIVE
C N            IS THE MAXIMUM DEGREE FOR WHICH VALUES OF THE POLYNOMIALS
C              ARE TO BE CALCULATED
C
C     NOTE : VALUES OF THE K-TH DEGREE POLYNOMIALS WILL BE PLACED IN
C A(K+1) OR B(K+1) AS APPROPRIATE.
C-------------------------------------------------------------------------
         R(N1)=A(N1)/B(N1)
         DO 50 J=1,N
            J1=N1-J
            R(J1)=A(J1)/B(J1)
            D(J1)=R(J1+1)-R(J1)
 50         E(J1)=R(N1)-R(J1)
         WRITE(6,3) N,AP,BP,Z
         DO 60 J=1,N1
            J1=J-1
 60         WRITE(6,4) J1,A(J),B(J)
         WRITE(6,5)
         DO 70 J=1,N1
            J1=J-1
 70         WRITE(6,6) J1,R(J),D(J),E(J)
 100  CONTINUE
      GOTO 10
 999  STOP
 1    FORMAT(2I2)
 2    FORMAT(Q39.32)
 3    FORMAT('1','VALUES OF THE POLYNOMIALS IN THE RATIONAL APPROXIMATIO
     :N OF (Z**AP)*U(AP;1+AP-BP;Z)'//' ','N = ',I2,T20,'AP = ',Q39.32/
     :' ',T20,'BP = ',Q39.32//' ','     Z = ',Q39.32//
     :' ','  I',T22,'A(I)',T65,'B(I)'/)
 4    FORMAT(' ',I2,2X,Q39.32,2X,Q39.32)
 5    FORMAT('0','VALUES OF THE APPROXIMATION, 1ST DIFFERENCES AND APPRO
     :XIMATE ERRORS'//' ','  I',T12,'I-TH APPROXIMATION -- F(I)',T47,
     :'1ST DIFF''S.',T60,'F(N)-F(I)'/)
 6    FORMAT(' ',I2,2X,Q39.32,2X,Q10.3,2X,Q10.3)
      END
```

```
      SUBROUTINE RU2G2(AP,BP,Z,A,B,N)
C     **********************************************************************
C     *         THIS SUBROUTINE RETURNS VALUES A(I) AND B(I), I=1,...,N+1 *
C     * OF THE NUMERATOR AND DENOMINATOR POLYNOMIALS IN THE RATIONAL      *
C     * APPROXIMATION OF (Z**AP)*U( AP ; 1+AP-BP ; Z ) .                  *
C     *                                                                   *
C     *         NO OTHER SUBROUTINES ARE CALLED BY THIS ONE.              *
C     **********************************************************************
      IMPLICIT REAL*16(A-H,O-Z)
      DIMENSION A(1),B(1)
      DATA ONE/1.Q0/,TWO/2.Q0/,THREE/3.Q0/,FIVE/5.Q0/,SIX/6.Q0/
C
C  INITIALIZATION :
C
      AB=AP*BP
      SZZ=Z+Z
      CT2=SZZ-(AB+AB)
      SAB=AP+BP
      B(1)=ONE
      A(1)=ONE
      CT3=SAB+ONE+AB
      B(2)=ONE+SZZ/CT3
      A(2)=ONE+CT2/CT3
      ANBN=CT3+SAB+THREE
      CT1=ONE+SZZ/ANBN
      B(3)=ONE+CT1/CT3*SZZ*THREE
      A(3)=ONE+AB/ANBN*SIX+CT1/CT3*CT2*THREE
      X2I1=FIVE
C
C  FOR I=3,...,N , THE VALUES OF A(I+1) AND B(I+1) ARE CALCULATED
C  USING THE RECURRENCE RELATIONS BELOW
C
      DO 100 I=3,N
C
C  CALCULATION OF THE MULTIPLIERS FOR THE RECURRENCE RELATIONS
C
      CT1=X2I1/(X2I1-TWO)
      ANBN=ANBN+X2I1+SAB
      CT2=(X2I1-ONE)/ANBN
      C2=X2I1*CT2-ONE
      D1Z=X2I1*SZZ/ANBN
      CT3=SAB*CT2
      G1=D1Z+CT1*(C2+CT3)
      G2=D1Z-C2
      G3=CT1*(ONE-(CT3+CT2+CT2))
C----------------------------------------------------------------------
C  THE RECURRENCE RELATIONS FOR A(I+1) AND B(I+1) ARE AS FOLLOWS
C----------------------------------------------------------------------
C
      B(I+1)=G1*B(I)+G2*B(I-1)+G3*B(I-2)
      A(I+1)=G1*A(I)+G2*A(I-1)+G3*A(I-2)
C
100   X2I1=X2I1+TWO
      RETURN
      END
```

VALUES OF THE POLYNOMIALS IN THE RATIONAL APPROXIMATION OF (Z**AP)*U(AP;1+AP-BP;Z)

N = 12   AP = 0.833333333333333333333333333333333Q+00
          BP = 0.166666666666666666666666666666667Q+00

Z = 0.200000000000000000000000000000000Q+01

I	A(I)	B(I)
0	0.100000000000000000000000000000000Q+01	0.100000000000000000000000000000000Q+01
1	0.274025974025974025974025974025970Q+01	0.287012987012987012987012987012990Q+01
2	0.975830052300640535934653581712400Q+01	0.102660208095575013222072045601460Q+02
3	0.342287438428148779474646451975800Q+02	0.365219915365291593586115056030250Q+02
4	0.118930561096648030405917716132578Q+03	0.125071089902286684133862717756430Q+03
5	0.390832939041419962248698767670620Q+03	0.411013105173107107623720285131510Q+03
6	0.123560794589049560968521433823000Q+04	0.129947207201076143388972670974814Q+04
7	0.377141893040226330702642330474910Q+04	0.396611909086684945370848113449049Q+04
8	0.111523100943671244682898504137420Q+05	0.117281470419824901914982732904320Q+05
9	0.320460042872718468544662367761100Q+05	0.337027668201887659537237004980213Q+05
10	0.897383615557140080637894457450770Q+05	0.943719005865792491305316224425144Q+05
11	0.245451551569323518481696887051151Q+06	0.258087268904322371161077997642130Q+06
12	0.656129720020809895381979276761340Q+06	0.690726768625464558463290788775230Q+06

VALUES OF THE APPROXIMATION, 1ST DIFFERENCES AND APPROXIMATE ERRORS

I	I-TH APPROXIMATION -- F(I)	1ST DIFF'S.	F(N)-F(I)
0	0.100000000000000000000000000000000Q+01	-0.452Q-01	-0.491Q-01
1	0.954751312217194570135746606334480Q+00	-0.421Q-02	-0.385Q-02
2	0.950542940061935808857622054185250Q+00	0.357Q-03	0.358Q-03
3	0.950899469694743495360820838233290Q+00	0.422Q-05	0.182Q-05
4	0.950903691430849370325915445508630Q+00	-0.229Q-05	-0.241Q-05
5	0.950901404656701081232958567394530Q+00	-0.144Q-06	-0.119Q-06
6	0.950901260687279607518811476223827Q+00	0.199Q-07	0.243Q-07
7	0.950901280664480733861709321823230Q+00	0.446Q-08	0.453Q-08
8	0.950901285125938545514699925643730Q+00	0.168Q-09	0.666Q-10
9	0.950901285129371432379185081318352Q+00	-0.813Q-10	-0.101Q-09
10	0.950901285212388922854014857608200Q+00	-0.188Q-10	-0.199Q-10
11	0.950901285193634115459243270131200Q+00	-0.114Q-11	-0.114Q-11
12	0.950901285192489900126948196252520Q+00	0.0	0.0

```
C     ............................................................
C     .           THIS MAINLINE PROGRAM ILLUSTRATES THE SYNTAX OF THE SUB-  .
C     . ROUTINE 'CU2G2' FOR GENERATING COEFFICIENTS IN THE POLYNOMIALS .
C     . FOR THE RATIONAL APPROXIMATION OF THE FUNCTION                      .
C     .                (Z**AP) * U( AP ; 1 + AP - BP ; Z )                  .
C     ............................................................
      IMPLICIT REAL*16(A-H,O-Z)
      DIMENSION CA(26),CB(26),NO(25)
   10 READ(5,2,END=999) AP,BP
      WRITE(6,3) AP,BP
      READ(5,1) M,(NO(J),J=1,M)
      DO 100 I=1,M
      N=NO(I)
      N1=N+1
C----------------------------------------------------------------
      CALL CU2G2(AP,BP,CA,CB,N)
C----------------------------------------------------------------
C     IN THE ABOVE :
C
C AP AND BP   ARE PARAMETERS IN THE U-FUNCTION
C N           IS THE DEGREE OF THE POLYNOMIALS IN THE RATIONAL
C             APPROXIMATION
C CA AND CB   WILL CONTAIN THE COEFFICIENTS OF THE NUMERATOR AND
C             DENOMINATOR POLYNOMIALS RESPECTIVELY
C
C     NOTE : THE COEFFICIENTS OF THE K-TH POWER OF Z WILL BE PLACED
C IN CA(K+1) OR CB(K+1) AS APPROPRIATE
C----------------------------------------------------------------
  100 WRITE(6,4) N,(CA(J),CB(J),J=1,N1)
      GOTO 10
  999 STOP
    1 FORMAT(26I2)
    2 FORMAT(Q39.32)
    3 FORMAT('1','COEFFICIENTS FOR THE RATIONAL APPROXIMATION OF (Z**AP)
     : * U( AP ; 1+AP-BP ; Z )'//' ',T20,'AP = ',Q39.32/' ',T20,'BP = ',
     :Q39.32/)
    4 FORMAT(' ','N = ',I2,T18,'CA(I)',T58,'CB(I)'//
     :26(1X,Q39.32,2X,Q39.32/)/)
      END
```

```
      SUBROUTINE CU2G2(AP,BP,A,B,N)
C     *****************************************************************
C     *          THIS SUBROUTINE RETURNS COEFFICIENTS A(I) AND B(I) ,   *
C     *    I = 1,2,...,N+1 , OF THE NUMERATOR AND DENOMINATOR POLYNOMIALS *
C     *    RESPECTIVELY, IN THE RATIONAL APPROXIMATION OF ORDER  N  FOR  *
C     *    (Z**AP) * U( AP ; 1 + AP - BP ; Z ) .                        *
C     *                                                                 *
C     *          NO OTHER SUBROUTINES ARE CALLED BY THIS ONE.           *
C     *****************************************************************
      IMPLICIT REAL*16(A-H,O-Z)
      DIMENSION A(1),B(1)
      DATA ONE/1.Q0/
C
C   INITIALIZATION :
C
      B(1)=ONE
      A(1)=ONE
      XI=ONE
      AP1=AP-ONE
      BP1=BP-ONE
      XN=N
      XN1I=XN
      DO 100 I=1,N
      I1=I+1
C----------------------------------------------------------------------
C  FOR I = 1,2,...,N , B(I+1) IS CALCULATED AS FOLLOWS
C----------------------------------------------------------------------
C
      B(I1)=B(I)/(AP+XI)*(XN+XI)/(BP+XI)*XN1I/XI
C
      CT1=B(I1)
      A(I1)=CT1
      XJ=ONE
      DO 50 J=1,I
C----------------------------------------------------------------------
C  THE NUMERATOR COEFFICIENTS ARE OBTAINED FROM THOSE OF THE
C  DENOMINATOR BY A SUMMATION PROCESS WHICH IS DESCRIBED IN THE MAIN
C  TEXT.  THE EQUATIONS FOR DOING THIS ARE AS FOLLOWS.
C----------------------------------------------------------------------
C
      CT1=-CT1/XJ*(AP1+XJ)*(BP1+XJ)
      A(I1-J)=A(I1-J)+CT1
C
 50       XJ=XJ+ONE
      XN1I=XN-XI
 100      XI=XI+ONE
      RETURN
      END
```

COEFFICIENTS FOR THE RATIONAL APPROXIMATION OF (Z**AP) * U(AP ; 1+AP-BP ; Z)

$$AP = 0.83333333333333333333333333333333333Q+00$$
$$BP = 0.16666666666666666666666666666666667Q+00$$

N = 3   CA(I)   CB(I)

0.67067930050834883248397853324492Q+00   0.10000000000000000000000000000000000Q+01
0.50875580201196031052000899894291Q+01   0.56103896103896103896103896103896Q+01
0.44649817809058622086395876300031Q+01   0.45695480989598636657460186871952Q+01
0.75287748998881048955083145418318Q+00   0.75287748998881048955083145418318Q+00

N = 4   CA(I)   CB(I)

0.61842575339542026477196663001592Q+00   0.10000000000000000000000000000000000Q+01
0.80703876959557168988057259092020Q+01   0.93506493506493506493506493506494Q+01
0.13054419719372072916662850797950Q+02   0.13708644296879590997238056061585Q+02
0.51974508101986158623474640388784Q+01   0.52701424299216734268558201792823Q+01
0.52337966200601446446016421090803Q+00   0.52337966200601446446016421090803Q+00

N = 5   CA(I)   CB(I)

0.57929383283151778913136225092451Q+00   0.10000000000000000000000000000000000Q+01
0.11571153276299584599544277150795Q+02   0.14025974025974025974025974025974Q+02
0.29663629461693717291874470151406Q+02   0.31986836692719045660222130810366Q+02
0.20472773983163580135792122786655Q+02   0.21080569719686693707423280717129Q+02
0.46670029768739077821678237239956Q+01   0.47104169580541301801414778981723Q+01
0.31258066449760126541031005407226Q+00   0.31258066449760126541031005407226Q+00

N = 6   CA(I)   CB(I)

0.54845064432598809554651335222318Q+00   0.10000000000000000000000000000000000Q+01
0.15551380038592784814958394687421Q+02   0.19636363636363636363636363636364Q+02
0.57793835222560527851750333939000Q+02   0.63973673385438091320444261620732Q+02
0.60431701890870787665067727599958Q+02   0.63241709159060081122269842151387Q+02
0.23098770708599772335181560803347Q+02   0.23552084790270650900707389490861Q+02
0.34157216053900699912371191604192Q+01   0.34383873094736139195134105947949Q+01
0.16319306940151628358929832750510Q+00   0.16319306940151628358929832750510Q+00

XXI PADÉ APPROXIMATIONS FOR zU(1;2-B;z)

We consider rational approximations for function in the title which is a form of the incomplete gamma function. See 2.2(41). The rational approximations of this section lie on the main diagonal of the Padé table. Except for a change in normalization, they follow from the rational approximations of 14(1), if there we replace z by c/z and let $c \to \infty$. Then put $b = 1-\nu$ to conform with the usual notation for the incomplete gamma function. This confluence principle does not hold for the asymptotic analysis of the error. The rational approximations also follow from 1.5(13-16). There, put $a = 0$, $f = g = 1$, $c_1 = 2$, $d_1 = 1$, $p = 2$, $q = 0$, $\alpha_1 = 1$, $\alpha_2 = 1-\nu$, $\lambda = \beta+1$ and let $\beta \to \infty$.

The incomplete gamma function can be defined by

$$\Gamma(\nu,z) = \int_z^\infty t^{\nu-1} e^{-t} dt \tag{1}$$

or

$$\Gamma(\nu,z) = z^\nu e^{-z} \int_0^{\infty e^{i\theta}} e^{-zt}(1+t)^{\nu-1} dt \, ,$$

$$z \neq 0 \, , \quad -\pi < \theta < \pi \, , \quad |\theta + \arg z| < \pi/2$$

$$\text{or} \quad |\theta + \arg z| = \pi/2 \quad \text{if} \quad R(\nu) < 1 \, . \tag{2}$$

We have

$$\Gamma(\nu,z) = z^\nu e^{-z} U(1;1+\nu;z) \, , \tag{3}$$

and the asymptotic expansion

$$\Gamma(\nu,z) \sim z^{\nu-1} e^{-z} {}_2F_0(1,1-\nu;-1/z) \, ,$$

265

$$|z| \to \infty , \quad |\arg z| \le 3\pi/2 - \varepsilon , \quad \varepsilon > 0 . \qquad (4)$$

We write

$$z^{1-\nu}e^z\Gamma(\nu,z) = \{E_n(z)/F_n(z)\} + T_n(\nu,z) , \qquad (5)$$

$$F_n(z) = {}_1F_1(-n;2-\nu;-z) \qquad (6)$$

$$E_n(z) = \sum_{k=0}^{\infty} \frac{(-n)_k(1-\nu)_k}{(2-\nu)_k k!} \; {}_2F_2\left(\begin{matrix} -n+k, \; 1 \\ 2-\nu+k, 1+k \end{matrix}\bigg| -z\right) . \qquad (7)$$

Here $T_n(\nu,z)$ is the error which we discuss later. For the polynomials $F_n(z)$ and $E_n(z)$, we have

$$F_0(z) = 1 , \quad F_1(z) = 1 + (2-\nu)^{-1}z ,$$

$$F_2(z) = 1 + 2(2-\nu)^{-1}z + [(2-\nu)(3-\nu)]^{-1}z^2 ,$$

$$F_3(z) = 1 + 3(2-\nu)^{-1}z + 3[(2-\nu)(3-\nu)]^{-1}z^2$$

$$+ [(2-\nu)(3-\nu)(4-\nu)]^{-1}z^3 , \qquad (8)$$

$$E_0(z) = 1 , \quad E_1(z) = (2-\nu)^{-1}(1+z) , \quad E_2(z) = [(2-\nu)(3-\nu)]^{-1}$$

$$\times \; [2 + (5-\nu)z + z^2] , \quad E_3(z) = [(2-\nu)(3-\nu)(4-\nu)]^{-1}$$

$$\times \; [6 + (26-9\nu+\nu^2)z + (11-2\nu)z^2 + z^3] . \qquad (9)$$

Both $E_n(z)$ and $F_n(z)$ satisfy the same recurrence formula

$$(n+1-\nu)F_n(z) = (z+2n-\nu)F_{n-1}(z) - (n-1)F_{n-2}(z) . \qquad (10)$$

This recurrence formula is stable in the forward direction.

For analysis of the error, it is convenient to have the

notation

$$m = \tfrac{1}{2}(1-\nu) \ , \ k = n+1-\tfrac{1}{2}\nu = n+m+\tfrac{1}{2} \ . \tag{11}$$

We present uniform asymptotic estimates of the error, that is, we give results for n large uniformly in z, z bounded away from the origin. If z and k are real, say $z = x$, there are four cases to consider. They are $0 < x/4k < \infty$, $-1 < x/4k < 0$, $x/4k \sim -1$ and $x/4k < -1$. More generally, for z and k complex, we study the regions $|\arg z/k| < \pi$ and $|\arg z/k| = \pi$ where the latter region is divided into three subregions. $x/4k \sim -1$ is sometimes called the transition region.

The Error $T_n(\nu,z)$ for $|\arg z/k| < \pi$

Let

$$z/4k = \sinh^2\alpha \ . \tag{12}$$

Because z and k may be complex, in the determination of $\sinh \alpha$ and α we choose the branch such that $\sinh \alpha$ is real and positive if $z/4k$ is real and positive. To be explicit, let

$$\sinh^2\alpha = z/4k = \rho e^{i\theta} \ , \ \rho > 0 \ , \ |\theta| < \pi \ ,$$

$$\alpha = \beta+i\delta \ , \ \beta \ \text{and} \ \delta \ \text{real,} \ \beta > 0 \ , \ |\delta| < \pi/2 \ , \tag{13}$$

so that

$$\cosh \beta = (\tfrac{1}{2}[1 + \rho + \{(1+\rho)^2 - 4\rho \sin^2\theta/2\}^{\frac{1}{2}}])^{\frac{1}{2}} \ ,$$

$$\sin \delta \cosh \beta = \rho^{\frac{1}{2}}\sin \theta/2 \ , \ \cos \delta \sinh \beta = \rho^{\frac{1}{2}}\cos \theta/2 \ . \tag{14}$$

Then

$$T_n(\nu,z) = - \frac{2\pi e^z z^{1-\nu} \exp[-2k(2\alpha + \sinh 2\alpha)][A(\alpha,k) + O(k^{-3})]}{\begin{pmatrix} \Gamma(1-\nu)[A(\alpha,-k) + O(k^{-3}) + \exp\{\pm i\pi(2m+\frac{1}{2})\} \\ \times \exp\{-2k(2\alpha + \sinh 2\alpha)\}\{A(\alpha,k) + O(k^{-3})\}] \end{pmatrix}} ,$$

$$(15)$$

where

$$A(\alpha,k) = 1 + P_1(\alpha)k^{-1} + P_2(\alpha)k^{-2} ,$$

$$P_1(\alpha) = (96t^3)^{-1}\{3(16m^2-1)t^4 - 6t^2 + 5\} ,$$

$$P_2(\alpha) = (18432t^6)^{-1}\{9(16m^2-9)(16m^2-1)t^8 + 36(48m^2-7)t^6$$

$$+ 6(121-112m^2)t^4 - 924t^2 + 385\} ,$$

$$t = \coth \alpha = (1 + 4k/z)^{\frac{1}{2}} .$$

$$(16)$$

Clearly, if ν is fixed and $n \to \infty$, arg $k \to 0$ and

$$\lim_{n \to \infty} T_n(\nu,z) = 0 , \quad |\arg z| < \pi$$

$$(17)$$

uniformly in z , z bounded away from the origin.

If $|z|$ is small with respect to n so that $|z/4k\alpha^2| \sim 1$, then from (15) we have

$$T_n(\nu,z) = - \frac{2\pi z^{1-\nu}}{\Gamma(1-\nu)} \exp[z - 4(kz)^{\frac{1}{2}}][1 + O(k^{-\frac{1}{2}})] .$$

$$(18)$$

If n is large but fixed, and $|z|$ is large so that $|z/4k| \gg 1$, then

$$T_n(\nu,z) \sim - \frac{n^{1-\nu}(n!)^2}{\Gamma(1-\nu)z^{2n+1}} , \quad |\arg k| < \pi/2 .$$

$$(19)$$

which shows that for n fixed but sufficiently large, the

Padé approximation has the same behavior as the asymptotic expansion for $\Gamma(\nu,z)$. Indeed, if $z = x$ with $x > 0$, then the error committed using $2n+1$ terms of the asymptotic expansion of $z^{1-\nu}e^z\Gamma(\nu,z)$ [see (4)] is negative provided $\nu < 1$ and does not exceed $x^{-2n-1}(1-\nu)_{2n+1}$ in magnitude. For n sufficiently large, the ratio of the right-hand side of (19) with sign omitted, to this last number is approximately $2^{-2n-1+\nu}(\pi/n)^{\frac{1}{2}}$. Under these conditions, it follows that for virtually the same number of operations, the Padé approximation is superior to the asymptotic expansion.

In the table on the following page, we illustrate evaluation of two cases of the incomplete gamma function and use of the error formula (15). The latter is remarkably realistic even for small n .

To facilitate use of this section, tables of a function closely related to $|T_n(\nu,z)|$, $\nu = 0,\frac{1}{2}$ are provided in Luke (1969,1975a) for a wide range of z-values in the complex plane. For ν sufficiently small, tables for $\nu = 0$ suffice since it can be shown that

$$\Gamma(1-\nu)z^\nu T_n(\nu,z) \sim T_n(0,z)\exp[2\alpha_0\nu + \frac{\nu^2\tanh\alpha_0}{4k_0} + O(k_0^{-2})]$$

$$\times\ [1 - \frac{\nu(2-\nu)\cosh\alpha_0}{2(k_0z)^{\frac{1}{2}}} + O(k_0^{-3/2})]\ ,\qquad (20)$$

where α_0 and k_0 are the values of α and k when $\nu = 0$. In illustration, let $z = 2$. Then the ratio of $z^{-\frac{1}{2}}e^{-z}T_4(\frac{1}{2},z)$ to $z^{-1}e^{-z}T_4(0,z)$ is 0.685 , while the corresponding number deduced from (20) with order terms omitted is 0.672 .

$z = 2e^{i\varphi}$ φ	$\int_z^\infty t^{-1}e^{-t}\,dt$	$z^{-1}e^{-z}E_4(0, z)/F_4(0, z)$
0	$4.89005 \cdot 10^{-2}$	$4.89190 \cdot 10^{-2}$
$\pi/4$	$-3.95846 \cdot 10^{-2} - 8.22921i \cdot 10^{-2}$	$-3.95652 \cdot 10^{-2} - 8.22414i \cdot 10^{-2}$
$\pi/2$	$-4.22981 \cdot 10^{-1} + 3.46167i \cdot 10^{-2}$	$-4.23980 \cdot 10^{-1} + 3.43682i \cdot 10^{-2}$
$3\pi/4$	$-2.16947 \quad + 0.31777i$	$-2.12827 \quad + 0.36462i$

$z = 2e^{i\varphi}$	$z^{-1}e^{-z}T_4(0, z)$	
φ	Exact	Approximate
0	$-0.185 \cdot 10^{-4}$	$-0.184 \cdot 10^{-4}$
$\pi/4$	$-0.194 \cdot 10^{-4} - 0.507i \cdot 10^{-4}$	$-0.194 \cdot 10^{-4} - 0.507i \cdot 10^{-4}$
$\pi/2$	$0.999 \cdot 10^{-3} + 0.248i \cdot 10^{-3}$	$0.996 \cdot 10^{-3} + 0.249i \cdot 10^{-3}$
$3\pi/4$	$-0.412 \cdot 10^{-1} - 0.468i \cdot 10^{-1}$	$-0.412 \cdot 10^{-1} - 0.473i \cdot 10^{-1}$

$z = 2e^{i\varphi}$ φ	$\int_z^\infty t^{-1/2}e^{-t}\,dt$	$z^{-1/2}e^{-z}E_4(\frac{1}{2}, z)/F_4(\frac{1}{2}, z)$
0	$8.06471 \cdot 10^{-2}$	$8.06597 \cdot 10^{-2}$
$\pi/4$	$-2.03962 \cdot 10^{-2} - 0.146768i$	$-2.03888 \cdot 10^{-2} - 0.146732i$
$\pi/2$	$-0.560363 \quad - 0.337570i$	$-0.560936 \quad - 0.337918i$
$3\pi/4$	$-2.14095 \quad - 2.19684i$	$-2.12898 \quad - 2.16019i$

$z = 2e^{i\varphi}$	$z^{-1/2}e^{-z}T_4(\frac{1}{2}, z)$	
φ	Exact	Approximate
0	$-0.126 \cdot 10^{-4}$	$-0.126 \cdot 10^{-4}$
$\pi/4$	$-0.074 \cdot 10^{-4} - 0.36i \cdot 10^{-4}$	$-0.074 \cdot 10^{-4} - 0.36i \cdot 10^{-4}$
$\pi/2$	$0.573 \cdot 10^{-3} + 0.348i \cdot 10^{-3}$	$0.572 \cdot 10^{-3} + 0.348i \cdot 10^{-3}$
$3\pi/4$	$-0.120 \cdot 10^{-1} - 0.366i \cdot 10^{-1}$	$-0.119 \cdot 10^{-1} - 0.369i \cdot 10^{-1}$

The Error $T_n(\nu,z)$ for $|\arg z/k| = \pi$

Here we consider uniform asymptotic formulas for $T_n(\nu,ze^{\pm i\pi})$. If ν is real, it is known that the denominator polynomial of the rational approximation, $F_n(-z)$, has zeros in the segment $0 < z/4k < 1$ where we must now understand that our previous z has been replaced by $ze^{\pm i\pi}$. The largest zero does not exceed $3k$ if n is large enough. Asymptotic developments for the zeros of $F_n(-z)$ and $T_n(\nu,ze^{\pm i\pi})$ to embrace the region $0 < z/4k < 1$ have been given by Luke (1969,1975a). We omit the details here. However, we do give results which include the transition region $z/4k \sim 1$ and the situation when $|z/4k| > 1$.

For the transition region, let

$$z - 4k = 2\varepsilon = O[(z/6)^{1/3}] . \tag{21}$$

Then

$$T_n(\nu,ze^{\pm i\pi}) \sim \frac{3^{\frac{1}{2}}\pi e^{-z}z^{1-\nu}(f+g)}{\Gamma(1-\nu)(f-g)} \{1 \pm \frac{i(f-g)}{3^{\frac{1}{2}}}\} , \quad f = 1 + \frac{\varepsilon(\varepsilon^2+\frac{1}{5})}{3z} ,$$

$$g = \frac{\Gamma(\frac{5}{6})}{\pi^{\frac{1}{2}}(z/3)^{1/3}} \{\varepsilon + \frac{(\varepsilon^4+\varepsilon^2+12m^2-33/35)}{6z} + O[\frac{\varepsilon^6}{(z/3)^2}]\} ,$$

$$|\arg z| \leq \frac{\pi}{2} . \tag{22}$$

If $z = x$ is real and positive, we take

$$2T_n(\nu,-x) = T_n(\nu,xe^{i\pi}) + T_n(\nu,xe^{-i\pi}) .$$

Thus,

$$T_n(\nu,-x) \sim \frac{3^{\frac{1}{2}}\pi e^{-x}x^{1-\nu}(f+g)}{\Gamma(1-\nu)(f-g)} . \tag{23}$$

Next we consider the asymptotic formula for $T_n(\nu,ze^{\pm i\pi})$

when $|z/4k| > 1$.   Put

$$z/4k = \cosh^2 \alpha \ . \tag{24}$$

Then

$$T_n(\nu, ze^{\pm i\pi}) \sim \frac{2\pi e^{-z} z^{1-\nu} \exp[2k(\sinh 2\alpha - 2\alpha)]}{\Gamma(1-\nu)}$$

$$\times \ \frac{[1 + Q_1(\alpha)k^{-1} + Q_2(\alpha)k^{-2} + O(k^{-3})]}{[1 - Q_1(\alpha)k^{-1} + Q_2(\alpha)k^{-2} + O(k^{-3})]} \ ,$$

$$k \to \infty \ , \ |z/4k| > 1 \ , \ |\arg z/k| \le \pi/2 \ . \tag{25}$$

Here, $Q_j(\alpha) = P_j(\alpha + i\pi/2)$, $j = 1,2$.   That is, $Q_j(\alpha)$ is
of the same form as $P_j(\alpha)$ [see (16)] if in the latter,
coth α is replaced by tanh α .   When   z   and   k   are complex, for the determination of α , we choose the branch such
that α is real and positive when   z   and   k   are real and
positive.   Specifically,

$$z/4k = \cosh^2 \alpha = \eta e^{i\tau}, \ \eta > 0, \ \alpha = \mu + i\xi, \ \mu \text{ and } \xi \text{ real}, \ \mu > 0 \ , \tag{26}$$

$$\cosh \mu = (\tfrac{1}{2}[\eta + 1 + \{(\eta-1)^2 + 4\eta \sin^2 \tfrac{1}{2}\tau\}^{\frac{1}{2}}])^{\frac{1}{2}} \ , \tag{27}$$

$$\cos \xi \cosh \mu = \eta^{\frac{1}{2}} \cos \tau/2 \ , \ \sin \xi \sinh \mu = \eta^{\frac{1}{2}} \sin \tau/2 \ . \tag{28}$$

Now suppose $|z/4k| \gg 1$.   Then, as in the development
of (19),

$$T_n(\nu, ze^{\pm i\pi}) \sim [n^{1-\nu-a}(n!)^2]/[\Gamma(1-\nu)z^{2n+1-a}] \ ,$$

$$|\arg k| < \pi/2 \ , \tag{29}$$

and so,

$$\lim_{z \to \infty} z^{2n} T_n(\nu, -z) = 0 \ . \tag{30}$$

Thus, if $z = x$, $x > 0$, and we fix the order of the rational approximate, then the accuracy of the approximation increases as x increases, and thus the rational approximations and asymptotic expansion for $\Gamma(\nu, -x)$ [see (4)] exhibit the same behavior. However, no efficient estimate is available to assess the error when using the truncated asymptotic expansion. If we truncate the asymptotic expansion for $x^{1-\nu}e^{-x}\Gamma(\nu, -x)$ after $2n+1$ terms, then the error certainly exceeds $x^{-2n-1}(1-\nu)_{2n+1}$. The ratio of the right-hand side of (29) to this last quantity is $2^{-2n-1+\nu}(\pi n)^{\frac{1}{2}}$ for n sufficiently large, and the superiority of the Padé approximations is manifest.

The following table illustrates use of formulas (22) and (25).

x	$(2x)^{-1}E_2(\frac{1}{2}, -x^2)/F_2(\frac{1}{2}, -x^2)$	Error	Approx. Error
3	0.17808 219(0)	0.740(-4)	0.741(-4) [a]
4	0.12934 631(0)	0.169(-5)	0.158(-5) [b]
5	0.10213 40744(0)	0.794(-7)	0.793(-7) [b]

a) Deduced from equation (22)

b) Deduced from equation (25)

Further numerical examples are given by Luke (1969, 1975a).

```
C       ...............................................................
C       .        THIS MAINLINE PROGRAM ILLUSTRATES THE SYNTAX OF THE SUB-  .
C       .  ROUTINE 'RU2G2P' FOR GENERATING VALUES OF THE NUMERATOR AND     .
C       .  DENOMINATOR POLYNOMIALS IN THE PADE APPROXIMATION OF            .
C       .  Z * U( 1 ; 1+V ; Z ) .                                          .
C       ...............................................................
        IMPLICIT REAL*16(A-H,O-Z)
        DIMENSION A(26),B(26),R(26),D(26),E(26)
        DATA ZERO/0.Q0/
   10   READ(5,1,END=999) N,M
        READ(5,2) V
        N1=N+1
        D(N1)=ZERO
        E(N1)=ZERO
        DO 100 I=1,M
          READ(5,2) Z
C-----------------------------------------------------------------------
        CALL RU2G2P(V,Z,A,B,N)
C-----------------------------------------------------------------------
C       IN THE ABOVE :
C
C   V             IS THE PARAMETER IN THE U-FUNCTION
C   Z             IS THE VALUE OF THE ARGUMENT
C   A AND B       WILL CONTAIN THE VALUES OF THE NUMERATOR AND DENOMINATOR
C                 POLYNOMIALS RESPECTIVELY FOR ALL DEGREES FROM 0 TO
C                 N INCLUSIVE
C   N             IS THE MAXIMUM DEGREE FOR WHICH VALUES OF THE POLYNOMIALS
C                 ARE TO BE CALCULATED
C
C   NOTE : VALUES OF THE K-TH DEGREE POLYNOMIALS WILL BE PLACED IN
C A(K+1) OR B(K+1) AS APPROPRIATE.
C-----------------------------------------------------------------------
        R(N1)=A(N1)/B(N1)
        DO 50 J=1,N
          J1=N1-J
          R(J1)=A(J1)/B(J1)
          D(J1)=R(J1+1)-R(J1)
   50     E(J1)=R(N1)-R(J1)
        WRITE(6,3) N,V,Z
        DO 60 J=1,N1
          J1=J-1
   60     WRITE(6,4) J1,A(J),B(J)
        WRITE(6,5)
        DO 70 J=1,N1
          J1=J-1
   70     WRITE(6,6) J1,R(J),D(J),E(J)
  100   CONTINUE
        GOTO 10
  999   STOP
    1   FORMAT(2I2)
    2   FORMAT(Q39.32)
    3   FORMAT('1','VALUES OF THE POLYNOMIALS IN THE PADE APPROXIMATION OF
       : Z * U( 1 ; 1+V ; Z )'//' ','N = ',I2,T20,'V = ',Q39.32//
       :' ',' ',       Z = ',Q39.32//' ',' I',T24,'A(I)',T65,'B(I)'/)
    4   FORMAT(' ',I2,2X,Q39.32,2X,Q39.32)
    5   FORMAT('0','VALUES OF THE APPROXIMATION, 1ST DIFFERENCES AND APPRO
       :XIMATE ERRORS'//' ',' I',T12,'I-TH APPROXIMATION -- F(I)',T47,
       :'1ST DIFF''S.',T60,'F(N)-F(I)'/)
    6   FORMAT(' ',I2,2X,Q39.32,2X,Q10.3,2X,Q10.3)
        END
```

```
      SUBROUTINE RU2G2P(V,Z,A,B,N)
C     ***********************************************************
C     *          THIS SUBROUTINE RETURNS VALUES A(I) AND B(I), I=1,...,N+1 *
C     * OF THE NUMERATOR AND DENOMINATOR POLYNOMIALS IN THE PADE    *
C     * APPROXIMATION OF Z * U( 1 ; 1+V ; Z ) .                     *
C     *                                                             *
C     *          NO OTHER SUBROUTINES ARE CALLED BY THIS ONE.       *
C     ***********************************************************
      IMPLICIT REAL*16(A-H,O-Z)
      DIMENSION A(1),B(1)
      DATA ONE/1.Q0/,TWO/2.Q0/
C
C  INITIALIZATION :
C
      XI=TWO
      V2=TWO-V
      ZV=Z-V
      A(1)=ONE
      B(1)=ONE
      XI1=ONE
      A(2)=(ONE+Z)/V2
      B(2)=ONE+Z/V2
      DO 100 I=2,N
          I1=I+1
          I2=I-1
C
C  FOR I = 2,...,N , THE VALUES A(I+1) AND B(I+1) ARE CALCULATED
C  USING THE RECURRENCE RELATIONS BELOW
C
      CT1=XI1+V2
      G1=(XI+XI+ZV)/CT1
      G2=-XI1/CT1
C-----------------------------------------------------------------
C  THE RECURRENCE RELATIONS FOR A(I+1) AND B(I+1) ARE AS FOLLOWS
C-----------------------------------------------------------------
C
      A(I1)=G1*A(I)+G2*A(I2)
      B(I1)=G1*B(I)+G2*B(I2)
C
      XI1=XI
 100  XI=XI+ONE
      RETURN
      END
```

VALUES OF THE POLYNOMIALS IN THE PADE APPROXIMATION OF Z * U(1 ; 1+V ; Z)

N = 12   V = 0.50000000000000000000000000000Q+00

Z = 0.80000000000000000000000000000Q+01

I	A(I)	B(I)
0	0.100000000000000000000000000000Q+01	0.100000000000000000000000000000000000Q+01
1	0.600000000000000000000000000000Q+01	0.633333333333333333333333333333330Q+01
2	0.272000000000000000000000000000Q+02	0.287333333333333333333333333333330Q+02
3	0.101485714285714285714285710Q+03	0.107209523809523809523809523809520Q+03
4	0.331428571428571428571428570Q+03	0.350121693121693121693121693121690Q+03
5	0.980373662337662337662337650Q+03	0.103605300062530062530062530062530Q+04
6	0.268726793206793206793206790Q+04	0.283883463943463943463943463943460Q+04
7	0.691891127539127539127539120Q+04	0.730915022804356137689471022280435Q+04
8	0.169157105232022879031702611114370Q+05	0.178697868097624960370054092740900Q+05
9	0.395788766461608566871724766461600Q+05	0.418111959815363324135253959915360Q+05
10	0.891597821962479016879724531120616Q+05	0.941885316232254067037103058529Q+05
11	0.194297800941980039416725761744318Q+06	0.205256552910708370122340384739700Q+06
12	0.411698714051988398160734372106706Q+06	0.434360586552142434348093270036Q+06

VALUES OF THE APPROXIMATION, 1ST DIFFERENCES AND APPROXIMATE ERRORS

I	I-TH APPROXIMATION -- F(I)	1ST DIFF'S.	F(N)-F(I)
0	0.100000000000000000000000000000Q+01	-0.526Q-01	-0.534Q-01
1	0.947368421052631578947368421052630Q+00	-0.733Q-03	-0.759Q-03
2	0.946635730858466677494199535962880Q+00	-0.247Q-04	-0.262Q-04
3	0.946610997601492404725948298836280Q+00	-0.135Q-05	-0.147Q-05
4	0.946609644416907197799706820969260Q+00	-0.102Q-06	-0.113Q-06
5	0.946609542579847752843166303164700Q+00	-0.966Q-08	-0.109Q-07
6	0.946609532918447045805578115799840Q+00	-0.110Q-08	-0.126Q-08
7	0.946609531822861279502426231936740Q+00	-0.143Q-09	-0.169Q-09
8	0.946609531679528269836823758052130Q+00	-0.211Q-10	-0.253Q-10
9	0.946609531658426053909348520074100Q+00	-0.343Q-11	-0.416Q-11
10	0.946609531654996732850789112083100Q+00	-0.608Q-12	-0.724Q-12
11	0.946609531654388935119645484902900Q+00	-0.116Q-12	-0.116Q-12
12	0.946609531654272953421848994925310Q+00	0.0	0.0

```
C     .................................................................
C     .          THIS MAINLINE PROGRAM ILLUSTRATES THE SYNTAX OF THE SUB- .
C     .  ROUTINE 'CU2G2P' FOR GENERATING COEFFICIENTS IN THE POLYNOMIALS.
C     .  FOR THE PADE APPROXIMATION OF Z*U(1;1+V;Z)                       .
C     .................................................................
      IMPLICIT REAL*16(A-H,O-Z)
      DIMENSION CA(26),CB(26),NO(25)
10    READ(5,2,END=999) V
      WRITE(6,3) V
      READ(5,1) M,(NO(J),J=1,M)
      DO 100 I=1,M
      N=NO(I)
      N1=N+1
C-----------------------------------------------------------------------
      CALL CU2G2P(V,CA,CB,N)
C-----------------------------------------------------------------------
C     IN THE ABOVE :
C
C  V            IS THE PARAMETER OF THE U-FUNCTION
C  N            IS THE DEGREE OF THE POLYNOMIALS IN THE RATIONAL
C               APPROXIMATION
C  CA AND CB    WILL CONTAIN THE COEFFICIENTS OF THE NUMERATOR AND
C               DENOMINATOR POLYNOMIALS RESPECTIVELY
C
C     NOTE : THE COEFFICIENTS OF THE K-TH POWER OF Z WILL BE PLACED
C IN CA(K+1) OR CB(K+1) AS APPROPRIATE.
C-----------------------------------------------------------------------
100   WRITE(6,4) N,(CA(J),CB(J),J=1,N1)
      GOTO 10
999   STOP
1     FORMAT(26I2)
2     FORMAT(Q39.32)
3     FORMAT('1','COEFFICIENTS FOR THE PADE APPROXIMATION OF Z * U( 1 ;
     :1+V ; Z )'//' ',T20,'V = ',Q39.32/)
4     FORMAT(' ','N = ',I2,T18,'CA(I)',T58,'CB(I)'//
     :26(1X,Q39.32,2X,Q39.32/)/)
      END
```

```
      SUBROUTINE CU2G2P(V,A,B,N)
C     *********************************************************************
C     *          THIS SUBROUTINE RETURNS COEFFICIENTS A(I) AND B(I) ,     *
C     *  I = 1,2,...,N+1 , OF THE NUMERATOR AND DENOMINATOR POLYNOMIALS *
C     *  RESPECTIVELY, IN THE PADE APPROXIMATION OF ORDER  N   FOR        *
C     *  Z * U(1;V+1;Z) .                                                 *
C     *                                                                   *
C     *          NO OTHER SUBROUTINES ARE CALLED BY THIS ONE.             *
C     *********************************************************************
      IMPLICIT REAL*16(A-H,O-Z)
      DIMENSION A(1),B(1)
      DATA ONE/1.Q0/
C
C   INITIALIZATION :
C
      B(1)=ONE
      A(1)=ONE
      XI=ONE
      XN=N
      XN1I=XN
      V1=ONE-V
      DO 100 I=1,N
        I1=I+1
C-------------------------------------------------------------------------
C  FOR I = 1,2,....,N , B(I+1) IS CALCULATED AS FOLLOWS
C-------------------------------------------------------------------------
C
        B(I1)=B(I)/(V1+XI)*XN1I/XI
C
        CT1=B(I1)
        A(I1)=CT1
        XJ=ONE
        DO 50 J=1,I
C-------------------------------------------------------------------------
C  THE NUMERATOR COEFFICIENTS ARE OBTAINED FROM THOSE OF THE
C  DENOMINATOR BY A PROCESS OF SUMMATION WHICH IS DESCRIBED IN THE MAIN
C  TEXT.  THE EQUATIONS FOR DOING THIS ARE AS FOLLOWS.
C-------------------------------------------------------------------------
C
          CT1=(V-XJ)*CT1
          A(I1-J)=A(I1-J)+CT1
C
 50       XJ=XJ+ONE
        XN1I=XN-XI
 100    XI=XI+ONE
      RETURN
      END
```

COEFFICIENTS FOR THE PADE APPROXIMATION OF Z * U(1 ; 1+V ; Z)

V =   0.50000000000000000000000000000000000Q+00

N =   3   CA(I)   CB(I)

0.45714285714285714285714285714286Q+00   0.10000000000000000000000000000000000Q+01
0.16571428571428571428571428571429Q+01   0.20000000000000000000000000000000000Q+01
0.76190476190476190476190476190476Q+00   0.80000000000000000000000000000000000Q+00
0.76190476190476190476190476190476Q-01   0.76190476190476190476190476190476Q-01

N =   4   CA(I)   CB(I)

0.40634920634920634920634920634921Q+00   0.10000000000000000000000000000000000Q+01
0.20634920634920634920634920634921Q+01   0.26666666666666666666666666666667Q+01
0.14603174603174603174603174603175Q+01   0.16000000000000000000000000000000Q+01
0.29629629629629629629629629629630Q+00   0.30476190476190476190476190476190Q+00
0.16931216931216931216931216931217Q-01   0.16931216931216931216931216931217Q-01

N =   5   CA(I)   CB(I)

0.36940836940836940836940836940837Q+00   0.10000000000000000000000000000000000Q+01
0.24329004329004329004329004329004Q+01   0.33333333333333333333333333333333Q+01
0.23434343434343434343434343434343Q+01   0.26666666666666666666666666666667Q+01
0.72188552138552188552188552188552Q+00   0.76190476190476190476190476190476Q+00
0.83116883116883116883116883116883Q-01   0.84656084656084656084656084656084Q-01
0.30784030784030784030784030784031Q-02   0.30784030784030784030784030784031Q-02

N =   6   CA(I)   CB(I)

0.34099234099234099234099234099234Q+00   0.10000000000000000000000000000000000Q+01
0.27738927738927738927738927738928Q+01   0.40000000000000000000000000000000000Q+01
0.33970473970473970473970473970474Q+01   0.40000000000000000000000000000000000Q+01
0.14097902097902097902097902097902Q+01   0.15238095238095238095238095238095Q+01
0.24508824508824508824508824508824Q+00   0.25396825396825396825396825396825Q+00
0.18233618233618233618233618233618Q-01   0.18470418470418470418470418470418Q-01
0.47360047360047360047360047360047Q-03   0.47360047360047360047360047360047Q-03

BIBLIOGRAPHY

Abramowitz, M., and Stegun, I. (eds.)(1964). "Handbook of Mathematical Functions with Formulas, Graphs and Mathematical Tables." Appl. Math. Ser. 55, U.S. Govt. Printing Office, Washington, D.C.

Chipman, D.M. (1972). Math. Comp. 26, 241-249.

Erdélyi, A., Magnus, W., Oberhettinger, F., and Tricomi, F.G. (1953). "Higher Transcendental Functions," Vols. I,II,III, McGraw-Hill, New York.

Fields, J.L. (1972). J. Approx. Theory 6, 161-175.

Fields, J.L. (1973). SIAM J. Math. Anal. 4, 482-507.

Fields, J.L. (1976). Written communication.

Luke, Y.L. (1962). "Integrals of Bessel Functions," McGraw-Hill, New York.

Luke, Y.L. (1969). "The Special Functions and Their Approximations," Vols. I,II, Academic Press, New York.

Luke, Y.L. (1970). SIAM J. Math. Anal. 1, 266-281.

Luke, Y.L. (1971-1972). Math. Comp. 25, 323-330, 789-795, and Math. Comp. 26, 237-240.

Luke, Y.L. (1972). On generating Bessel functions by use of the backward recurrence formula. Rept. ARL 72-0030, Aerospace Research Labs., Wright-Patterson Air Force Base, Ohio.

Luke, Y.L. (1975a). "Mathematical Functions and Their Approximations," Academic Press, New York

Luke, Y.L. (1975b). SIAM J. Math. Anal. 6, 829-839.

Luke, Y.L. (1976a). On the expansion of exponential type integrals in series of Chebyshev polynomials. See Law, A.G., and Sahney, B.N. (eds.). "Theory of Approximation with Applications," pp. 180-199, Academic Press, New York.

Luke, Y.L. (1976b). J. Comput. Appl. Math. 2, 85-93.

Luke, Y.L. (1977a). Algorithms for rational approximations for a confluent hypergeometric function. Utilitas Math. (In press).

Luke, Y.L. (1977b). Algorithms for rational approximations for the Gaussian hypergeometric function. "Proceedings of a Conference on Rational Approximation with Emphasis on Applications of Padé Approximations," (E.B. Saff and R.S. Varga, eds.), Tampa, Florida, December, 1976, Academic Press, New York. (In press).

Wimp, J. (1969). On recursive computation. Rept. ARL 69-0186, Aerospace Research Labs., Wright-Patterson Air Force Base, Ohio.

Wimp, J. (1970). Recent developments in recursive computation. "SIAM Studies in Applied Mathematics," Vol. VI, pp. 110-123, Philadelphia. Also published under the same title, Rept. ARL 69-0104, Aerospace Research Labs., Wright-Patterson Air Force Base, Ohio.

Wimp, J. (1974). Computing 13, 195-203.

Wimp, J., and Luke, Y.L. (1969). Rend. Circ. Mat. Palermo 18, 251-275.

NOTATION INDEX

The system of notation employed to locate specific material in the text lists, in the following order, chapter, section (if any) and sub-section (if any).

EXAMPLE. 19.2 means Chapter XIX, second section. Only 19.4 has subsections, 19.4.1 and 19.4.2. A number in parenthesis refers to an equation.

EXAMPLE. 19.2(6) refers to the sixth equation in 19.2. An equation number standing by itself refers to the equation of the particular section in which the reference occurs.

EXAMPLE. A reference to (6) in 19.2 means the sixth equation in 19.2.

There is much ad hoc notation which is explained in the text near where it occurs. Data of this kind are excluded in the index.

In the listings below, the numbers refer to the page on which the notation is defined.

281

J

$J_\nu(z)$, Bessel function of the first kind, 44

K

$K_\nu(z)$, modified Bessel function of the second kind, 46

$K(k)$, complete elliptic integral of the first kind, 42

L

$L_\nu(z)$, modified Struve function, 44

P

$P_\nu^\mu(z)$, associated Legendre function of the first kind, 42

Q

$Q_\nu^\mu(z)$, associated Legendre function of the second kind, 42

S

$S_{\mu,\nu}(z)$, Lommel function, 47

$s_{\mu,\nu}(z)$, Lommel function, 44

T

$T_n(x)$, Chebyshev polynomial of the first kind, 5

$T_n^*(x)$, shifted Chebyshev polynomial of the first kind, 5

U

$U(a;c;z)$, confluent hypergeometric function, 45

Y

$Y_\nu(z)$, Bessel function of the second kind, 46

Greek Letters

$\Gamma(a,z)$, complementary incomplete gamma function, 45

$\gamma(a,z)$, incomplete gamma function, 43

$\Gamma(a) = \Gamma(a,z) + \gamma(a,z)$, gamma function

γ, Euler-Mascheroni constant, often referred to as Euler's constant. To 5D, $\gamma = 0.57722$

ε_n, a parameter such that $\varepsilon_0 = 1$ and $\varepsilon_n = 2$ if $n > 0$, 5

$\lambda = \alpha + \beta + 1$, 21

$\psi(z)$, logarithmic derivative of the gamma function

Miscellaneous Notations

$z = x + iy$, $i = (-1)^{\frac{1}{2}}$, is a complex number, x and y real

$R(z) = x =$ real part of z

$I(z) = y =$ imaginary part of z

$|z|$ = absolute value of z, $|z| = (x^2+y^2)^{\frac{1}{2}}$

arg z = argument of z, $\tan(\arg z) = y/x$

ln z = principal value of the natural logarithm of z

$\ln z = \ln|z| + i \arg z$, $-\pi < \arg z \leq \pi$

$z^\alpha = e^{\alpha \ln z}$ with ln z defined as above

$k! = 1 \cdot 2 \cdots k$, factorial function

$f'(z) = df(z)/dz$

$[x]$ = largest integer contained in x, x > 0

$(-)^n = (-1)^n$, n an integer or zero

P.V.$\int$ means Cauchy principal value of an integral. See p. 43, 2.2(43)

~ means asymptotic equality

$\alpha_p = \Pi_{j=1}^p \alpha_j$ and $(\alpha_p)_k = \Pi_{j=1}^p (\alpha_j)_k$ are frequently used in connection with generalized hypergeometric and G-functions. See p. 2, 1.2(2)

D as in 6D means six decimals

The notation x = 0(0.2)2.0, for example is used in connection with the description of tabular data and means that data are given for x from 0 to 2.0 in gaps of 0.2

A number in parentheses following a base number indicates the power of ten by which the base number is to be multiplied. In illustration, 2.35786(-3) means $2.35786 \cdot 10^{-3}$

SUBJECT INDEX

A

Algorithms, general remarks, 49

Approximation, see Rational and Padé approximation

Arc sine, 42

Arc tangent, 42

Asymptotic expansion for large n for coefficients in expansion of functions in series of Chebyshev polynomials of the first kind and for denominators and remainders in Padé and rational approximations, see under pertinent functions

n for gamma function ratio, 7

z for $G_{p,q+1}^{1,p}(z)$, 4,7,10

$U(a;c;z)$, 252

$\Gamma(\nu,z)$, 265

μ for $I_\mu(z)$, $K_\mu(z)$, 234

$J_\mu(z)$, $Y_\mu(z)$, 240

B

Basic series, 31, 154

Bessel functions and their integrals

$J_\nu(z)$, $I_\nu(z)$, 6,30-33,43-48,70,71, 77,82,88,91,92,101,105,182,194, 203,220,230,233,239

$Y_\nu(z)$, $K_\nu(z)$, 30-33,46-48,88,92, 101,104,105,233,239,252

$H_\nu^{(r)}(z)$, $r = 0,1$, 46,101

Binomial function, 42,53,57,160,176

C

Chebyshev polynomials, first kind,5, 17,20,154; differential and integral properties of, 116,126

Complex arithmetic, 49,56

Computation and checks on coefficients and tables, 15,23,29

Computation by use of recurrence formulas, 5,8,10-14,17,31,230

Computer programs, 49

Confluent hypergeometric functions, see under Hypergeometric functions

Conversion of power series into Chebyshev series, 154

Cosine and hyperbolic cosine, 32,44, 78

Cotangent, 32

Coulomb wave functions, 46

D

Dilogarithm function, 33

E

Elliptic integrals, 42,52

Error functions, 43,46,70,72,88,192

Evaluation of series of Chebyshev polynomials, 17,20

Expansion-To simplify this index, we do not cite pages where functions are represented by infinite series of Chebyshev polynomials of the first kind or where functions are represented by Padé and rational approximations. To get such information, use this index, the Table of Contents and the listings in Chapter 2. Even this might not be sufficient for named functions which are special cases of more generalized functions treated here. To fill out this index, see also my references Luke (1962; 1969, Ch. 6; 1975a, Ch. 8). In illustration, note that Airy functions and their derivatives are related to Bessel functions of order 1/3 and 2/3. More precise data are not given here, but will be found in the sources cited.

Exponential functions, 7,30,35,40,43, 52,54,71,135,175,195,234

Exponential and exponential type integrals, 30,45,46,48,70,88,126,182, 192

G

G-function, 3,7,9,25,34,41,47

Gamma function and derivatives, 31,32

Gamma function, ratio of, 1,7

Gaussian hypergeometric function, see under Hypergeometric function